MECHANICS OF DRILLSTRINGS AND MARINE RISERS

DON W. DAREING

ASME PRESS

Library of Congress Cataloging-in-Publication Data

Dareing, Don W.
Mechanics of drillstrings and marine risers / by Don W. Dareing.
p. cm.
Includes bibliographical references and index.
ISBN 978-0-7918-5999-5
1. Oil well drilling. 2. Offshore oil well drilling. I. Title.
TN871.2.D332 2012
622'.33819–dc23

2012016752

Cover Art by David E. Dareing.

TABLE OF CONTENTS

PREFACE

Mechanics of Materials typically addresses bending of beam columns where axial forces are assumed to be constant over the length of the beam. This assumption is valid for most engineering structures, and there are many good reference books that cover this subject. To my knowledge, there is no reference book that covers the mechanics of long beam columns that are several hundred or even thousands of feet long.

Since Spindle Top, much has been learned about the performance of drillstrings through field experience and since the 1950s a considerable amount through analysis and testing. Drilling is now a mature technology with the capability to drill along a specified well path to reach a target within a few feet using Measurements While Drilling (MWD) and Rotary Steerable Tools. Marine riser technology has also advanced through better understanding of environmental loads and riser responses.

In light of these advances, there is a need to document the fundamental mechanics of these long tubulars. Writing this book offered the challenge of bringing together the works of many who have contributed to this technology. Even though both areas of application are different, there are similarities in their mechanics. It is, therefore, appropriate to cover both topics together.

Field operations of drillstrings and marine risers are complex and require the skill of experts with years of experience, especially in offshore drilling and production. Also, there are many excellent computer software programs that predict responses of these tubulars under different operating conditions. Software design tools are typically based on numerical methods. It is not the purpose of this book to explain practices, hardware, or computer software but to focus on the fundamental mechanics of long beam columns. Closed form solutions will be emphasized throughout.

Chapter 1 gives historical background on the evolution of the mechanics of drilling tubulars and marine risers. Chapter 2 explains the mechanics and mathematics of long tubulars. These concepts are utilized throughout. Chapter 3 covers the mechanics of drillstrings including both static and dynamic responses. Sources of excitation are explained along with methods of vibration control. Chapter 4 applies tubular mechanics to marine risers. Chapter 5 covers the application of drilling mechanics, while Chapter 6 covers selected topics in marine riser design.

Each of us has special friends for different areas of our life, whether it be community, church, sports, or profession. It has been my pleasure to work with truly outstanding engineers and scientists who have expanded my understanding of this subject.

Dr. Tseng (Leo) Huang, Professor of Civil Engineering at the University of Texas at Arlington, was brilliant in his understanding of engineering mechanics and its application to long tubulars. Much of the material in this book is extrapolated from several papers which we wrote together. He would have had much to add to this manuscript. Selected papers by Dr. Huang have been added to the list of references at the end of Chapter 2.

Dr. Jiri (George) Tlusty, Professor of Mechanical Engineering at the University of Florida, explained the self exciting mechanism of machine tool chatter. He discarded the stick-slip theory, which was commonly accepted at the time. Dr. Tlusty's pioneering work set the engineering baseline for high speed machining, allowing machine tools to produce smooth surfaces at high cutting rates. This technology has been advanced by Dr. K. Scott Smith, Mechanical Engineering Department Chair at the University of North Carolina, Charlotte.

Out of our discussions of Polycrystalline Diamond Compact (PDC) drill bits, Dr. Tlusty saw a direct correlation between machine tool chatter and PDC drill bit chatter. This technology transfer is covered in Chapter 3. Dr. Mostafa A. Elsayed, Professor of Mechanical at the University of Louisiana, Lafayette, expanded this theory and verified its validity through extensive testing at the Sandia National Laboratories.

It is my pleasure to collaborate with Dr. Thomas G. Thundat, Professor, Canada Research Chair in Oil Sands Molecular Engineering, University of Alberta (formerly Corporate Fellow at the Oak Ridge National Laboratory), renowned for his research in biomechanics and micro sensors. Our research collaboration merges engineering mechanics and bioscience with a focus on microcantilever sensors, a beam of much smaller dimensions.

This book does not address all of the issues of these technologies. Offshore drilling and production requires field experience and engineering insight, which are difficult to capture in writing. It is my hope this book will serve as a reference for designers of offshore drilling and production systems and provide a fundamental background on the mechanics of long beam columns for those entering the petroleum industry.

I am grateful to Dr. Kirk E. Boatright, Dr. Mostafa A. Elsayed, and Dr. Nick S. Winowich for their valuable feedback. I would like to especially thank members of the ASME book review committee and the editorial staff for evaluating and processing this manuscript.

Don W. Dareing
Life Fellow Member, ASME
Professor Emeritus, University of Tennessee

NOMENCLATURE

$\alpha = \dfrac{(w - A_0\gamma_0 + A_i\gamma_i)L^3}{\text{EI}}$, dimensionless

$\beta = \dfrac{(F_B + LA_0\gamma_0 - LA_i\gamma_i)L^2}{\text{EI}}$, dimensionless

$\lambda^4 = \dfrac{m\omega^2 L^4}{\text{EI}}$, dimensionless

$\zeta = \dfrac{x}{L}$

C_d, C_m — Morison drag and inertia coefficients

$F_n G_n H_n I_n$ — Constants that relate $a_0 a_1 a_2 a_3$ to a_n (see Eq. (2-109))

$Y = Y(x)$ or $Y(\zeta)$

$y = y(x, t)$

F_B — internal force at bottom end of tubular, lbs

$F_{B,cr}$ — critical internal force (used to predict riser buckling)

F_{cr} — critical buckling force applied to drill pipe (force above hydrostatic pressure force)

BF — buoyancy factor

L — Length of tubular, ft

A_0 — cross section area based on OD, in.^{2}

A_i — cross section area based on ID, in.^{2}

γ_0 — density of fluid outside of pipe, ppg (lb/ft^3)

γ_i — density of fluid inside of pipe, ppg (lb/ft^3)

γ_m — density of drilling mud, ppg (lb/ft^3)

EI — bending stiffness, lb-in.^{2}

m — mass per unit length of pipe, slugs/ft

$a = w - A_o\gamma_o + A_i\gamma_i$

$b = F_B + LA_o\gamma_o - LA_i\gamma_i$

c — damping coefficient, lb/fps/ft

$\psi^2 = m\omega^2 L^2$

M_t — applied torque

H_a — angular momentum with respect to point a

G_x — x component of linear momentum

x, y, z — coordinate system, (fixed or rotating)

$u = u(\varsigma)$, complex displacement

U — complex function of x and y; $U = x(z,t) + iy(z,t)$

$P_{Euler} = \dfrac{\pi^2 EI}{L^2}$ (Euler buckling force)

$\omega_{crit} = \left(\dfrac{\pi}{L}\right)^2 \sqrt{\dfrac{EI}{m}}$ (critical synchronous whirl velocity)

$(M_t)_{crit} = \dfrac{2\pi EI}{L}$ (Greenhill critical torque)

T — applied torque

$T_{\text{eff}} = [(F_B + wx) + (L - x)(A_0\gamma_0 - A_i\gamma_i) - A_i p]$ (effective tension)

A, B, C — constants used by Southwell and Dough

w — weight per unit length of pipe (air weight)

$\Theta = \dfrac{TL}{EI}$ (dimensionless torque)

$\gamma = R_\gamma + iI_\gamma$ (complex eigenvalue)

$\Omega^2 = \dfrac{m\omega^2 L^4}{EI}$

$\eta = c\left(\dfrac{L^4}{E\,\mathrm{Im}}\right)^{\frac{1}{2}}$

$\eta_{cr} = \dfrac{|I_\gamma|}{\sqrt{R_\gamma}}$ (critical damping factor)

$\Omega_{oc} = R_\gamma^{\frac{1}{2}}$ (critical angular velocity parameter)

p — rotation of whirl relative to rotating reference frame, xyz

λ_1, λ_2 — complex constants (see Eq. (2-218))

λ_i — frequency eigenvalues

X_i — mode shape

$\eta_i(t)$ — time history of mode shape

$M_i C_i K_i Q_i$ — modal mass, damping, stiffness, force

$m^3 = \dfrac{EI}{p}$ (dimensionless unit, defined by Arthur Lubinski)

$p = (w - w_m)$ (Lubinski's notation)
q — unit side loading (positive in +y direction)
Q — compression within drill collars (also used as side load on buoys)
V — shear force, lbs
M — bending moment, ft-lb

$$\lambda^2 = \frac{Q}{EI}$$ (used in Stabilizer placement derivation)

$\Phi = \omega + p$ (absolute whirl velocity)
ρ — radius of curvature
B_x, B_y — bending moment vector components (used by Vandiver)
AE — axial stiffness of pipe
u(x.t) — axial displacement
$G(\omega)$ — complex transfer function
D_1, D_2 — exponents in tapered flex joint solution

$$I(x) = I_p \left(\frac{h - x}{a} \right)^2$$ (moment of inertia of parabolic flex joint)

$$h = \frac{L}{1 - \sqrt{\frac{I_p}{I_{SJ}}}}$$ (tapered flex joint parameter)

$$a = h \sqrt{\frac{I_p}{I_{sJ}}}$$ (tapered flex joint parameter)

$$\gamma^2 = \frac{a^2 F_B}{EI_p}$$ (tapered flex joint parameter)

Chapter 1

INTRODUCTION

From 1860 to 1900, the cable tool method was used extensively. This method uses a heavy steel bar and bit attached to the bottom of a cable. The bottom assembly is dropped rhythmically against the bottom of the hole to crush and chip away at rock formations. Periodically, drilling is stopped and special tools are lowered to the bottom to clean out the cuttings. Prior to 1900, several wells exceeded 2000 ft, and in 1953, one well was drilled to 11,000 ft using the cable tool method.

Rotary drilling uses a rotating string of pipe to power drill bits designed to chip and grind rock. Drilling mud is pumped through the pipe to clean cuttings from around drill bits and carry the cuttings to the surface. Joints of pipe are added as the hole gets deeper to lengthen the drillstring and advance the drill bit. Weight applied to the drill bits is controlled by a top tension or hook load supported by the derrick.

Rotary rigs were patented and envisioned as early as 1833; however, portable steam engines had not been fully developed, and rotaries that were used before 1900 were often powered by mules walking in circles. Before 1900, more than 100 wells had been drilled in Texas by the rotary method.

All elements of rotary drilling, as known today, were first brought together at Spindletop in 1901. This initial drilling system included: steam engines, draw works, rotary table, mud pumps, swivel, drill, traveling block and crown block, and a special drill bit. Early drill bits were fish tail drag bits. One of the most significant advances of this time was the rolling cone drill bit, invented in 1909. Drilling rigs were usually built on site and left in place. Mobile derricks were introduced in the 1950s.

Drilling muds have greatly improved since Spindletop where cattle pond water was muddied to add weight for well bore stability and well control. Modern drilling fluids (still called drilling mud) provide better well bore stability, mud weight refinement for well control and drilling pressure balance, and environmentally friendly properties.

The growth in drilling technology over the 20th century was enormous [1]. Technology has advanced from crude drill bits and drilling equipment to current tools that drill along specified well paths and with great accuracy, which is accomplished through Measurement While Drilling (MWD), Rotary Steerable Tools, and advanced drill bits. Yergin [2] gives an excellent overview of the evolution of the petroleum industry, discussing key milestones in history and their political impact on international affairs.

Methods for offshore drilling were greatly expanded in the 1950s. Floating drill ships, initially called semisubmersibles, were anchored over a well site for exploratory drilling. These drilling vessels were designed for stability in rough seas. Drilling proceeded as on land with the aid of motion compensators, which maintained top tension in risers and drillstrings.

Marine risers are an essential component to offshore drilling and production. They connect the well head at the sea floor to drill ships or production platforms. The operation of marine risers requires special topside equipment and knowledge of vessel response to ocean currents and waves. Lateral motion of rises is controlled mainly by top tension. Structural connections at top and bottom ends are critical too. These areas can be subjected to high bending stresses. Dynamic stresses caused by inline wave action and vortex shedding can be detrimental.

The lateral motion of both drillstrings and marine risers is described by the same differential equation of bending so it is appropriate to discuss the mechanic of each within the same book even though drilling engineers and marine engineers may view the results quite differently. Both tubulars accommodate tension/compression and hydrostatic pressure which vary from top to bottom. These factors introduce a term in the differential equation having a variable coefficient, which makes the analysis different from classical beam theory.

This chapter gives an overview of drillstring and marine riser technology and provides a backdrop for more detailed discussions of the mechanics of these long vertical beam columns.

1.1 DRILLSTRINGS

The drillstring is central to rotary drilling. It transmits mechanical rotary power from the surface to drill bits, serves as a conduit for drilling fluid, applies force to drill bits, and affects hole direction. While the majority of the drillstring is made up of drill pipe, the bottom portion (roughly 150 ft) of drill collars and bottom hole assembly contains tools (drill bit, PDM, Turbine, MWD, and stabilizers) that affect rate of penetration, footage cost, and well bore direction.

Drillstrings are subjected to many operational loads such as direct pull, torsion, and bending. Each has a static and a dynamic component. They are typically designed from static load considerations following the state-of-the-art as given in API RP 7G [3]. Since 1960, much has been learned about the dynamic behavior of drillstrings through downhole measurements and analytical studies. Vibrations are known to affect the performance of downhole equipment. Well bore curvature and local dog-legs also affect fatigue life.

1.1.1 Drill Pipe [4]

Lap-welded pipe made for general industrial maritime and industrial applications was first used as a drill pipe. Pipe material was wrought iron having yield strength of about 25,000 psi and an ultimate strength of 40,000 psi. Joints of drill pipe were connected by threaded couplings called collars. The threads had a slight taper (3/4 in./ft). Failures typically occurred in the last engaged thread under the coupling. Sealing across the coupling was a problem and washouts were common. Casing was also used as drill pipe and left in place when the hole was finished.

The first tool joint was patented in 1910. It contained coarse tapered threads and mating shoulders to provide a seal across the pipe connection. By making tool joints separate from the tubular body, threads could be heat treated and this eliminated the galling problem. Tapered threads also allowed for quick makeup and breakout of connections. Early tool joints were threaded onto the pipe. This was followed by induction welding. Today tool joints are attached to the pipe body by friction or inertia welding.

The first seamless pipe was manufactured in the early 1920s. A steel billet is rolled into a long bar while simultaneously being pierced by a die to produce the pipe bore. Seamless pipe became an API standard for drill pipe in 1927. At this time, a drill pipe was made of mild steel. Grade C pipe had a yield strength of 45,000 psi and was introduced around 1926. Grade D pipe, which had a yield strength of 55,000 psi, was introduced in 1928. Both C and D grades are seldom used today. Currently, the industry uses four grades of steel (Grades E75, X95, G105, and S135). Grade E75 is considered a low-strength steel. The other three grades are high-strength steels.

1.1.2 Drill Collars [4]

During the early 1920s, the lower portion of drill pipe was put in compression to apply force to advance the drill bit. Short-threaded collars were used to connect drill bits to drill pipe. These collars were lengthened to about 5 to 6 ft in some cases to facilitate the recovery of collars and drill bits in the event of a twist-off and to strengthen this area against failure. They were eventually lengthened to help maintain hole direction. After some time, these collars became known as drill collars. The use of drill collars to control well bore direction was not a consideration at that time even though it was believed that they helped in drilling straighter well bores. Today, the drill collar section is much longer and enhances directional control and rate of penetration.

1.1.3 Rotary Shouldered Connections

Rotary shouldered connections are special threaded connections used to attach individual joints of pipe [3]. They also connect special drilling tools within drillstrings. They are designed to withstand structural loads while providing a mechanical seal, separating internal pressure (~2000 psi) from annulus pressure. Threads are tapered to a 1:6 pitch to allow quick make-up and break-out. These threads are machined directly into the collar collars, which are made of high-strength steel. Because of this, drill collar connections cannot be as strong as the drill collar body. Fortunately, drill collars are run in compression. Drill pipe is run in tension but can be subjected to dog-leg type bending.

1.2 STATIC ANALYSIS

A good starting point for the discussion of drillstring mechanics is the definition and location of the neutral point.

In practice, weight on bit (WOB) is established by slacking off the desirable bit force from the hook load at the surface. This puts drill collars in compression while drill pipe is in tension. A basic question is: how long should the drill collar section be to prevent drill pipe buckling? This is important because of the large difference in structural stiffness between drill pipe and drill collars. From a structural point of view, stiffness attracts bending moment and resulting stresses can be damaging, especially under stress reversals caused by drillstring rotation.

1.2.1 Neutral Point

The location and definition of the neutral point was not generally understood until the 1950s [5]. The confusion was due to a misunderstanding of buoyancy and how it is applied to submerged pipe. According to Archimedes' principle, the buoyant force acting on a submerged body is equal to the weight of fluid displaced by the body. It is important to remember that buoyancy is a surface force and not a body force. On the other hand, pipe weight is a body force.

Applying Archimedes' principle to the entire string hanging freely from the draw works predicts a hook load of

$$H = W - B \tag{1-1}$$

where H is hook load, W is air weight of total string and B is the weight of the drill mud displaced by the string as stated by Archimedes' principle. In this equation, W is a body force and B is a surface force. Equation (1-1) can also be written

$$H = W\left(1 - \frac{B}{W}\right) = W\left(1 - \frac{\gamma_m}{\gamma_{stl}}\right) \tag{1-2}$$

$$H = W \times BF \tag{1-3}$$

where BF is the buoyancy factor.

Assuming a 12-ppg mud, the buoyancy factor is 0.817 as calculated below.

$$\gamma_m = 12\text{lb/gal}\left|\frac{7.48\text{ gal}}{1\text{ft}^3}\right| = 89.76\frac{\text{lb}}{\text{ft}^3}$$

so

$$\text{BF} = \left(1 - \frac{89.76}{490}\right) = 0.817$$

The buoyed weight of a drillstring (including tools, etc.) is total air weight multiplied by the buoyancy factor.

The magnitude of buoyancy forces can be quite high. For example, consider 5 ½ in. (19.2 lb/ft) drill pipe having a cross section area of 4.9624 in². If 10,000 ft of 5 ½ in. drill pipe hangs freely in 12 ppg mud, hydrostatic pressure applied to the lower end is

$$p = \gamma \times L \tag{1-4}$$

$$p = 89.76 \text{ lb/ft}^3 \ (10{,}000 \text{ ft}) \left| \frac{1 \text{ ft}^2}{144 \text{ in}^2} \right|$$

$$p = 6{,}233 \text{ psi}$$

This pressure creates a force on the lower free end of

$$F = p \times A \tag{1-5}$$

$$F = 6233 \text{ lb/in}^2 \ (4.9624 \text{ in}^2)$$

$$F = 30{,}932 \text{ lbs}$$

It would appear that this force is great enough to buckle the drill pipe several times. However, this does not happen.

Assume the bottom end of an open-ended pipe (Fig. 1-1a) is given a sidewise displacement. The dashed line indicates the static equivalence of the hydrostatic pressure forces summed over the surface is below section a-a. The force B is equal to the weight of fluid displaced by the pipe up to line a-a. The solid vector, W, represents the air weight of the drill pipe below cross section a-a. W is truly a body force while B is a surface force. Note that the bottom end is exposed to hydrostatic pressure. Moments about point "0" show there is a restoring effect that always moves the pipe back toward the vertical, assuming the density of the pipe is greater than the density of the fluid. In other words, an

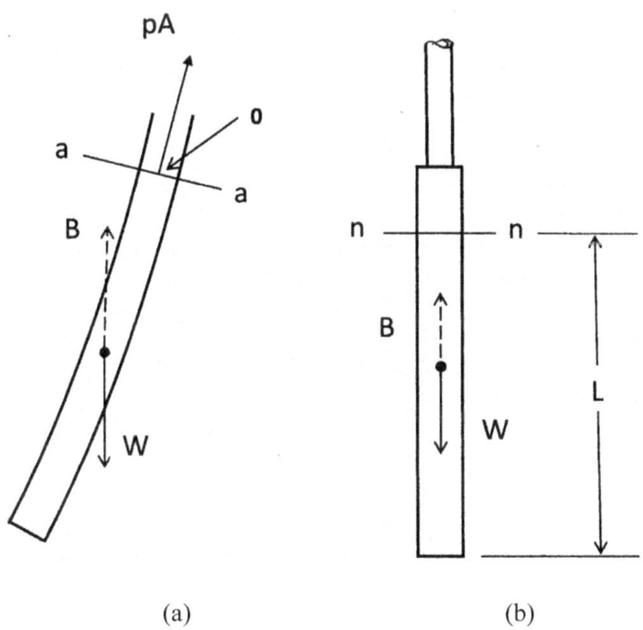

Figure 1-1. Stability of vertical pipe under hydrostatic loading.

open-ended drill pipe will not buckle from hydrostatic pressure alone regardless of well bore depth.

Now, assume the drill pipe is attached to the top of a drill collar section (Fig. 1-1b) and the compression at the top of the drill collars is equal to the local hydrostatic pressure. According to the argument above, the drill pipe will not buckle under this condition.

A good definition for the neutral point is the point in the drillstring where compressive stress is equal to the local hydrostatic pressure. If the hydrostatic pressure point is located within the drill collars, the drill pipe will not buckle. Drill pipe will buckle when the internal force at the lower end reaches a critical level, which is somewhat higher than the hydrostatic force level. Hydrostatic pressure does affect pipe bending, however, and this will be discussed in Chapter 2.

The location of the neutral point depends on bit force (WOB). If bit force is zero, the neutral is located at the drill bit. As bit force is increased, the neutral point moves up the drill collars.

In practice, the neutral point is kept within the drill collars as a safety measure to avoid buckling the drill pipe. The relation between WOB and distance to neutral point is determined by

$$\mathrm{WOB} = wL_{\mathrm{np}} - w_{\mathrm{m}}L_{\mathrm{np}} \tag{1-6}$$

or

$$\mathrm{WOB} = wL_{\mathrm{np}}\left(1 - \frac{w_{\mathrm{m}}}{w}\right) \tag{1-7}$$

$$\mathrm{WOB} = wL_{\mathrm{np}}\left(1 - \frac{\gamma_{\mathrm{m}}}{\gamma}\right) \tag{1-8}$$

where

γ — steel density (490 lb/ft^3 or 65.5 ppg)
γ_{m} — drilling mud density (ppg)

The distance to the neutral point (or point of hydrostatic compression) from the drill bit is determined by

$$L_{\mathrm{np}} = \frac{\mathrm{WOB}}{w\mathrm{BF}} \tag{1-9}$$

Drill collar length is generally computed by

$$L_{\mathrm{np}} = 0.85L_{\mathrm{c}} \tag{1-10}$$

which means that the neutral point is 85% of drill collar length. The extra length represents a safety factor against drill pipe buckling.

$$L_{\mathrm{c}} = \frac{\mathrm{WOB}}{0.85w\mathrm{BF}} \tag{1-11}$$

Bottom hole assemblies are generally made up of many components, such as motors, MWD, and stabilizers. In this case, the neutral point is located by

$$\mathrm{WOB} = \sum_{i=1} (W_i + xw_{i+1})\mathrm{BF} \tag{1-12}$$

where W_i is the air weight of each component up to the top of the ith component and x is the distance from the top of the ith component to the neutral point.

1.2.2 Drill Collar Buckling

Even though the neutral point can be kept within drill collars by adding more and more collars, the collars themselves do buckle at some level of WOB.

During the late 1950s, Lubinski [5] began analyzing drill collar buckling by mathematically modeling drill collars as a long vertical beam column under variable internal compression. Prior to this, little was understood about the effects of bit force, gravity and even hydrostatic pressure on the lateral stability of drill collars. His landmark study provided understanding and insight into drill collar buckling, bit side force, side loads against the well bore and bending stresses. In 1881, Greenhill [6–8] solved the buckling problem for a vertical column fixed at the lower end and free at the top. He also pointed out a variety of buckling problems that can be solved by Bessel functions. However, his solution did not apply directly to drill collars.

Lubinski defines the condition of drill collar buckling in terms of a dimensionless unit designated as m where

$$m = \sqrt[3]{\frac{EI}{p}} \tag{1-13}$$

and (using Lubinski's notation)

E — modulus of elasticity, lb/ft^2
I — moment of inertia, ft^4
p — weight in mud per unit length of the drilling string, lbs/ft ($p = w - w_m$)
m — length in feet of one dimensionless unit

His results show that buckling of the first and second modes occurs when the neutral point is located 1.94 and 3.75 dimensionless units, respectively, above the drill bit.

$$(L_{np})_1 = 1.94m$$

$$(L_{np})_2 = 3.75m$$

Relating this formula to critical WOB

$$L_{np} = \frac{\mathrm{WOB}}{w\mathrm{BF}} \tag{1-14}$$

$$\frac{\text{WOB}_{cr}}{w\text{BF}} = 1.94\left(\frac{EI}{w\text{BF}}\right)^{\frac{1}{3}} \tag{1-15}$$

$$\text{WOB}_{cr} = 1.94\left(w\text{BF}\right)^{\frac{2}{3}}\left(EI\right)^{\frac{1}{3}} \quad \text{(first mode)} \tag{1-16}$$

$$\text{WOB}_{cr} = 3.75\left(w\text{BF}\right)^{\frac{2}{3}}\left(EI\right)^{\frac{1}{3}} \quad \text{(second mode)} \tag{1-17}$$

Figure 1-2 gives critical bit weights for two collar sizes, based on Eqs. (1-16) and (1-17).

Lubinski derived the equation of bending and showed how the solution is expressed in terms of a power series and Bessel functions. The assumption is made that the well bore is vertically straight. His analysis locates and quantifies the contact force between drill collars and well bore and predicts transverse reaction forces at the drill bit. He concludes … "it is impossible to drill a straight vertical hole because of this side force on the drill bit". He explains well bore deviation in terms of drill bit side forces and shows that hole angle and the direction of the total bit force are, in general, different. This insightful

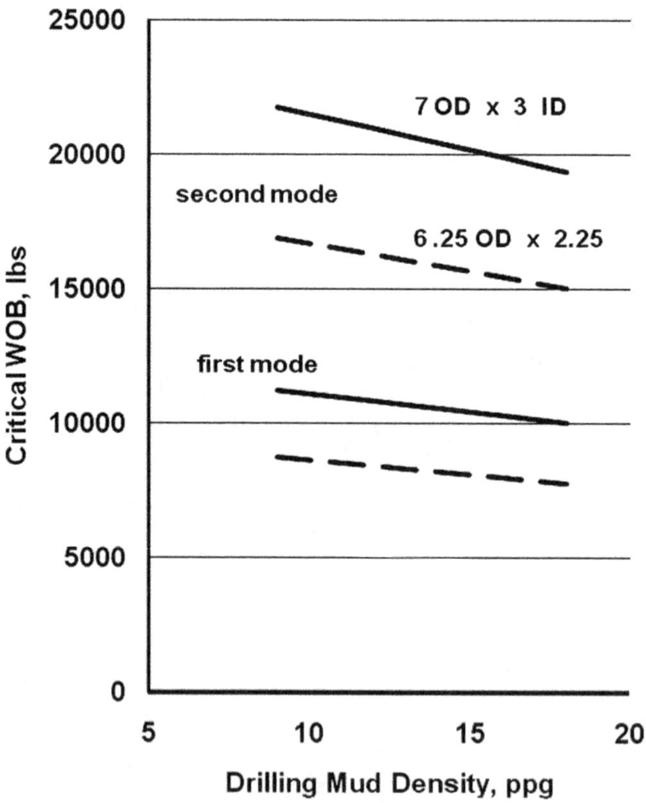

Figure 1-2. Critical WOB for two collar sizes.

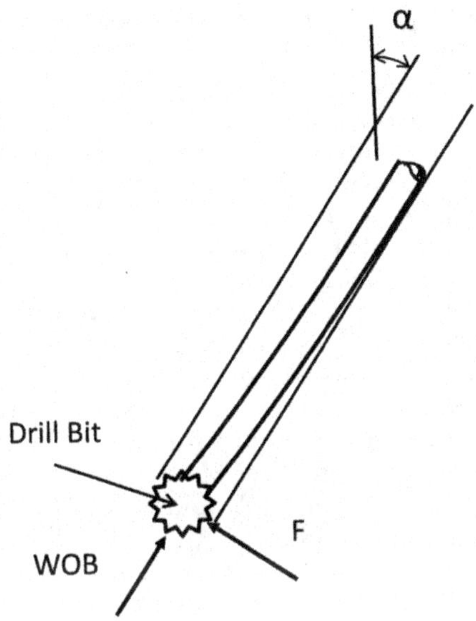

Figure 1-3. Resultant bit force.

conclusion provides the basis for directional control technology used throughout the industry today.

A short time later, this work was extended by Lubinski and Woods [9] to show the effects of well bore inclination (Fig. 1-3) on bit side forces. This work also predicts the effect of formation type (isotropic and non-isotropic) on hole deviation. Graphs are given to show conditions that make bit force angle and hole inclination the same, which would be required to drill a straight hole along a given hole inclination. The data defines drilling conditions that would either build or drop well bore angle. Stabilizers had not been introduced at this time.

1.2.3 Stabilized Drill Collars

Initially, drilling operations focused on vertically straight well bores. It became apparent that hole deviation depended on bit weight, collar size, and formation. The main criteria for an acceptable wellbore path was well bore deviation; the hole should not stray outside of a 3- to 5-deg imaginary cone with apex at the rig. During the 1950s, the industry felt that local well bore curvature or dogleg severity (DLS) was a better measure of an acceptable well path because of its potential negative effect on drilling operations [5].

Drill collar stabilizers were introduced in the 1950s to help keep the well bore move fairly straight and vertical. These tools were slightly under gauge in terms of drill bit diameter. Their purpose was to keep drill collars lined up with the direction of the hole. They were designed with open blades to allow drilling fluid to move vertically up the annular space unobstructed. At first, several stabilizers were placed near the drill bit. Later, a more scientific approach was made to determine stabilizer locations that would cause the bit to drill in a preferred direction.

In 1951, MacDonald and Lubinski [10] proposed using conventional stabilizers to allow greater WOB without collar buckling with the added benefit of eliminating high curvature or dog legs. Further work by Woods and Lubinski [11] showed that it is possible to drill a slightly inclined well bore by establishing an equilibrium angle or a result bit force angle that is in line with the centerline of the well bore. This work provided the bases for the use of multiple stabilizers to build angle, hold angle, or drop hole angle. Walker [12] gives an excellent review of the development of this technology. Miska [13] provides a comprehensive review of Arthur Lubinski's work.

Other methods of pushing drill bits forward were also being considered. In 1958, Kellner and Roberts [14] patented a special wall anchor drill collar for the purpose of applying force to drill bits regardless of hole inclination. Hydraulic pressure expanded anchor shoes to make contact with the well bore. Hydraulic pressure was then used to apply force to drill bits. The inner mandrel travelled downward for a given distance before the wall anchors were reset.

During the early 1960s, Garrett and Rollins [15,16] designed and tested a directional control tool that intentionally pushed drill bits sideways. The tool could also be used as a wall anchor type drill collar to push the drill bit forward. The purpose of the tool was to correct and control well bore direction. Its use required multiple wireline surveys. This tool was a forerunner to rotary steerable tools.

1.2.4 Drill Pipe Buckling

Earlier, the neutral point was identified as a location in the drillstring where internal compressive stress is equal to the local hydrostatic pressure. Furthermore, pipe above this point can be viewed as being exposed to the local hydrostatic pressure as if it were hanging freely and open ended in the fluid. Even though this compressive force can be quite high, pipe under hydrostatic end loading will not buckle. However, buckling can occur by applying an additional compressive force (F_{cr}) over and above the hydrostatic bottom end force. Drill pipe buckling under this condition is predicted by Huang and Dareing [17]. The boundary condition at the lower end was assumed to be fixed to simulate the structurally stiff interface between pipe and drill collars. The critical force for the first mode is be predicted by

$$F_{cr} = \phi_{cr}(EI)^{\frac{1}{3}}(w - w_m)^{\frac{2}{3}} \tag{1-18}$$

$$F_{cr} = 3.29(EI)^{\frac{1}{3}}(w\mathrm{BF})^{\frac{2}{3}} \tag{1-19}$$

This buckling force is independent of hole depth (drill pipe length). The 3.29 factor applies to the case where the lower end of the drill pipe is fixed and the upper end is pinned. This explains the difference between Lubinski's 1.94 (Eq. (1-16)), which applies to pinned-pinned boundary conditions. The fixed lower end condition more closely models the connection between drill collars and drill pipe. Using Eq. (1-18), the critical buckling force for 4 ½ in. drill pipe (assuming 10 ppg mud) is calculated to be 2200 lbs.

1.3 DYNAMIC ANALYSIS

During the early 1960s, drilling and production research focused on drilling efficiency. The goal was to identify areas that offered opportunity to increase rate of penetration (ROP) and reduce footage cost. Specific area of focus were

- drill bits
- weight on bit (WOB)
- rotational speed
- hydraulics
- pressure balance
- mud properties
- bottom hole assembly

Studies show that significant drilling improvements could be achieved by paying attention to these factors. Good drilling practices involve doing the very best under each category. The application of the results became known as drilling optimization. Drillstring vibration was one area of focus.

1.3.1 Drillstring Vibrations

There has always been much interest in understanding vibrations in drillstrings. They were suspected of causing drillstring and drill bit failures. Placido, Santoes, and Galeano [18] show that lateral vibrations also affect well bore stability. Kelly bounce was common, and it was believed to be symptoms of more severe vibrations down hole.

Finnie and Bailey [19] measured real time axial and torsion loads directly below the Kelly. At the same time, velocity sensors at the top of the Kelly give real time measurements of velocity and displacement in vertical and angular directions. Measurements were taken in two different wells having depths of 3445 ft and 3496 ft under a variety of drilling conditions. The measured data was somewhat random with no dominate frequency in each of the four output signals. Axial and torsion natural frequencies were calculated for each drillstring in order to check for resonance [20].

Paslay and Bogy [21] developed a mathematical model to predict forced vibration response of axial modes in drillstrings. Their model was based on the mobility method developed by Plunkett [22]. The force driving these vibrations was assumed to be generated by intermittent contact of the teeth on roller cones. Predictions of bit forces and drillstring motion are given for both damped and undamped cases. The objective of the study was to correlate bit vibrations with penetration rate.

Garrett [23] attached an accelerometer to a swivel and detected frequencies of three cycles per rotation of the drillstring. This was a clear indication that bit cones were driving the vibrations. It was not uncommon to find three lobed core patterns in core tops. The purpose was to test a new shock absorbing tool placed directly above drill bits. Measurements showed a marked reduction in Kelly bounce when shock absorbers were applied in this manner.

The core shown in Fig. 1-4 was taken from a very hard dolomitic limestone formation. The mechanism which generates these lobes is not clearly

Figure 1-4. Top of core taken from dolomite limestone formation (courtesy of A. Scovil Murray, Imperial Oil Company, Calgary, Alberta, Canada).

understood, but it is believed to be the result of bit motions coupled with frequency tuning. Spanos et al [24] explain analytically how these lobed patterns can be enhanced or reduced during drilling.

A field method used to eliminate Kelly bounce was to lift the bit off bottom, spin the string and touch back down for a few minutes [25]. It was envisioned that this would "clean off these high spots." The mechanics of this practice is scientifically sound because vibration motion is 180 deg out of phase with the force when operating on the high side of resonance.

1.3.2 Downhole Measurements of Drill Bit Forces and Motion

In the late 1950s Jersey Production Research Company in Tulsa (now EPRCo, Houston) designed a downhole recording tool [26]. The objective was to measure forces and motions at drill bits in real time. This had never been done. Several parameters were measured: axial force, torque, bending, accelerations (axial, torsion, lateral) as well as pressure and temperature. Near bit data was measured and recorded on a magnetic tape. About six minutes of data could be recorded; the recorder was activated by pump pressure. The tool was retrieved during a bit change and results displayed on a strip chart. This tool was the industry's first measurement-while-drilling (MWD) tool. Downhole data was obtained at several drilling location in West Texas, Oklahoma, Louisiana, and Alberta, Canada during the early 1960s [27]. The data was unique and special and much credit is due those who planned, designed, built and tested this tool over a 3 year period.

An example of real time bit force and torque, as measured by EPRCo's downhole recording tool, is shown in Fig. 1-5. Rotary speed was between 80 and 90 rpm. The drill bit was a roller cone bit. Well depth was 5282 ft. Note that the recorded frequency is 4 cps. This relates directly to the 80 rpm of the three cone bit.

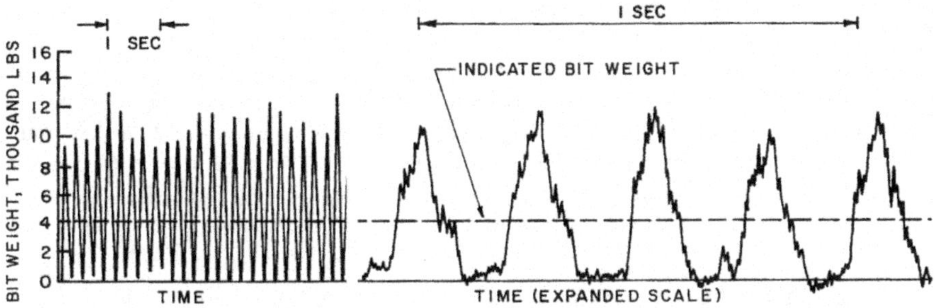

Figure 1-5. Resonant vibrations; West Texas, Midland Area [27].

It was not always possible to generate resonant vibrations during the six minute recording time frame. In most cases the vibration was random with relatively small amplitudes (Fig. 1-6). This particular data was obtained while drilling in a limestone/shale formation near Edmonton, Alberta. The drill bit was a roller cone bit. Well depth was 4365 ft and rotary speed was approximately 60 rpm.

As part of this study, a mathematical model was developed by Dareing and Livesay [28] to compliment the field data. During one test run, Kelly vibrations were measured simultaneously with downhole measurements. By using measured data at the bit, top side predictions were made with the model. There was reasonable agreement between predicted response and measured response

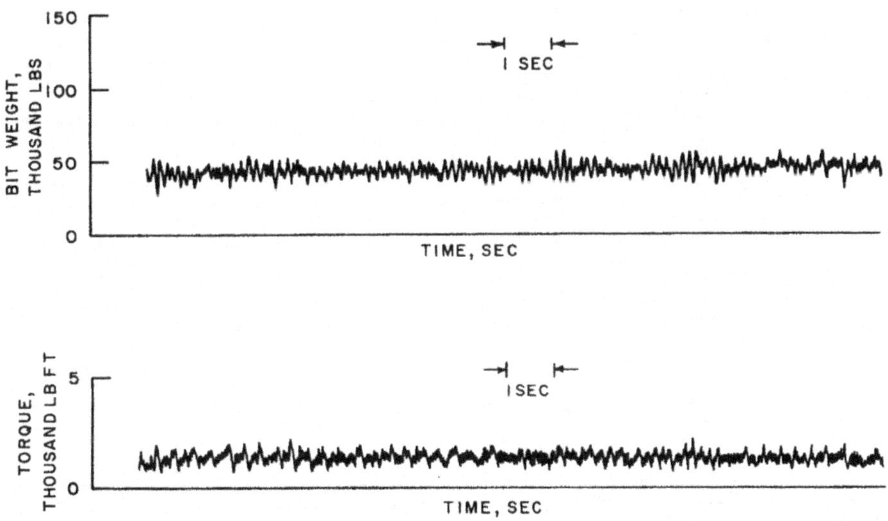

Figure 1-6. Typical drill bit force and torque; Edmonton, Alberta Area [27].

at the surface. The prediction depends to a large extent on the damping factor, which was extracted from field measurements.

The vision for the mathematical model was to make predictions of drillstring behavior using surface measurement as input. Surface measurement were easier to obtain on a real time basis while drilling The need for such a model is not as important today because MWD tools can feedback this type of information continuously. Later, the EPRCo data was made available to industry and additional interpretations were subsequently made by Cunningham [29].

1.4 DIRECTIONAL DRILLING

Directional drilling refers to the process of drilling along a prescribed well path to a target located at a given depth and some lateral distance from a vertical axis through the drill site. This type of drilling became important as offshore production drilling expanded during the 1960s and as many as 40 directional wells were typically drilled from one platform. Early methods of directional drilling required intervals of drilling, measurements of hole angle and inclination, hole correction, drilling, etc. Well bores were deflected off vertical by whipstocks at a kick-off point some 2000 ft below the mudline. The specified orientation of the well was established by multiple whipstock settings. After that, the straight inclined portion of the well was maintained using stabilized drill collars. When hole corrections were needed, adjustments were made to stabilizer locations. This process required numerous trips which were time consuming and costly.

1.4.1 Stabilized Bottom Hole Assemblies

There are basically three types of stabilized bottom-hole assemblies [12]:

- building assembly
- holding assembly
- dropping assembly

They differ primarily in the number and location of stabilizers.

The affect of stabilizer location on magnitude and direction of drill bit side force can be calculated by treating the stabilized section as a multi-spanned beam column (see Ch. 3). The goal is to produce the needed side force on drill bits to maintain or affect a desired well direction.

Prediction of bit side forces produced by various stabilizer assemblies is a good guide to predicting directional performance. Ultimately stabilizer assemblies have to be tried, tested, and usually modified in a given area until they perform as needed. A good data base of BHA performance in a given area is very useful in this regard.

1.4.2 Directional Control Using Downhole Motors

Downhole motors were introduced in the late 1960s as a means of deflecting a direction well away from vertical [30,31] without using whipstocks. A bent sub was placed on top of the motor to bring about lateral drilling (Fig. 1-7a). Tool face is within the bend. It was oriented in a desired plane of drilling to bring about a directional change. This arrangement began to replace whipstocks. Tool face, hole direction, and inclination were measured with a downhole sensor lowered by means of a wire line. Measurements were taken intermittently until the desired hole angle was achieved. Positive Displacement Motors (PDM) emerged as a viable directional tool during this time.

The advantage of using PDMs in directional drilling was huge. It replaced the time consuming, multiple steps of setting a whipstock and surveying to establish hole direction. The bent sub and PBM did the same thing faster. At first, surveys were made by placing a survey tool in a side pocket and using a hard wire to bring hole direction and inclination information back to the surface. Later survey instruments were inserted and retrieved from within the drill collars to track progress. Once the angle had been established, a stabilized bottom hole assembly was used to maintain hole direction to the target. When the bit drilled off course, a bent sub and PDM were put back in to bring the hole back on course.

Eventually, the bent sub was replaced by a bend within the motor housing (Fig. 1-7b). The bent housing configuration brought about a more responsive direction change. This configuration was subsequently replaced by a tilted bearing assembly (Fig. 1-7c) which located the tool face even closer to the drill bit. This configuration produced even a greater rate of directional change, which can be expressed by $\frac{\Delta\theta}{\Delta s} = \frac{\alpha}{L}$. By reducing the distance, L, the rate of directional

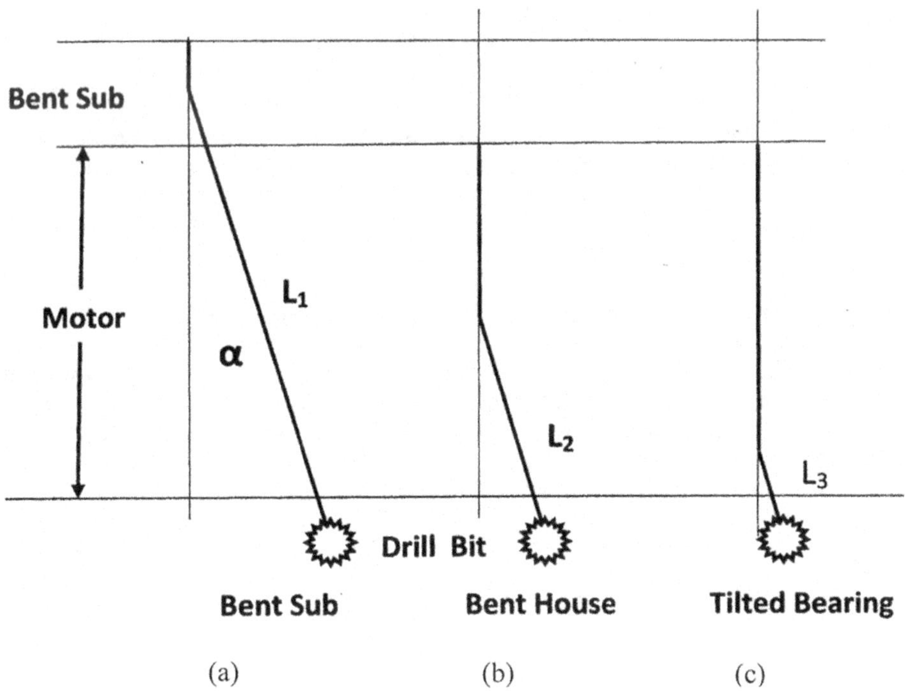

Figure 1-7. Effect of bend location on rate of build.

change is increased. By moving the bend angle close to the bit, the bend could be reduced and still achieve the same rate of build.

The realigned bearing or tilted bearing assembly gave another advantage too. The motor housing itself could be used to drill ahead as well as for making hole corrections. By rotating the motor housing, the drill bit would be continually disoriented and drilling could proceed with motor power. Rotation of the BHA at low speeds simply disoriented the drill bit. When hole corrections were needed, the tool face in the housing was set and the correction was made by motor power. This, along with MWD, eliminated costly tripping time.

This progression set the stage for Rotary Steerable Tools (RST), which do not require stopping for hole corrections. Drilling continues while well bore corrections are made. The drill bit is actually navigated along a predetermined well path.

1.4.2.1 Positive Displacement Motors (PDM)

It has long been recognized that only a small portion of total power (~8000 hp) at the surface is used at the drill bit for boring into a formation. For example, under normal rotary drilling operations (N = 100 rpm, T = 1500 ft-lbs), horsepower delivered to drill bits for boring into a formation is

$$P_{\text{bit}} = \frac{2\pi TN}{33{,}000} \tag{1-20}$$

$$P_{\text{bit}} = \frac{2\pi(1500)100}{33{,}000} = 28.6 \text{ hp},$$

a very small portion of total rig power. By using PDMs, bit rotary speed increases to, say 400 rpm, creating 4 times the power at the drill bit.

$$P_{\text{bit}} = \frac{2\pi(1500)400}{33{,}000} = 114.2 \text{ hp}$$

This increase in drilling power comes at a cost, which has to be weighed against rig rate ($/day) and increase in rate of penetration (ROP).

One of the issues facing the use of PDMs was the need for a companion drill bit which could operate at high rotary speeds. Roller cone drill bits operated successfully up to near 150 rpm. A companion drill bit was needed for the motor. The solution came with Polycrystalline Diamond Compact (PDC) drill bits, which were introduced during the 1970s.

1.4.2.2 Downhole Drilling Turbines

Downhole turbines were introduced into drilling programs before PDMs [30]. They generate power at a much higher rotary speed. Turbine speed is approximately double that of PDMs. Consider, again the power Eq. (1-20). Assuming an output speed of 800 rpm, then

$$P_{bit} = \frac{2\pi(1500)800}{33,000} = 228.4 \text{ hp}$$

a factor of 8 increase over normal rotary speeds of 100 rpm. However, maximum power of drilling turbines is developed at one particular rotational speed and this optimum speed is sometimes difficult to monitor and control.

1.5 MARINE DRILLING AND PRODUCTION RISERS

Marine risers provide a means of reentering well bores below the sea floor. Initially, when wells were drilled in shallow or marshlands, surface casing was extended to the surface of drilling or production platforms. Early platforms were built on piers extending into the sea or from barges towed to location and sunk in marshlands. During the 1930s platforms were erected several miles offshore in about 20 ft of water. Since then drilling and production has moved far beyond the continental shelf where oil is being produced in water depths of 5000 ft.

There are two types of marine risers: drilling risers and production risers. Drilling risers are used to connect drill ships to mudline equipment. They serve as a direct extension of the subterranean well bore back to drill ships. Offshore drilling practices are much the same as inland drilling. Drilling risers are used in exploratory drilling.

Production risers are tied back to offshore platforms, which may be fixed or compliant. These risers are also an extension of the well bore but serve a different purpose. Production risers accommodate production tubing and perform over the life span of reservoirs. As many as 40 production wells may be drilled from one platform and each well requires a production riser. Fatigue life is an important consideration as these tubulars are expected to perform over the reservoir production life (~20 years).

The structural mechanics of both drilling and production risers are the same. They must withstand current and wave forces and support its self. A critical design consideration of these special structures is how they are connected to topside equipment and the wellhead. These are locations of potentially high bending stresses.

Marine riser computer software has been developed over the years to make performance predictions under a variety of operating and environmental conditions. These software packages are based on either finite difference [32] or finite element [33] numerical techniques. These methods provide a means of accounting for stepwise changes in geometry and loads along both riser types. Numerical approaches involve many (often hundreds) of algebraic equations which have to be solved simultaneously. The higher the expected stress level, the higher the number of numerical elements. Reference [34] summarizes capabilities of various software packages.

Closed form solutions offer a more direct approach to riser response and are especially useful for areas of high stress. The challenge is to approximate the real situation with a mathematical model that gives reasonable engineering predictions. Closed form solutions also give a clear picture of the mechanics, which is sometimes lost in numerical approaches.

1.5.1 Hydrodynamic Forces on Cylinders

Fluid forces on cylinders in a flow stream are a major load consideration in the design of offshore structures. Analytical predictions of these forces, based on hydrodynamic theory, are limited to lower Reynold's numbers and laminar flow. At high Reynold's numbers, turbulence develops around pipes obstructing the flow and empirical models are usually used to predict fluid forces. In 1950 Morison, O'Brien, Johnson, and Schaff [35] proposed an empirical formula (Eq. (1-21)) to predict these forces. It has been commonly accepted as a useful formula in designing offshore structures. This equation includes effects of both fluid velocity and acceleration relative to cylinders.

$$F = \frac{1}{2}\rho D C_{\mathrm{d}} |U| U + \frac{1}{4}\pi\rho D^2 C_{\mathrm{m}} \frac{\mathrm{d}U}{\mathrm{d}t} \tag{1-21}$$

F is force per length of the cylinder. Equation (1-21) indicates that total fluid force on cylinders is the algebraic sum of drag and inertia effects. This equation applies directly to the design of fixed offshore structures. However, relative velocity and acceleration between currents (and waves) and marine risers is time dependent, so both terms in Eq. (1-21) have to account for relative motion. The inertia effect may be the result of time dependent wave action or acceleration of pipe relative to the fluid. Studies by Keulegan and Carpenter [36] and Sarpkaya [37,38] add further understanding to this formula.

Drag and inertia coefficients are key to the engineering application of Eq. (1-21). Experimental data [35] show that

$$C_{\mathrm{m}} = 1.508 \pm 0.197$$
$$C_{\mathrm{d}} = 1.626 \pm 0.414$$

Commonly used values of C_{d} and C_{m} are given in Table 1-1.

C_{d} and C_{m} are dimensionless. Reynold's number ranged from 0.22×10^4 to 1.11×10^4 based on maximum surface orbital velocity of waves. Values of C_{d} for cylinders in a steady stream within this range of Reynold's numbers varied from 0.65 to 1.2. Studies conducted by Vandiver [40], show that drag coefficient for long flexible cylinders can reach as high as 3.

By Eq. (1-21), fluid force (per unit length) on a fixed cylinder is

$$F = \frac{1}{2}\rho D C_{\mathrm{d}} U^2 \tag{1-22}$$

Table 1-1. Drag and mass coefficients [38] (based on the diameter of the main tube).

	C_d	C_m
$\mathrm{Re} \prec 10^5$	1.2–2.0	1.5–2.0
$10^5 \prec \mathrm{Re} \prec 10^6$	1.0–2.0	1.5–2.0
$\mathrm{Re} \succ 10^6$	1.0–1.5	1.5–2.0

The inertia is not included because current velocity is assumed to be constant with time and the cylinder is assumed to be stationary.

If, for example, a 20 in OD cylinder is stationary in a fluid moving at 1 knot (1 nautical mile = 6076 ft; 1 knot = 1.1508 mph = 1.688 ft/sec), assuming a drag coefficient of C_d = 1.2, the drag force on the cylinder, according to Eq. (1-22) is

$$F = 0.5\frac{64}{32.2}\frac{20}{12}1.2(1.688)^2 = 5.663\text{lb/ft}$$

This is a significant rate of loading, considering risers are sometimes several hundred feet long.

1.5.2 Vortex Shedding

Under certain conditions of fluid flow around submerged cylinders, vortices form on the sides of cylinders and create cyclic pressure variations transverse to fluid flow.

When a fixed cylinder interrupts fluid flow, boundary layers form and the flow pattern is modified and changes progressively as Reynold's increases. At some fluid, separating occurs. Eventually, with increased fluid velocity, vortices form and break away form the body in a periodic manner as illustrated in Fig. 1-8.

The vortices are characterized by a reduction in fluid pressure, which brings about a pressure difference across the cylinder in the transverse direction. The intensity of this (lift) force is expressed by

$$F_L = C_L\frac{\rho V^2}{2}D \tag{1-23}$$

The lift coefficient, C_L, varies with Reynold's number. Over a wide range of Reynold's numbers (10^2 to 10^7) $C_L \cong 1$; this lift coefficient is greatest at low Reynold's numbers. This means that the intensity of the side force is equal to the stagnation pressure of fluid flow times projected area; den Hartog [41].

Experimental studies show that the frequency of vortex shedding depends on the Strouhal number

$$St = \frac{fD}{V} \tag{1-24}$$

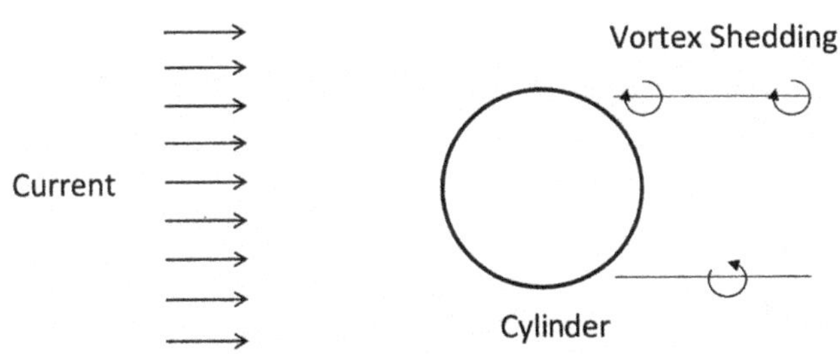

Figure 1-8. Formation of vortices.

which varies with Reynold's number. Over a wide range of Reynold's numbers, the Strouhal number is fairly constant at 0.22. Therefore, frequency of vortex shedding is commonly determined from

$$\frac{fD}{V} = 0.22 \tag{1-25}$$

For example, if V = 1 knot (1.688 ft/sec) and D = 20 in., the frequency of vortex shedding is f = 0.223 cps (T = 4.49 sec). According to Eq. (1-23) the magnitude of the lift force per unit length of the cylinder is

$$F_L = (1)\frac{\rho V^2}{2}D = \frac{(64)1.688^2}{2(32.2)}\frac{20}{12} = 4.718 lb/ft \tag{1-26}$$

This side force varies periodically and can be described mathematically by

$$F(t) = F_L \sin \omega t \tag{1-27}$$

where $\omega = 2\pi f$ and f is the frequency of vortex shedding. Note that the lift force formula is similar to the drag component in Eq. (1-22); only the coefficients differ.

If a cylinder is mounted on elastic supports, the vortex shedding interacts with the natural vibration of the cylinder [42]. Experiments show that when the frequency of vortex shedding is close to the natural frequency of the elastically mounded cylinder, the motion of the cylinder takes control of the vortex shedding, which become synchronized with the natural frequency in violation of the Strouhal number. This lock-in occurs over a range of "reduced velocity" defined by $U_r = \frac{U}{f_n D}$. This range is affected by Re and K-C numbers but in general varies between $4.5 < U_r < 10$ with maximum amplitude falling between $6.5 < U_r < 8$ [43].

In the case of multiple modes, such as occur in marine risers, each natural mode should be compared with the vortex shedding frequency as determined from the Strouhal number. If the shedding frequency is near a natural frequency, vortex shedding could lock-on to the frequency of the mode.

1.5.3 Boundary Conditions

Structurally, both drilling and production risers must be designed to withstand the same oceanographic loads. A fundamental difference is the boundary conditions at top and bottom (Fig. 1-9).

Drilling risers must accommodate the passage of drillstrings and downhole drilling equipment. Flex/ball joints are used at both top and bottom to reduce the angle of rotation at these boundaries to allow passage of drilling equipment. Recommended flex/ball joint stiffness [39] restricts this rotation while lowering bending moments attracted to these structural boundaries. Drilling risers are expected to last over a drilling span of time (~4 months).

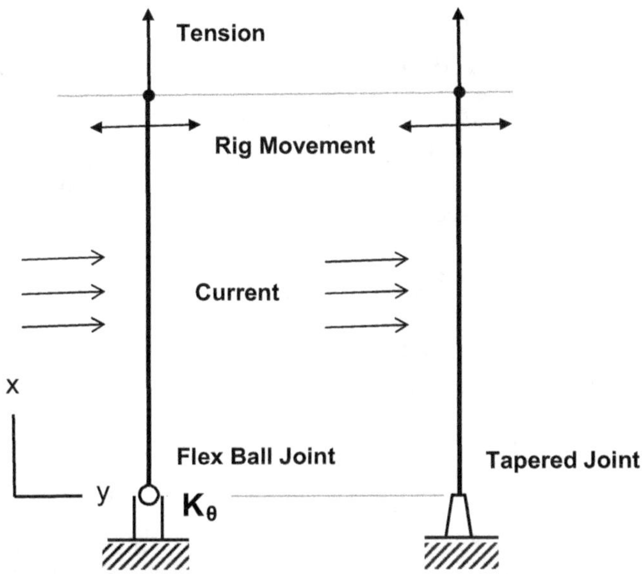

Figure 1-9. Boundary conditions for drilling and production risers.

Production risers must accommodate production tubing and other production equipment which require a smoother transition across the lower boundary. Tapered flex joints provide a good structural transition at a low bending stress level. Production risers are expected to last over the production life of the reservoir, which is typically 20 years or longer.

Structural boundary conditions are critical in mathematical models as local bending moments and angular rotation across these connections could impede tool passage and affect bending stress levels.

Much of the discussion of marine riser behavior in following chapters is based on analytical methods and closed form solutions. The following illustrates the basic equations of bending and one set of boundary conditions for a drilling riser.

1.5.4 Marine Riser Response Prediction

Closed form solutions are emphasized throughout to explain the mechanics of these long beam columns. Advantages to closed form solutions are:

- basic concepts and design tools are easily explained and visualized
- initial riser configurations can be easily scoped out
- solutions can be programmed for small computers

There are, however, practical limitations. A complete assessment of riser performance should be conducted by numerical based software.

1.5.4.1 Closed Form Method of Solution

A method of obtaining closed form solutions for bending of long pipes is presented by Huang and Dareing [44,45]. This approach is illustrated below. Petty [46] also gives a closed form solution for the riser bending problem.

The differential equation of bending for marine risers is developed in Ch. 2. Assuming motion in one plane, the differential equation defining lateral displacement, $y(x.t)$, is expressed by

$$EI\frac{\partial^4 y}{\partial x^4}-\frac{\partial}{\partial x}\left\{\left[(F_B+wx)+(L-x)(A_0\gamma_0-A_i\gamma_i)\right]\frac{\partial y}{\partial x}\right\}+c\frac{\partial y}{\partial t}+m\frac{\partial^2 y}{\partial t^2}=F(x,t) \tag{1-28}$$

This equation is being introduced here to point out the uniqueness of the mathematical problem. The term

$$T_{eff}=\left[(F_B+wx)+(L-x)(A_0\gamma_0-A_i\gamma_i)\right] \tag{1-29}$$

defines effective tension. The real tension $(F_B + wx)$ that is internal to the riser is only part of this expression. The other terms carry the effect of hydrostatic fluid pressure inside and outside of the riser. For drilling risers, the inside fluid is drilling mud while the outside fluid is sea water. This difference in fluid densities has a huge effect on the static and dynamic behavior of risers. Since this term occupies the space in the differential equation reserved for tension, the entire expression has been named, effective tension.

Only a static case of rig offset will be considered here. Other cases are covered in Ch. 4. For the sake of simplicity and numerical convenience, Eq. (1-28) can be casted in terms of dimensionless variables and parameters as

$$\frac{d^4Y}{d\zeta^4}-\frac{d}{d\zeta}\left[(\beta+\alpha\zeta)\frac{dY}{d\zeta}\right]=0 \tag{1-30}$$

where

$$\alpha=\frac{(w-A_0\gamma_0+A_i\gamma_i)L^3}{EI}$$

$$\beta=\frac{(F_B+LA_0\gamma_0-LA_i\gamma_i)\,L^2}{EI}$$

The solution to Eq. (1-30) cannot be expressed in terms of traditional trigonometric functions because of the variable coefficient $(\beta + \alpha\zeta)$. It will be shown that its solution takes the form of a power series

$$y(\varsigma)=\sum_0^\infty a_n\varsigma^n \tag{1-31}$$

The details of this approach are explained in Ch. 2.

1.5.4.2 Riser Displacements due to Rig Offset

The method of solution explained in Ch. 2 is used to calculate deflections and bending stresses in a particular riser having the following specifications.

water depth = 1000 ft
riser = 22 in. OD × 20.75 in. ID
$E = 29(10)^6$ psi
$I = 2399$ in.^{4}
mud weight = 12 ppg
sea water = 64 lbs/ft^3
flex/ball joint stiffness = 30,000 ft-lbs/deg
bottom tension = 100,000 lbs
rig offset = 50 ft.

Under these conditions

$\alpha = 382.2$
$\beta = 120.4$

Riser displacements (Fig. 1-10) are given for the lower 100 ft ($\zeta = 0.1$), where bending stresses (Fig. 1-11) are the largest. The first four coefficients in Eq. (1-31) are determined to be

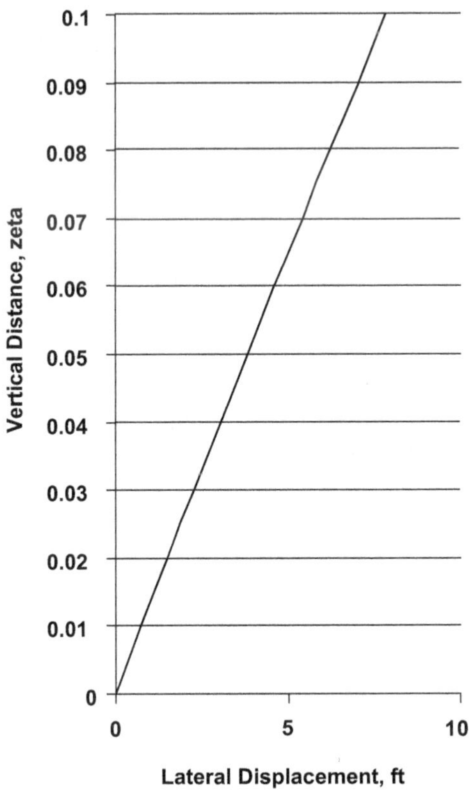

Figure 1-10. Predicted lateral displacement of drilling riser in 1000 ft of water (rig offset is 50 ft, $K_\phi = 30{,}000$ ft-lb/deg; $\zeta = 0.1$ corresponds to x = 100 ft).

$$\begin{aligned} a_0 &= 0 \\ a_1 &= 72.17 \\ a_2 &= 128.38 \\ a_3 &= -882.1 \end{aligned}$$

The remaining coefficients are related to these coefficients. These coefficients are used to determine slope, moment and shear at the lower end as explained below.

Similar calculations are made for ocean currents in Chapter 4. The slope of the deflection curve along the riser is determined by

$$\frac{dy}{dx} = \frac{1}{L}\frac{dy}{d\varsigma} = \frac{1}{L}\sum_{1}^{\infty} a_n n \varsigma^{n-1} \tag{1-32}$$

The amount of angular rotation across the flex/ball joint is

$$\frac{dy}{dx}(0) = \frac{a_1}{L} \tag{1-33}$$

$$\Delta\theta = \frac{72.17}{1000} = 0.0722 \qquad \text{radians or } (4.137 \text{ deg})$$

The bending moment across the flex joint is

$$M(0) = 30{,}000(4.135) = 124{,}050 \text{ ft-lb}$$

The bending moment distribution (Fig. 1-11) along the riser is determined from

$$M = \frac{EI}{L^2}\frac{d^2y}{d\varsigma^2} = \frac{EI}{L^2}\sum_{2}^{\infty} a_n n(n-1)\varsigma^{n-2} \tag{1-34}$$

Note the point of inflection near $\zeta = 0.073$ Applying Eq. (1-34) to the lower end ($x = 0$) gives

$$M = \frac{EI}{L^2}2a_2 = \frac{29(10)^6 2399}{1000^2}\left|\frac{1\text{ft}^2}{144\text{in}^2}\right| 2(128.38)$$

$$M = 124{,}050 \text{ ft-lb}$$

which agrees with the above calculation based on $\Delta\theta$ across the flex joint.

Bending stresses can be determined from

$$\sigma_b = \frac{Ec}{L^2}\sum_{2}^{\infty} a_n n(n-1)\varsigma^{n-2} \tag{1-35}$$

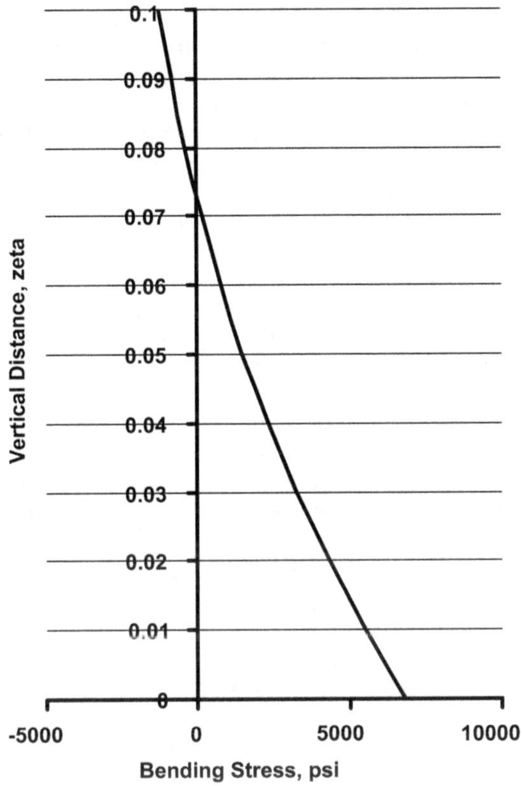

Figure 1-11. Bending stress over lower 100 ft of riser pipe. Water depth = 1000 ft, rig offset = 50 ft.

The shear force transmitted to the mudline equipment is determined from

$$V = \frac{dM}{dx}(0) - F_b \frac{dy}{dx}(0) \tag{1-36}$$

$$V = \frac{EI}{L^3}\frac{d^3y}{d\zeta^3}(0) - \frac{F_B}{L}\frac{dy}{d\zeta}(0)$$

$$V = \frac{EI}{L^3} 6a_3 - \frac{F_B}{L} a_1$$

$$V = \frac{29(10)^6 2399}{1000^3} \left| \frac{1\,\text{ft}^2}{144\,\text{in}^2} \right| 6(-882.1) - \frac{100{,}000}{1000}(72.17) = -9774\ \text{lbs}$$

The minus is consistent with the sign convention used throughout for shear. It means that the direction of the shear force applied to the bottom of the riser is to the left and the shear force applied to the well head is to the right. Since, in this particular example, there are no intermediate side loads, this same force is also applied to the motion compensation support equipment. Riser buckling, a separate topic, should also be considered during the early stages of design.

Production risers are slightly different in purpose and design. Operational loads are different even though environmental loads are the same as for drilling risers. Fluid densities are different and usually production tubulars are attached to them. They are expected to stay on location over a relatively longer life cycle so fatigue considerations are critical considerations. Dynamic loads and induced stresses became especially important.

1.5.5 Dynamic Response

Static displacements provide a baseline from which to reference riser performance. This is a reasonable start to any riser design. Real time-response to rig motion, currents, and waves are often accounted for by a quasi-static approach where real time response is determined on a time step basis. Basically, inertia is ignored in this approach. However, there are situations where inertia becomes important, such as when force frequencies are close to or tuned to a natural frequency.

There are two types of vibrations: forced vibrations and self excited vibrations. Both are possible in marine risers. Forces that are independent of system response cause forced vibrations; if the vibration stops, the force may continue. There are traditional methods for predicting response in this case. A good example is wave action against a riser. If the risers were fixed, the wave force against the riser would continue. The solution here, if possible, is to separate natural frequency from the frequency of the force.

A system is self-excited if driving force depends on vibration response. Vortex-induced vibration is a good example. The frequency of the driving force depends on the motion. As pointed out earlier, the frequency of vortex shedding (the driving frequency) can lock on to a natural frequency and feed energy into given modes. The solution here is to break up the feedback mechanism; fairings are often used for this purpose. Accurate response predictions depend on damping and effective mass. The drag and mass force components given by Eq. (1-21) are usually used to account for these two factors. If a cylinder is fixed, vortex shedding frequency follows the Strouhal number.

When lock-on occurs, large amplitudes and cyclic bending stresses are produced, sometimes near the lower end where effective tension is lowest. Higher modes with higher natural frequencies can accumulate fatigue cycles at a higher rate. Much work has been conducted to predict riser response under lock-on and shear flow conditions [47–51]. Useful software programs are available to predict riser response and fatigue life under VIV conditions [42,52].

1.6 MATURITY OF DRILLING AND PRODUCTION TECHNOLOGY

Four significant technologies, developed in recent years, greatly advanced oil exploration and production.

1. Mapping of underground geologic structures
2. Computer modeling of oil and gas reservoirs

3. Navigation drilling into complex geologic reservoirs with Rotary Steerable Tools
4. Deep Ocean Platforms

Geophysics technology is capable of creating an accurate picture in three-dimensional space of underground geologic formations using computer-based models. Input to these models comes from improved geophysical equipment, data collection, and formation mapping. This technology allows potential oil reservoirs to be mapped accurately regardless of the complexity of the formations. Furthermore, underground formation can be accurately mapped whether on land or offshore. This technology is currently used to find promising geologic structures below the floor of deep ocean waters.

The computer became a research and engineering tool during the 1960s. High-speed computers along with advanced software provide a means to develop production plans for reservoirs for maximum recovery. This information sets the technical specifications for offshore production design and reservoir development programs necessary for optimum oil and gas recovery.

The third technology relates to drilling laterally along a prescribed three-dimensional path in order to penetrate a reservoir at a particular location. These penetration points are prescribed by reservoir analysis and are critical for optimum oil and gas production. Rotary steerable tools, coupled with MWD, can navigate continuously along a specified path and intersect a target, several thousand feet away, within a few feet.

Offshore compliant production platforms are now in place in water depths of 5000 ft and greater. A high level of reliability has been established in the design, fabrication, and installation of these structures through testing and computer models. Marine production risers remain a critical part of these offshore systems. The technology is in place to produce oil from almost any remote geographic and geologic locations.

Ledgerwood [53] gives an excellent overview of international efforts to find new and improved drilling methods. Studies by Maurer [54] on novel drilling methods show that rotary drilling is still the best way to bore through rock. The drillstring is central to this boring method. Also, marine risers are a necessary and vital linkage between well bore and drill ship or production platform. Both of these tubulars are linked through mechanics. The next chapter gives the mathematical foundation for predicting lateral behavior of these tubulars.

REFERENCES

1. Bernt S. Andnoy, Iain Cooper, Stefan Z. Miska, Robert F. Mitchell, Michael L. Payne; *Advanced Drilling and Well Technology*, Society of Petroleum Engineers, 2009.
2. Yergin, D, *The Prize*, (1991) Simon & Schuster, NY.
3. API RP 7G; API Recommended Practice for Drill Stem Design and Operating Limits.
4. Brantly, J.E., 1971, *History of Oil Well Drilling*, Gulf Publishing Company; also see *History of Petroleum Engineering*, API, 1961.

5. Lubinski A., "A study of the Buckling of Rotary Drilling Strings", Drilling and Production Practice (API), **17**, 1950, pp. 178–214.
6. Todhunter I. and Pearson, K.; *History of the Theory of Elasticity*, Vol. 1, pp. 39–50, Cambridge, 1886.
7. Stephen P. Timoshenko and James M. Gere; *Theory of Elastic Stability*, McGraw-Hill Book Company, Inc., NY 1961, pp. 101–103.
8. Greenhill G., *Proc. Cambridge Phil. Soc.*, Vol. 4, 1881.
9. Lubinski A. and Woods, H. B. "Factors Affecting the Angle of Inclination and dog-legging in rotary bore holes," Drilling and Production Practice (API), **20**, 1953, pp. 222–250.
10. MacDonald, G. C. and Lubinski, A.: "Straight-Hole Drilling in Crooked-Hole Country", Drill. And Prod. Prac., API (1951) p. 80.
11. Woods, H. and Lubinski, A.: "Use of Stabilizers in Controlling Hole Deviation", Drill. And Prod. Prac., API (1955) p. 165.
12. Bruce H. Walker; "Some Technical and Economic Aspects of Stabilizer Placement", *Journal of Petroleum Technology*, June 1973.
13. Development in Petroleum Engineering; Collected works of Arthur Lubinski, Edited by Stefan Miska, Gulf Publishing Company, Houston, 1987.
14. Kellner, J. M. and Roberts, A. P.; "Hydraulic Wall-Anchored Drill Collar Promises Lower Drilling Costs," *Oil & Gas Journal*, Vol. 58, No. 40, October 3, 1960.
15. Garrett, W. R. and Rollins, H. M.; "New Deviation Control Looks Good in Field Tests," *World Oil*, October, 1964.
16. Rollins, H. M.; "Deviation Control Using Fluid-Operated Bit-Guiding Tools," API Division of Production, April 1966.
17. Huang T. and Dareing D. W.; "Buckling and Lateral Vibration of Drill Pipe," Trans. *ASME, Journal of Engineering for Industry*, Vol. 90, Series B, No. 4, 1968, pp. 613–619.
18. Placido, Joao Carlos R., Helio M. R. Santos and Yadira Diaz Galeano, "Drillstring Vibration and Wellbore Instability," *Journal of Energy Resources Technology*, December 2002, Vol. 124, pp. 217–222.
19. Finnie I. and Bailey J. J.; "An Experimental Study of Drill-String Vibration," *Transactions of the ASME, Journal of Engineering for Industry*, No. 129, May 1960.
20. Bailey J. J. and Finnie I.; "An Analytical Study of Drill-String Vibration," *Transactions of ASME, Journal of Engineering for Industry*, No. 129, May 1960.
21. Paslay, P. R. and Bogy, D. B.; "Drill String Vibrations Due to Intermittent Contact of Bit Teeth," *ASME Journal of Engineering for Industry*, Vol. 85, No. 2, May 1963.
22. Plunkett, R. "Mechanical Impedance Methods", American Institute of Mining Met. And Petroleum Engr. Colloquium Pub. 1958.
23. Garrett, W. R.; "Effect of a Downhole Shock Absorber on Drill Bit and Drill Stem Performance, Paper ASME 62-Pet-21 presented at the ASME 1962 Petroleum and Mechanical Engineering Conference, Dallas, Sept 23–26, 1962.
24. Spanos, P. D., Payne, M. L., and Secora, C. K.; "Bottom-Hole Assembly Modeling and Dynamic response Determination", *Journal of Energy Resources Technology*, Sept. 1997, Vol. 119, pp. 153–158.

25. Personal discussions with Bill Garrett (1980).
26. Deily, F. H., Dareing, D. W., Paff, G. H., Ortloff, J. E., and Lynn, R. D.; "Downhole Measurements of Drill String Forces and Motions," *Transactions of ASME Journal of Engineering for Industry*, Vol. 90, Series B., No. 2, May 1968, 217–225.
27. Dareing, D. W.: "Downhole Measurements of Drill-String Forces and Motions", Esso Production Research Report, No. EPR65-PR12, October 28, 1965.
28. Dareing, D. W. and Livesay, B. J.; "Longitudinal and Angular Drillstring Vibration with Damping," *Transactions of the ASME Journal of Engineering for Industry*, Vol. 90, Series B, No. 4, 1968, pp. 671–679.
29. Cunningham, R. A.; "Analysis of Downhole Measurements of Drill String Forces and Motions", *ASME Journal of Engineering for Industry*, May 1968, pp. 208–216.
30. Garrett, W. R. and Rollins, H. M.: "A Steering Wheel for Rock Bits", Paper presented at *19th Annual Petroleum Mechanical Engineering Conference*, Los Angeles, Calif. (Sept. 20, 1964).
31. C. E. Murphey, and J. B. Cheatham, Jr.; "Hole Deviation and Drill String Behavior," *Society of Petroleum Engineers Journal* (March 1966) 44–54; Trans., AIME, 237.
32. Burke, B. G.; "An Analysis of Marine risers for Deep Water", *Journal of Petroleum Technology*, Apr. 1974.
33. Gardner, T. N. and Kotch, M. A., "Dynamic Analysis of risers and Caissons by the Element Method', *Offshore Technology Conference*, OTC 2651, 1976.
34. API Bulletin 16J, "Comparison of Marine Drilling Riser Analysis."
35. Morison, J.R., O'Brien, M.P., Johnson, J.W. and Schaaf, S.A. (1950), "The Forces Exerted by Surface Waves on Piles," *Petroleum Trans., AIME*, **189**, 149–157.
36. Keulegan, G. H. & Carpenter, L. H. (1958), "Forces on Cylinders and Plates in an Oscillating Fluid," *J. Res. Nat. Bur. Standards*, **60**, 423–440.
37. Sarpkaya, T. (1976) Vortex Shedding and resistance in harmonic flow about smooth and rough circular cylinder at high Reynold's number Tech. Rep. No. NPS-59SL76021, Naval Post Graduate School, Monterey, CA.
38. Sarpkaya, T. (1986) "Force on a circular cylinder in viscous oscillatory flow at low Keulegan-Carpenter numbers, J. Fluid Mech. Vol. 165, pp. 61–71.
39. API Recommended Practice 16Q (RP 16Q), Aug 2001.
40. J. Kim Vandiver, "Drag Coefficients of Long flexible Cylinders," Offshore Technology Conference paper, OTC 4490, 1983.
41. Den Hartog, J. P., *Mechanical Vibrations*, 4th Edition, McGraw-Hill 1956.
42. Le Cunff, C., Biolley, F., Fontaine E., Etienne, S., and Facchinetti M. L.; "Vortex-Induced Vibrations of Risers: Theoretical, Numerical and Experimental Investigation", Oil & Gas Science and Technology–Rev. IFP, Vol. 57 (2002), No. 1, pp. 59–69.
43. Sarpkaya, T.; "Vortex-Induced Oscillations", *Journal of Applied Mechanics*, June 1979, Vol. 46, pp. 241–258.
44. Huang, T. and Dareing, D. W.; "Buckling and Lateral Vibration of Drill Pipe," *Transactions of the ASME, Journal of Engineering for Industry*, Vol. 90, Series B., No. 4, 1968, pp. 613–619.

45. Dareing, D. W. and Huang, T.; "Natural Frequencies of Marine Drilling Risers," *Journal of Petroleum Technology,* July 1976, pp. 813–818.
46. Petty, T. D.; "Analysis of Marine risers Used with floating Drilling Vessels", MS Thesis, University of Oklahoma, 1970.
47. Stansby, P. K.; "The locking-on of vortex shedding due to the cross-stream vibration of circular cylinders in uniform and shear flows", *J. Fluid Mech.* (1976) Vol. 74, part 4, pp. 641–665.
48. Griffin, O. M. and Ramberg S. E.; "Some Recent Studies of Vortex shedding with Application to Marine Tubulars and Risers", ASME, JERT, March 1982, Vol. 104, pp. 2–13.
49. Sarpkaya, Turgut; "Force on a circular cylinder in viscous oscillatory flow at low Keulegan-Carpenter numbers", *Journal of Fluid Mechanics* (1986), Vol. 165, pp. 61–71.
50. Vandiver, J. K.; "Dimensionless Parameters Important to the Prediction of Vortex-Induced Vibrations of Long, Flexible Cylinder in Ocean Currents," *J Fluids and Structures* (1993), pp. 423–455.
51. Vandiver, J. K., Swithenbank, S., Jaiswal, W., and Jhingran, W.; "Fatigue from High Mode Number Vortex-Induced Vibration", *Proceedings of OMAE 2006: 25the International Conference on Offshore Mechanics and Arctic Engineering*, Paper No. OMAE2006-9240, June 4–9, 2006, Hamburg, Germany.
52. Vandiver, J. K.; "A Universal Reduced Damping Parameter for Prediction of Vortex Induced Vibration, *21st International Conference on OMAE*, Paper 28292, June 23–28, 2002, Oslo.
53. Ledgerwood, Jr., L. W.; "Efforts to Develop Improved Oilwell Drilling Methods", J. of Petroleum Technology, 1960.
54. William C. Maurer; *Advanced Drilling Techniques*, Petroleum Publishing Co., Tulsa, OK, 1980.

Chapter 2

MECHANICS OF LONG BEAM COLUMNS

Beam columns are defined as structural beams subjected to axial and transverse loads. Buckling of columns is often analyzed separately in consideration of boundary conditions. Axial forces are assumed to be constant throughout. Beam response to these loads, whether considered separately or combined, is well documented [1,2].

Mechanics of long vertical beam columns is similar but complicated by linear changes in internal axial forces (caused by gravity) and variations in fluid pressure (internal and external) applied along the beam (pipe). Load distribution, due to the weight and hydrostatic pressure, poses a unique mathematical problem not usually addressed in classical beam column mechanics.

Drillstrings and marine risers, whose lengths are measured in terms of hundreds and thousands of feet, are good examples of long beam columns. Drillstrings, for example, bend in well bores from a combination of gravity, fluid pressure and forces applied directly to a drill bit. Side loads may be applied from the well bore through different tools which make up part of drillstrings. Responses to each of these loads may be either static or dynamic.

Marine risers are subjected to wave and current action and are held in place by top tension, which also controls bending. Internal tension varies from a maximum at the rig to a smaller value at the well head. Marine risers respond to these loads with static and dynamic lateral motion. Motion compensators maintain top tension to offset heave motion of the drill ship.

The differential equation of bending is the same for both drillstrings and marine risers, and for this reason, it is appropriate to discuss their mechanics together. The differential equation of bending is developed in this chapter. Closed form solutions to several situations are given. Specific applications of this technology to drillstrings and marine risers are discussed in following chapters.

2.1 BUCKLING DUE TO WEIGHT OF VERTICAL COLUMN

Buckling of vertical beams caused by gravity was first considered by Greenhill [3]. Euler recognized the potential of buckling due to gravity and formulated the differential equation of bending, but did not obtain a satisfactory solution [4]. Greenhill solved the differential equation of bending and was able to predict critical lengths that cause buckling. Willers [5] expanded the analysis by including different boundary conditions along with an externally applied force. Greenhill's solution is summarized below [25].

Figure 2-1 illustrates the problem and gives the coordinate system used in the analysis. Summing forces and moment on the differential element gives

$$EI\frac{d^3y}{dx^3} = -w(L-x)\frac{dy}{dx} \tag{2-1}$$

By changing variables using,

$$z = \frac{2}{3}\left[\frac{w}{EI}(L-x)^3\right]^{\frac{1}{2}} \tag{2-2}$$

and

$$\frac{dy}{dz} = u \tag{2-3}$$

The basic differential equation converts to

$$\frac{d^2u}{dz^2} + \frac{1}{z}\frac{du}{dz} + \left(1 - \frac{1}{9z^2}\right)u = 0 \tag{2-4}$$

The solution to Eq. (2-4) is expressed in terms of Bessel functions [6] of order one third. Applying boundary conditions gives a critical value of z = 1.866.

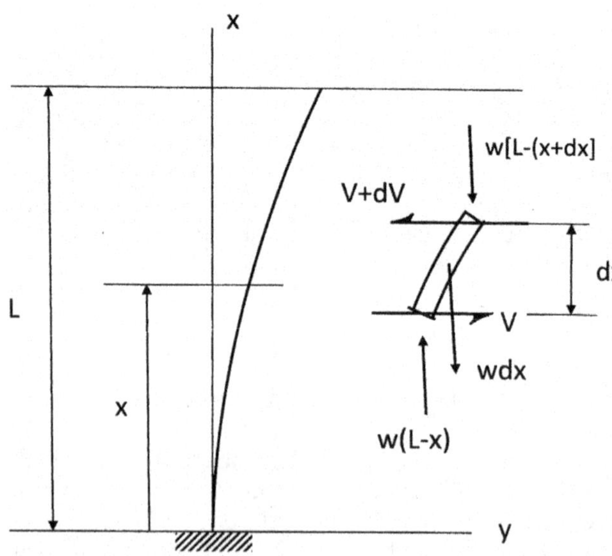

Figure 2-1. Vertically cantilevered beam loaded by its own weight.

This number converts to a lowest critical buckling load of

$$(wL)_{cr} = \frac{7.837EI}{L^2} \tag{2-5}$$

or a critical length of the column

$$L_{cr} = \sqrt[3]{\frac{7.837EI}{w}} \tag{2-6}$$

where w is weight per length.

Timoshenko [4] gives an approximate method based on the principle of minimum potential energy where stability is determined from the second derivative of the total potential energy expression.

$$V = U + \Omega \tag{2-7}$$

where

V is total potential energy
U is elastic potential
Ω is potential of external forces (in this case, gravity potential)

For the particular boundary condition shown in Fig. 2-1, the lateral deflections can be approximated by

$$y = \delta\left(1 - \cos\frac{\pi x}{2L}\right) \tag{2-8}$$

The elastic potential energy is

$$U = \int_0^L \frac{M^2}{2EI}dx \tag{2-9}$$

where

$$M(x) = \int_x^L w\,(\eta - y)d\xi \tag{2-10}$$

and

$$\eta(\xi) = \delta\left(1 - \cos\frac{\pi\xi}{2L}\right) \tag{2-11}$$

The integration of Eq. (2-10) gives

$$M = 2w\left[(L - x)\cos\frac{\pi x}{2L} - \frac{2L}{\pi}\left(1 - \sin\frac{\pi x}{2L}\right)\right] \tag{2-12}$$

After substitution and integration, Eq. (2-9) gives

$$U = \frac{\delta^2 w^2 L^3}{2EI}\left(\frac{1}{6} + \frac{9}{\pi^2} - \frac{32}{\pi^3}\right) \tag{2-13}$$

The reduction in gravity potential is

$$\Omega = \frac{1}{2} w \int_0^L \left(\frac{dy}{dx} \right)^2 (L - x) dx \tag{2-14}$$

$$\Omega = \frac{\pi^2 \delta^2 w}{8} \left(\frac{1}{4} - \frac{1}{\pi^2} \right) \tag{2-15}$$

For stable equilibrium $\frac{dV}{d\delta} = 0$ and $\frac{d^2V}{d\delta^2} \geq 0$. Therefore wL is critical when

$$(wL)_{cr} = \frac{7.89EI}{L^2} \tag{2-16}$$

which is very close to the solution (Eq. (2-5)) developed by Greenhill. The energy method offers a somewhat simpler approach for other boundary conditions as well.

2.2 OSCILLATIONS OF A HANGING CHAIN

This topic provides an introduction to the more practical subject of the vibration of steel catenaries. The model deals only with the effects of gravity and introduces the complexity of predicting responses of very long pipe. Hydrostatic fluid pressure is covered later.

2.2.1 Natural Frequency of a Freely Hanging Chain

In this model (Fig. 2-2) tension is greatest at the top and zero at the lower end. The analysis shows the effect of gravity and inertia on the differential equation of motion and vibration response.

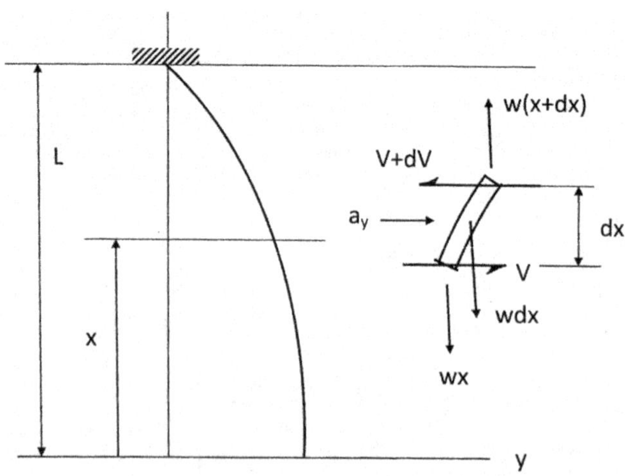

Figure 2-2. Oscillation of a hanging chain.

The differential equation of motion is determined from the freebody diagram shown in Fig. 2-2. Summation of forces in the y direction is equal to the product of the differential mass and acceleration in the y direction ($mdxa_y$). Combining the force components gives

$$\frac{\partial}{\partial x}\left(wx\frac{\partial y}{\partial x}\right)=\frac{w}{g}\frac{\partial^2 y}{\partial t^2} \tag{2-17}$$

which is rewritten as

$$x\frac{\partial^2 y}{\partial x^2}+\frac{\partial y}{\partial x}=\frac{1}{g}\frac{\partial^2 y}{\partial t^2} \tag{2-18}$$

The solution proceeds by assuming a solution of the form [7]

$$y(x,t)=e^{i\omega t}v(x) \tag{2-19}$$

giving

$$x\frac{d^2 v}{dx^2}+\frac{dv}{dx}+\frac{\omega^2}{g}v=0 \tag{2-20}$$

This equation can be reformatted into a Bessel equation of order zero. Applying boundary conditions, the eigenvalues are the roots to

$$J_o\left(2\omega\sqrt{\frac{L}{g}}\right)=0 \tag{2-21}$$

which are 2.4048, 5.5201, 8.6537, 11.7915, 14.9309, etc. It follows that the natural frequencies are determined from

$$\omega_1=\frac{2.405}{2}\sqrt{\frac{g}{L}} \tag{2-22a}$$

$$\omega_2=\frac{5.52}{2}\sqrt{\frac{g}{L}} \tag{2-22b}$$

$$\omega_3=\frac{8.654}{2}\sqrt{\frac{g}{L}},\ \text{etc.} \tag{2-22c}$$

Table 2-1 gives natural frequencies of hanging chains of different lengths for the first vibration mode based on Eq. (2-22a).

Table 2-1. Fundamental natural frequencies of hanging chains.

L (ft)	ω (rad/sec)	*f* (cps)
500	0.305	0.0485
1000	0.216	0.0343
1500	0.176	0.028
2000	0.153	0.0244

2.2.2 Natural Frequencies of a Hanging Chain Constrained at the Lower End

A hanging chain with constraints shown in Fig. 2-3 was considered by Huang and Dareing [8]. This model incorporates pinned-pinned boundary conditions with bottom tension. The objective was to determine the natural frequencies of lateral vibration corresponding to different values of d and L. Equation (2-18) still applies. This equation can be reorganized into a Bessel type differential equation as shown above. Applying boundary conditions, the solution can be expressed as a combination of both Bessel functions: (J_o) of order zero of the first kind and (K_o) of order zero of the second kind.

Because of these particular boundary conditions and the availability of high speed computers, it is convenient to assume a solution in the form of a power series. The power series approach is useful for solving beam problems where effective tension varies linearly with depth. This approach will be used later to predict natural frequencies and buckling of elastically flexible pipes.

Assuming a solution of the form

$$y(x,t) = v(x)\sin\omega t \tag{2-23}$$

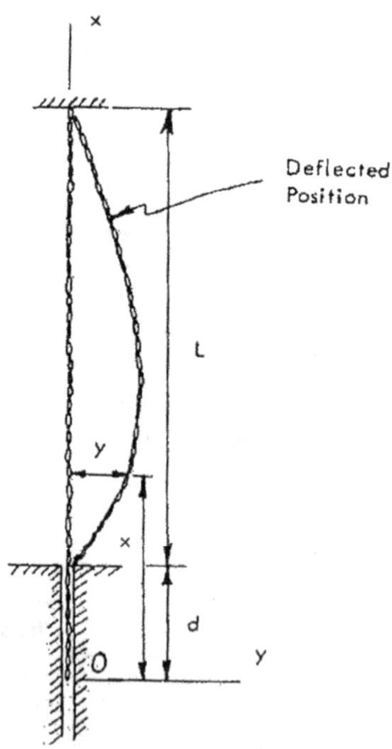

Figure 2-3. Hanging chain with lower boundary constraint [8]. (Reprinted with permission from the Acoustical Society of America).

gives

$$v\frac{d^2v}{dx^2}+\frac{dv}{dx}+\frac{\omega^2}{g}v=0 \tag{2-24}$$

By letting

$$\lambda=\omega\left(\frac{L}{g}\right)^{\frac{1}{2}} \tag{2-25}$$

and

$$\xi=\frac{x}{(L+d)} \tag{2-26}$$

Eq. 2-24 becomes

$$\xi\frac{d^2v}{d\xi^2}+\frac{dv}{d\xi}+\lambda^2\left(1+\frac{d}{L}\right)v=0 \tag{2-27}$$

The solution to this equation can be represented by the power series

$$v=\sum_{n=0}^{\infty}a_n(1-\xi)^n,\ \text{for } \xi>0 \tag{2-28}$$

where a_0 and a_1 are two arbitrary constants and

$$a_n=\frac{n-1}{n}a_{n-1}-\frac{\lambda^2(1+d/L)}{n(n-1)}a_{n-2} \tag{2-29}$$

Boundary conditions imposed on the power series are

$$v=0 \text{ at } \xi=1$$

and

$$v=0 \text{ at } \xi=\frac{d}{L+d}$$

These two boundary conditions are satisfied by $a_0 = 0$ and

$$\sum a_n\left[\frac{L}{(L+d)}\right]^n=0,\quad \text{with } \frac{d}{L}>0 \tag{2-30}$$

Let the arbitrary constant, $a_1 = 1$. Incorporating Eq. (2-29) into Eq. (2-30) gives the characteristic equation, whose solution gives the eigenvalues, λ. Values for λ for the first three modes are given in Table 2-2.

When $d/L = 1$, the bottom tension is equal to the weight of the chain between top and bottom confinement. When $d/L = 0.5$, the bottom tension is equal to 1/2 the weight of the chain between top and bottom confinement points, etc.

Figure 2-4 gives the first three eigenvalues, λ, plotted against d/L. The natural frequencies for a uniform tensioned string are determined from ref. [12]

Table 2-2. Lowest three natural frequencies.

d/L	λ_1	λ_2	λ_3
0.01	1.648	3.410	5.160
0.02	1.737	3.570	5.391
0.05	1.911	3.891	5.861
0.1	2.108	4.268	6.418
0.2	2.400	4.834	7.261
0.5	3.023	6.063	9.100
1.0	3.787	7.582	11.375
2.0	4.940	9.883	14.826
5.0	7.359	14.720	22.080
10.0	10.177	20.354	30.531
20.0	14.223	28.446	42.669
50.0	22.325	44.650	66.975
100.0	31.494	62.989	94.483

$$\omega = \frac{n\pi}{L}\sqrt{\frac{T}{m}}, \quad \text{n} = 1,2,3... \tag{2-31}$$

and plotted in Fig. 2-4 for comparison.

When $d/L = 1$ (Fig. 2-4), eigenvalues for a chain of distributed weight and a chain under uniform tension are essentially the same. Uniform tension is taken as the average tension between top and bottom tension.

For example, if $L = 1000$ ft and assuming a d/L ratio of 1, Table 2-2 shows

$\lambda_1 = 3.787$
$\lambda_2 = 7.582$
$\lambda_3 = 11.375$

Corresponding values of frequency (using Eq. (2-25)) are

$\omega_1 = 0.68$ rad/sec
$\omega_2 = 1.36$ rad/sec
$\omega_3 = 2.04$ rad/sec

2.3 SHORT BEAMS UNDER UNIFORM TENSION

The previous discussion of chains did not include bending stiffness (EI) or hydrostatic pressure. These factors will now be included starting with a discussion of short beams.

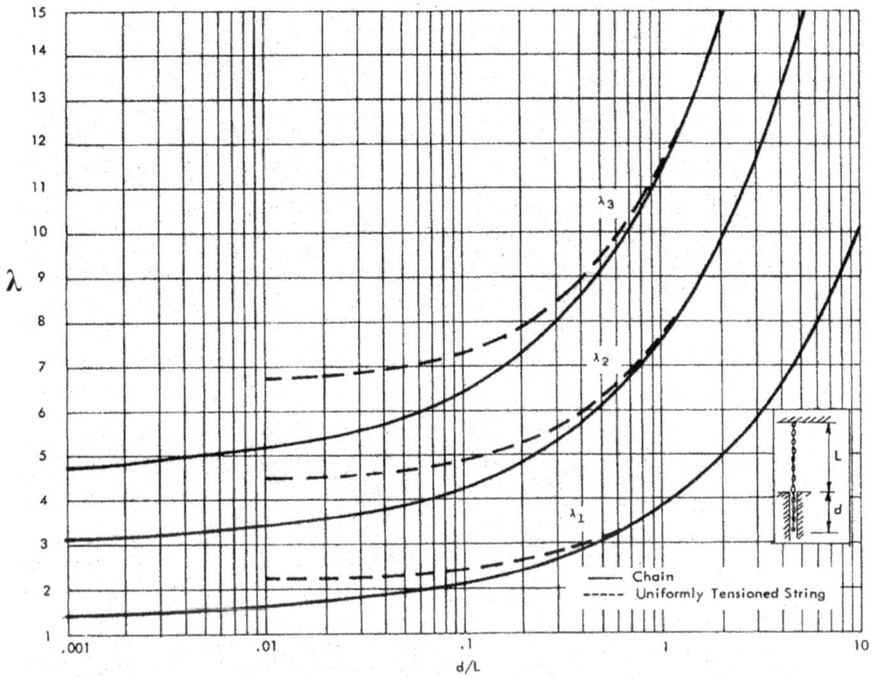

Figure 2-4. Comparison of natural frequencies for variable and uniform tension [8]. (Reprinted with permission from the Acoustical Society of America).

A short beam is defined as having constant internal tension over beam length. Also, in the case of pipe, fluid pressures, both inside and out, are constant throughout. In some cases, the short beam model may be adequate for reasonable engineering results. Possible applications are drill pipe within marine risers and drill collar sections between stabilizers. The mathematics is greatly simplified and solutions can be found in the literature [1]. However, effects of internal and external pressure are usually not included.

2.3.1 Constant Inside and Outside Pressure with Uniform Tension

The general equation of motion for short beams, including internal and external pressures, is developed below.

Consider the freebody diagram in Fig. 2-5. Equating force components in the y direction to $ma_y dx$ gives

$$\frac{dV}{dx} = -ma_y + q + P\frac{d^2y}{dx^2} \tag{2-32}$$

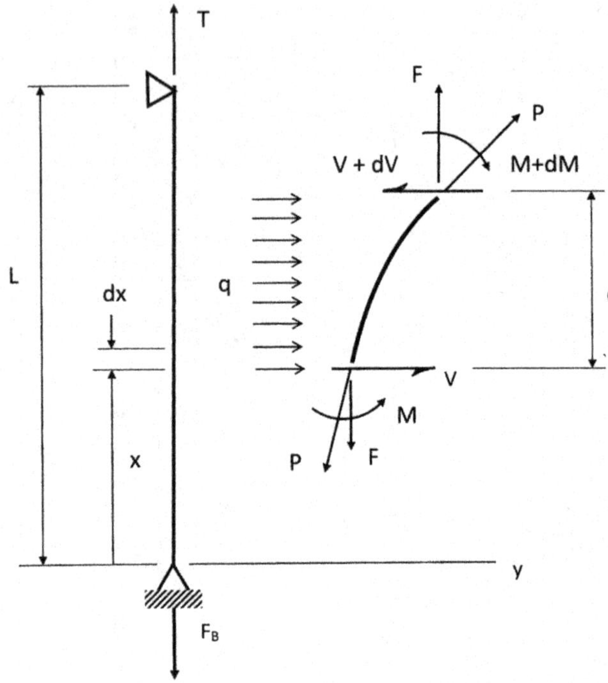

Figure 2-5. Freebody of beam element.

where

q = force per unit length
$P = p_0 A_0 - p_i A_i$
F = axial internal force (actual) within pipe

Summation of moments gives

$$\frac{dM}{dx} - F\frac{dy}{dx} - V = 0 \tag{2-33}$$

Combining Eqs. (2-32) and (2-33) gives

$$\frac{d^2M}{dx^2} - (P+F)\frac{d^2y}{dx^2} + ma_y = q(x) \tag{2-34}$$

Assuming Euler bending, $EI\dfrac{d^2y}{dx^2} = M$, Eq. (2-34) produces

$$EI\frac{\partial^4 y}{\partial x^4}-(P+F)\frac{\partial^2 y}{\partial x^2}+c\frac{\partial y}{\partial t}+m\frac{\partial^2 y}{\partial t^2}=q(x,t) \tag{2-35}$$

For static bending, Eq. (2-35) simplifies to

$$EI\frac{d^4y}{dx^4}-(P+F)\frac{d^2y}{dx^2}=q(x) \tag{2-36}$$

The term $F + P$ is effective tension where F is the actual internal force and P accounts for pressure effects. Effective tension is discussed further in this chapter.

2.3.2 Static Analysis of Short Pipe

The following examples feature the effects of internal and external pressures on the bending and stability of short pipe.

2.3.2.1 Buckling of Short Pipe

Applied loads in this case are: axial force applied at both ends and internal and external fluid pressure. Internal force and fluid pressures are assumed to be constant along the pipe. Equation (2-36) simplifies to

$$EI\frac{d^4y}{dx^4}-(P+F)\frac{d^2y}{dx^2}=0 \tag{2-37}$$

The force F was assumed to be tension in the derivation. Its sign will now be changed and Eq. (2-37) rewritten as

$$EI\frac{d^4y}{dx^4}+(F-P)\frac{d^2y}{dx^2}=0 \tag{2-38}$$

where the force F is now a compressive force and is a positive number. Also, $P = p_0A_0 - p_iA_i$.

Direct integration gives

$$\frac{d^2y}{dx^2}+\beta y=C_1x+C_2 \tag{2-39}$$

where

$$\beta=\frac{F-P}{EI}$$

The symbol, β, is chosen to be consistent with an effective tension term, which will be introduced later.

The particular solution to Eq. (2-39) is

$$y_p = \frac{1}{\beta}(C_1 x + C_2) \tag{2-40}$$

The complimentary solution is

$$y_c(x) = C_3 \sin\sqrt{\beta}x + C_4 \cos\sqrt{\beta}x \tag{2-41}$$

The total solution to Eq. (2-38) is therefore

$$y(x) = C_1 x + C_2 + C_3 \sin\sqrt{\beta}x + C_4 \cos\sqrt{\beta}x \tag{2-42}$$

Assuming the pipe is structurally pinned at top and bottom end, boundary conditions, are

$$y(0) = 0$$
$$y(L) = 0$$
$$\frac{d^2y}{dx^2}(0) = 0$$
$$\frac{d^2y}{dx^2}(L) = 0$$

Applying these boundary conditions to Eq. (2-42) gives

$$C_3 \sin\sqrt{\beta}L = 0 \tag{2-43}$$

From which

$$\sqrt{\beta}L = \pi \tag{2-44}$$

so

$$\frac{(F-P)}{EI} = \left(\frac{\pi}{L}\right)^2 \tag{2-45}$$

or

$$\left[F - (p_0 A_0 - p_i A_i)\right]_{cr} = \left(\frac{\pi}{L}\right)^2 EI \tag{2-46}$$

This is Euler's buckling force for simply supported columns. Note that external pressure stabilizes the beam while internal pressure contributes to instability.

2.3.2.2 *Buckling from Internal Pressure*

A thorough discussion of the effects of internal fluid pressure on pipe buckling is given by Lubinski et al [9]. A brief review is given here as a background for the formulation of effective tension for long pipe.

Consider the buckling of pipe with internal pressure acts alone as illustrated in Fig. 2-6. The surface pressure force is shown in the left drawing. The right element represents the statically equivalent force system which is obtained by adding and subtracting pressure forces over the inside area across an imaginary plane. The net effect of the pressure within an enclosed volume is zero, leaving the end force vectors as shown. The equivalent force system is easier to incorporate into the bending equation.

This is a special case of Eq. (2-38). The solution, using Eq. (2-46) and noting that $P = -p_i A_i$ is

$$p_{cr} A_i = \frac{\pi^2 EI}{L^2} \tag{2-47}$$

The pipe is assumed to be pinned-pinned as shown in Fig. 2-7 and not capped.

If the pipe is capped at the both ends and free of the supports nothing happens because the pressure forces at both ends cancel out the effects of the side pressure. However, if the tubular is supported as shown, the inside pressure can cause the pipe buckling. In this case, forces outside of the supports are not a factor in the lateral buckling of the supported section.

2.3.2.3 *Uniformly Distributed Side Load*

Now consider a short beam column with side loading. In this case Eq. (2-36) becomes

$$EI \frac{d^4 y}{dx^4} - (F + P)\frac{d^2 y}{dx^2} = q_0 \tag{2-48}$$

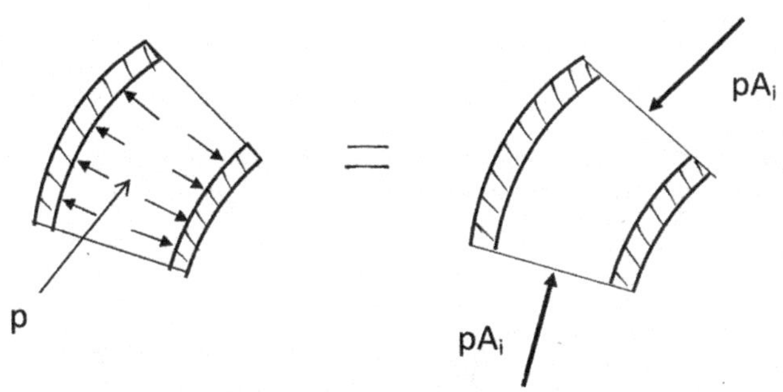

Figure 2-6. Statically equivalent constant pressure effects.

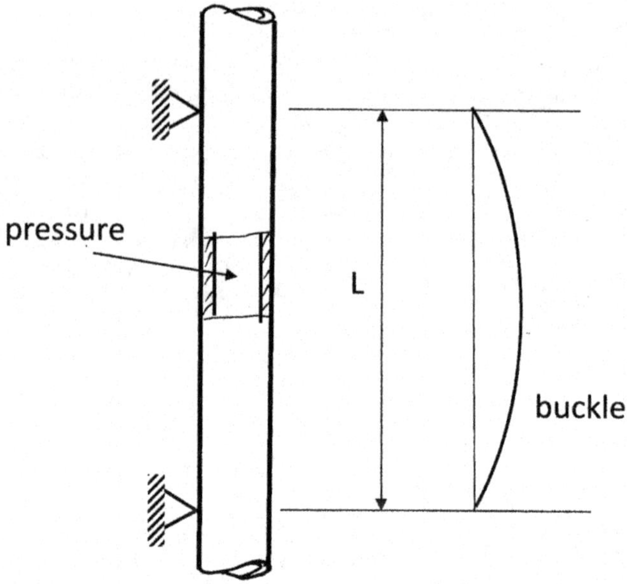

Figure 2-7. Buckled mode from internal pressure.

where q_0 is a constant, which could be caused by fluid flow described by Eq. (1-22). The force, F, is assumed to be tension and a plus number.

$$\frac{d^4y}{dx^4} - \beta\frac{d^2y}{dx^2} = \frac{q_0}{EI} \tag{2-49}$$

where

$$\beta = \frac{F+P}{EI}$$

The assumption is that $F + P$ is a positive number. Integrating of Eq. (2-49) gives

$$\frac{d^2y}{dx^2} - \beta y = \frac{1}{2}\frac{q_0}{EI}x^2 + C_1x + C_2 \tag{2-50}$$

The particular solution is determined using (D) operators, where D stands for differentiation and its inverse stands for integration [10].

$$\left(D^2 - \beta\right)y = \frac{1}{2}\frac{q_0}{EI}x^2 + C_1x + C_2 \tag{2-51}$$

$$y = -\left(\frac{1}{\beta} + \frac{D^2}{\beta^2}\right)\left[\frac{1}{2}\frac{q_0}{EI}x^2 + C_1x + C_2\right] \tag{2-52}$$

The particular solution is

$$y_p = -\frac{1}{\beta}\left[\frac{1}{2}\frac{q_0}{EI}x^2 + C_1 x + C_2\right] - \frac{1}{\beta^2}\frac{q_0}{EI} \tag{2-53a}$$

The complimentary solution is

$$y_c = C_3 \sinh\sqrt{\beta}x + C_4 \cosh\sqrt{\beta}x \tag{2-53b}$$

The total solution to Eq. (2-49) is

$$y = y_c + y_p \tag{2-54}$$

Applying the pinned-pinned set of boundary conditions as illustrated in Fig. 2-5,

$$y(0) = 0 \tag{2-55a}$$

$$\frac{d^2y}{dx^2}(0) = 0 \tag{2-55b}$$

$$y(L) = 0 \tag{2-55c}$$

$$\frac{d^2y}{dx^2}(L) = 0 \tag{2-55d}$$

gives

$$C_1 = -\frac{q_0}{EI}\frac{L}{2} \tag{2-56a}$$

$$C_2 = 0 \tag{2-56b}$$

$$C_3 = \frac{1}{\beta \sinh\sqrt{\beta}L}\frac{1}{\beta}\frac{q_0}{EI}(1 - \cosh\sqrt{\beta}L) \tag{2-56c}$$

$$C_4 = \frac{1}{\beta^2}\frac{q_0}{EI} \tag{2-56d}$$

Consider, for example, pipe and loading as follows:

OD = 2.5 in.
ID = 2.25 in.
A_i = 3.9761 in.^{2}
I = 0.6594 in.^{4}
q_0 = 2 lb/ft
p = 1000 and 2000 psi (internal)
L = 30 ft
F = 10,000 lb (tension)

In this case, $F + P = F - p_iA_i$ is positive so Eq. (2-48) applies. Figure 2-8 clearly shows the effect of internal pressure on beam deflection.

Now consider a slightly different case where F = 0 and $P = -p_iA_i$. Here, $F + P$ is negative and Eq. (2-48) becomes

$$EI\frac{d^4y}{dx^4} + p_iA_i\frac{d^2y}{dx^2} = q_0 \tag{2-57}$$

and

$$\frac{d^4y}{dx^4} + \beta\frac{d^2y}{dx^2} = \frac{q_0}{EI} \tag{2-58}$$

where

$$\beta = \frac{p_iA_i}{EI}$$

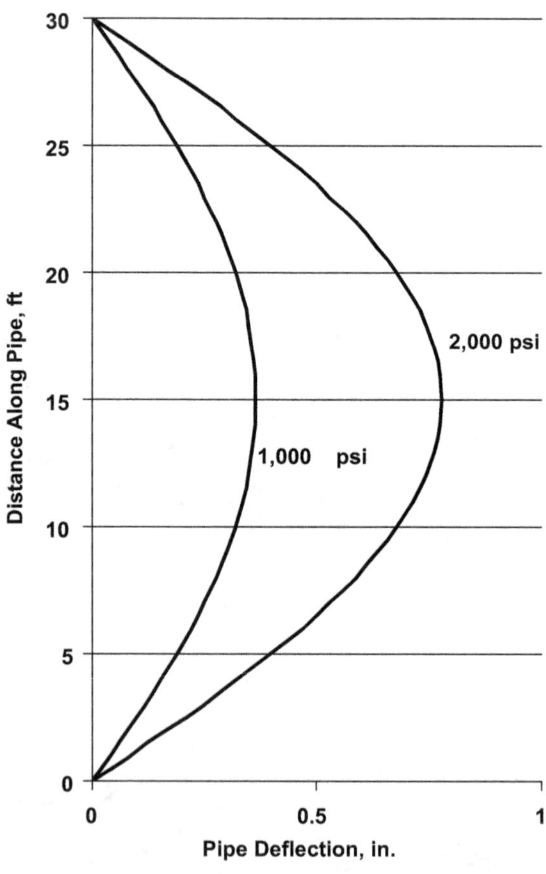

Figure 2-8. Pipe deflection caused by side load, direct pull and internal pressure.

Integrating Eq. (2-58) twice gives

$$\frac{d^2y}{dx^2}+\beta y=\frac{1}{2}\frac{q_0}{EI}x^2+C_1x+C_2 \tag{2-59}$$

Proceeding as before,

$$(D^2+\beta)y=\frac{1}{2}\frac{q_0}{EI}x^2+C_1x+C_2 \tag{2-60}$$

$$y=\left(\frac{1}{\beta}-\frac{D^2}{\beta^2}\right)\left[\frac{1}{2}\frac{q_0}{EI}x^2+C_1x+C_2\right] \tag{2-61}$$

The particular solution becomes

$$y_p=\frac{1}{\beta}\left[\frac{1}{2}\frac{q_0}{EI}x^2+C_1x+C_2\right]-\frac{1}{\beta^2}\frac{q_0}{EI} \tag{2-62a}$$

The complimentary solution is

$$y_c=C_3\sin\sqrt{\beta}x+C_4\cos\sqrt{\beta}x \tag{2-62b}$$

The total solution is

$$y=y_c+y_p \tag{2-63}$$

Applying the boundary conditions gives

$$C_1=-\frac{q_0}{EI}\frac{L}{2} \tag{2-64a}$$

$$C_2=0 \tag{2-64b}$$

$$C_3=\frac{1}{\beta\sin\sqrt{\beta}L}\frac{1}{\beta}\frac{q_0}{EI}(1-\cos\sqrt{\beta}L) \tag{2-64c}$$

$$C_4=\frac{1}{\beta^2}\frac{q_0}{EI} \tag{2-64d}$$

Figure 2-9 shows deflections for three values of internal pressure. Side loading is q_0 = 2 lb/ft.

From Eq. (2-47), the critical bucking pressure is

$$p_{cr}=\frac{\pi^2EI}{A_iL^2} \tag{2-65}$$

$$p_{cr}=\frac{\pi^2 30(10^6)0.6594}{(3.976)360^2}=378.9\text{ psi}$$

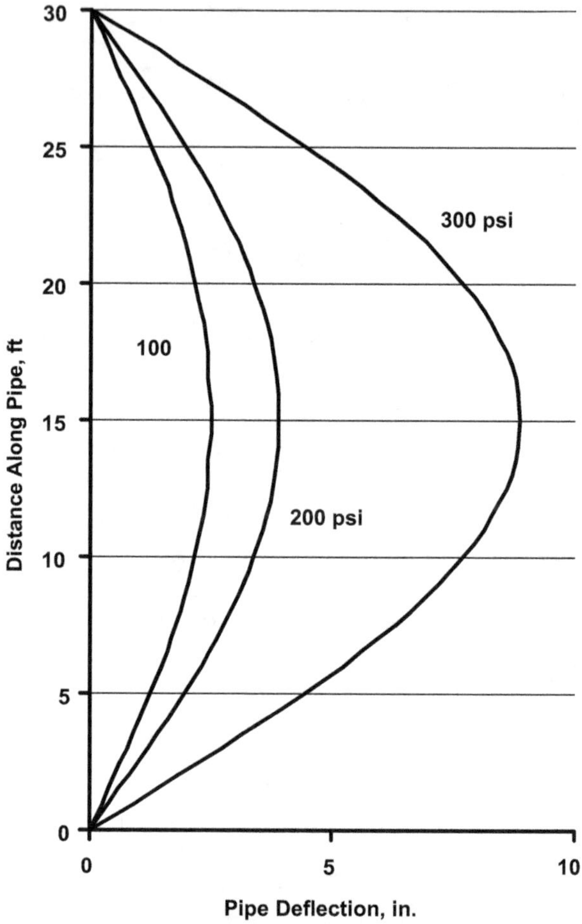

Figure 2-9. Pipe deflection caused by side load and internal pressure.

In this example, the pipe becomes unstable as internal pressure approaches the critical buckling pressure. Already, at a pressure of 300 psi amplitudes starts to increase at a rapid rate.

2.3.2.4 Non-Uniformly Distributed Side Load

Non-uniform load distributions can be determined similarly. The differential equation for general side loading is

$$EI\frac{d^4y}{dx^4} - (F + P)\frac{d^2y}{dx^2} = q(x) \qquad (2\text{-}66)$$

Integrating twice as before and making use of the (D) operator gives

$$(D^2 - \beta)y = \int\int q(x)dxdx + C_1x + C_2 \qquad (2\text{-}67)$$

The particular solution is determined as before. The complimentary solution is the same as Eq. (2-53b).

2.3.3 Dynamic Analysis of Short Pipe

Two cases are considered: 1) natural frequencies of short beams and 2) forced vibration of short beams.

2.3.3.1 *Natural Frequencies and Mode Shapes*

With reference to Fig. 2-5 and Eq. (2-35), the differential equation for this case is

$$EI\frac{\partial^4 y}{\partial x^4} - (F+P)\frac{\partial^2 y}{\partial x^2} + m\frac{\partial^2 y}{\partial t^2} = 0 \tag{2-68}$$

Its solution is assumed to be of the form

$$y(x,t) = Y(x)\sin\omega t \tag{2-69}$$

After substituting and simplifying

$$\frac{d^4 Y}{\partial x^4} - \beta\frac{\partial^2 Y}{\partial x^2} - \lambda^4 Y = 0 \tag{2-70}$$

where

$$\beta = \frac{F+P}{EI}$$

$$\lambda^4 = \frac{m\omega^2}{EI}$$

Let the solution to Eq. (2-70) be expressed by

$$Y(x) = Ce^{px} \tag{2-71}$$

Substituting Eq. (2-71) into Eq. (2-70) gives

$$p^4 - \beta p^2 - \lambda^4 = 0 \tag{2-72}$$

The four roots of Eq. (2-72) are

$$p^2 = \frac{1}{2}\left[\beta \pm \sqrt{\beta^2 + 4\lambda^4}\right] \tag{2-73}$$

Using all four roots gives

$$Y(x) = C_1 e^{p_1 x} + C_2 e^{p_2 x} + C_3 e^{p_3 x} + C_4 e^{p_4 x} \tag{2-74}$$

Applying pinned-pinned boundary conditions and working through the algebra, the eigenvalues are

$$-\frac{\beta}{2}+\sqrt{\left(\frac{\beta}{2}\right)^2+\lambda_n^4}=\left(\frac{n\pi}{L}\right)^2 \quad \text{n = 1, 2, 3. .....} \tag{2-75}$$

which is rewritten

$$\lambda_n^4=\left[\left(\frac{n\pi}{L}\right)^2+\frac{\beta}{2}\right]^2-\left(\frac{\beta}{2}\right)^2 \tag{2-76}$$

Note, λ increases with tension as expected. If the tension term, β, in Eq. (2-76) is dropped, the resulting equation is $\lambda_n=\frac{n\pi}{L}$, giving

$$\omega_n=\left(\frac{n\pi}{L}\right)^2\sqrt{\frac{EI}{m}} \tag{2-77}$$

which agrees with the natural frequency of simply supported beams [11,12].

2.3.3.2 *Instability of Rotating Shafts with End Loads*

A similar solution is given by den Hartog [13] for a rotating beam (pipe) subjected to compressive end loading, *P*. If the shaft rotates slowly and jarred sidewise, the elastic stiffness will restore the shaft to its original straight position. At some higher rotational speed the shaft, if jarred, will not return. Under this condition, the shaft has buckled from the rotation. This scenario is different from the whirling of unbalanced rotating shafts, which experience radial displacement once the shaft begins to rotate. It is a dynamic stability problem.

The differential equation of bending for this case is similar to the previous case except the applied end force is compressive and the side loading inertia is due to centrifugal forces based on synchronous whirl.

$$\frac{d^4y}{dx^4}+\frac{P}{EI}\frac{d^2y}{dx^2}-\frac{m\Omega^2}{EI}y=0 \tag{2-78}$$

This equation has the same form as Eq. (2-70). The end force, *P*, is compressive and carries a (+) sign. The notation for the speed of rotation has been changed from ω to Ω in order to avoid confusion with the natural frequency discussion.

The solution is the same as above and the condition for dynamic instability is defined by

$$\frac{P}{2EI}+\sqrt{\left(\frac{P}{2EI}\right)^2+\frac{m\,\Omega^2}{EI}}=\frac{\pi^2}{L^2} \tag{2-79}$$

This equation is comparable to Eq. (2-75). Equation (2-79) is put into the form

$$\frac{P}{P_{cr}} + \left(\frac{\Omega}{\Omega_{cr}}\right)^2 = 1 \tag{2-80}$$

where

$$P_{cr} = \frac{\pi^2 EI}{L^2}$$

$$\Omega_{cr} = \left(\frac{\pi}{L}\right)^2 \sqrt{\frac{EI}{m}}$$

If P is tension instead of compression, P in Eq. (2-80) changes signs and the rotary speed required to create whirling motion by the centrifugal force increases as a result. If the left side of Eq. (2-80) is greater than one (1), the rotating column (pipe) is dynamically unstable.

Consider, for example, a 60-ft span of drill collars between two stabilizers. Assuming pinned support at each stabilizer and

OD = 7 in.
ID = 2 in.
I = 117.07 in.^{4}
E = 29 $(10)^6$ psi
w = 120 lb/ft
m = 3.73 slugs/ft
N = 100 rpm
Ω = 10.47 rad/sec; $\Omega = \frac{2\pi N}{60}$
WOB = 50,000 lbs

Then

$$P_{cr} = \frac{\pi^2 29(10^6)117.07}{[12(60)]^2} = 64{,}636 \text{ lbs}$$

and

$$\Omega_{cr} = \left(\frac{\pi}{60}\right)^2 \sqrt{\frac{29(10^6)117.07}{3.73}\left|\frac{1ft^2}{144in^2}\right|} = 6.89 \text{ rad/sec}$$

Applying Eq. (2-80)

$$\frac{50{,}000}{64{,}636} + \left(\frac{10.47}{6.89}\right)^2 = 0.774 + 2.31 = 3.08$$

Based on these numbers the collar section between the two stabilizers is unstable and will experience synchronous whirl. The compressive force contributes to this instability but collar rotary speed is the main culprit.

2.3.3.3 *Modal Analysis of Short Pipe*

Modal analysis is a useful tool for predicting vibration response to time-dependent forces [12]. Modal analysis expresses solutions in terms of mode participation.

This method gives solutions directly but usually requires assumptions on damping [12]. Modal damping may not be uncoupled; however, cross-terms in the modal damping matrix are often ignored. Damping is usually assumed to be proportional. The method also is amenable to experimental analysis, which can give modal mass, damping, and stiffness directly. The solution to Eq. (2-35) is assumed to be of the form

$$y(x,t) = \sum X_i(x)\eta_i(t) \tag{2-81}$$

$X_i(x)$ represents the ith mode shape and is determined from an undamped free vibration analysis. $\eta_i(t)$ is the time history of each mode. Substituting Eq. (2-81) into Eq. (2-35) gives

$$\sum_1^\infty \left[EI\frac{d^4X_i}{dx^4} - (F+P)\frac{d^2X_i}{dx} \right]\eta + \sum_1^\infty cX_i\frac{d\eta}{dt} + \sum_1^\infty mX_i\frac{d^2\eta}{dt^2} = F(x,t) \tag{2-82}$$

Next, multiply through by X_j and integrating over the length of the pipe.

$$\sum_1^\infty\left\{\int_0^L\left[EIX_i''''X_j - (F+P)X_i''X_j\right]dx\,\eta + c\int_0^L X_iX_j dx\,\dot{\eta} + m\int_0^L X_iX_j dx\,\ddot{\eta}\right\}$$
$$= \sum_1^\infty \int_0^L X_j F(x,t)dx \tag{2-83}$$

Based on the orthogonal property of mode vectors, the cross products are zero leaving

$$M_i\frac{d^2\eta_i}{dt^2} + C_i\frac{d\eta_i}{dt} + K_i\eta_i = Q_i(t) \tag{2-84}$$

where

$$M_i = m\int_0^L X_iX_i dx \tag{2-85}$$

$$C_i = c\int_0^L X_iX_i dx \text{ (assumming proportional damping)} \tag{2-86}$$

$$K_i = \int_0^L\left[EIX_i''''X_i - (F+P)X_i''X_i\right]dx \tag{2-87}$$

$$Q_i = \int_0^L X_iF(x,t)dx \tag{2-88}$$

The solution to Eq. (2-84) is well documented [12,14]. The complex frequency response of each mode is

$$\overline{\eta_i}(t) = \Xi_i e^{i\omega t} \tag{2-89}$$

where

$$\Xi_i = \frac{Q_i / K_i}{(1 - r^2) + i2\varsigma_i r}$$

Beam response is completely defined in terms of the participation of each mode. A key aspect of the modal analysis approach is the determination of the natural modes of vibration.

2.4 BENDING EQUATION FOR LONG BEAM COLUMNS

Long beam columns are defined as having variable effective tension as a result of gravity and hydrostatic pressure. These forces have a significant effect on drillstring and riser bending. The differential equation of bending containing these forces is developed below. A method of solution will be presented and used to solve a number of drillstring and marine riser problems. Small deformation theory is applicable since riser displacement slopes are assumed to be less than 10 deg.

2.4.1 Hydrostatic Effects of Surrounding Fluids of Equal Density

Consider the equilibrium of a differential pipe element (Fig. 2-10). The left element shows fluid of the same density, γ, applying pressure to both the inside and outside of the pipe. The right element shows the statically equivalent force system. The right element shows that the net effect of inside and outside pressure is a buoyancy force plus end effects. This pressure effect exists in drillstrings that are surrounded by drill mud both inside and outside. The pressure at points 1 and 2 are different because of elevation difference.

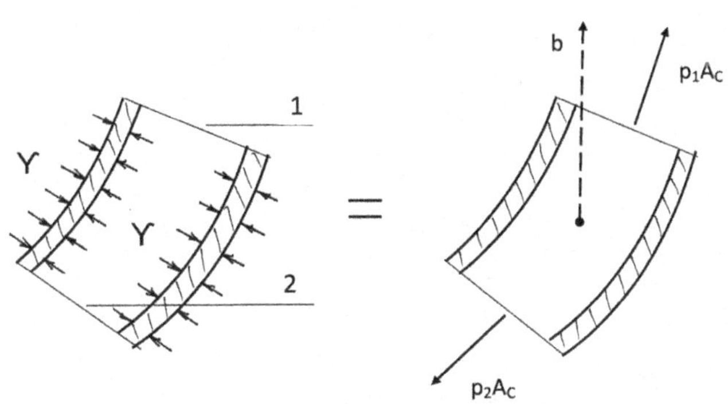

Figure 2-10. Statically equivalent force systems for hydrostatic pressure.

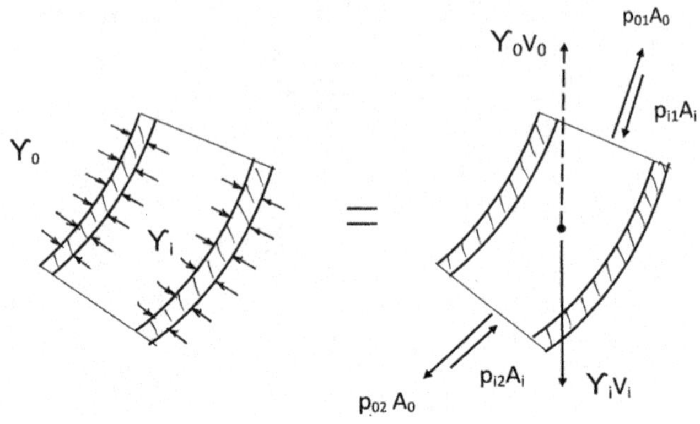

Figure 2-11. Statically equivalent force systems for unequal hydrostatic pressures.

2.4.2 Hydrostatic Effects of Surrounding Fluids of Different Densities

Now consider the case where densities of fluids inside and outside of pipes are different. Such is the case for marine drilling risers where drilling mud density inside is greater than the density of ocean water (Fig. 2-11). The left element depicts a difference in fluid pressure. The right element gives the statically equivalent force system, which is useful for a force analysis. This force system reverts back to the equal fluid density case of Fig. 2-10.

2.4.3 Unique Features of the Differential Equation of Bending

The differential equation of bending is developed from the differential element shown in Fig. 2-12. The solid vectors represent the actual applied forces and body force of gravity, while the dashed vectors represent the statically equivalent hydrostatic forces both inside and outside the riser pipe.

The parameters in the drawing are defined below.

F_x = actual internal force in pipe, lb
F_B = internal force generated at $x = 0$, lb
V = shear force, lb
M = bending moment, lb-ft
θ = slope of deflection at x location, rad
$P = A_0 p_0 - A_i p_i = (A_0 \gamma_0 - A_i \gamma_i)(L - x)$ lb
$b_0 = A_0 \gamma_0$ lb/ft
$b_i = A_i \gamma_i$ lb/ft
γ_0 = weight density of external fluid, lb/ft^3
γ_i = weight density of internal fluid, lb/ft^3
γ = weight density of pipe material, lb/ft^3
w = air weight of pipe per length (γA_c)

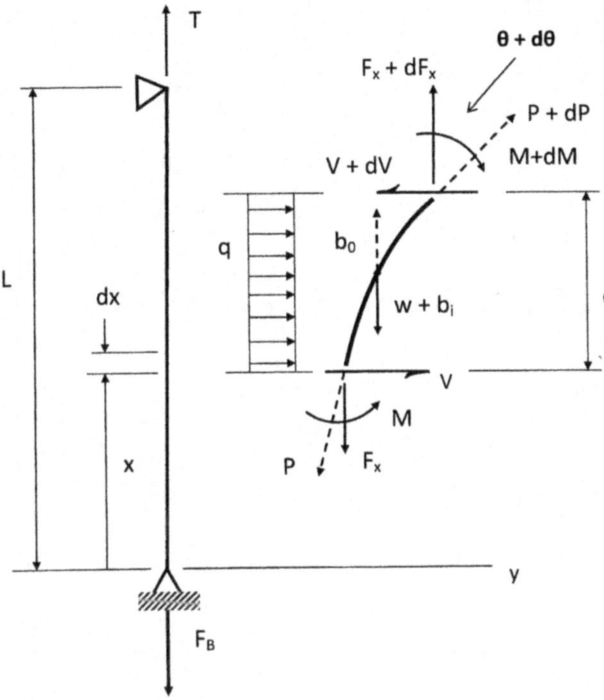

Figure 2-12. Freebody diagram of a differential marine riser section.

Summation of forces in the x direction gives

$$\frac{dF_x}{dx} = w \tag{2-90}$$

$$F_x = wx + F_B \tag{2-91}$$

Summation of forces in the y direction gives

$$\frac{dV}{dx} = \frac{d}{dx}\left[\left(A_o\gamma_o - A_i\gamma_i\right)\left(L - x\right)\frac{dy}{dx}\right] + q \tag{2-92}$$

where q represents transverse loads applied over length, dx. The affect of an additional uniform constant pressure, p, throughout the interior of the pipe (Fig. 2-6) can be included by adding $-pA_i\frac{d^2y}{dx^2}$ to the right side of Eq. (2-92). This term is important in bending analysis of production tubing. Summation of moments gives

$$\frac{dM}{dx} - F_x\frac{dy}{dx} - V = 0 \tag{2-93}$$

Combining Eqs. (2-91), (2-92), and (2-93) gives

$$\frac{d^2M}{dx^2} - \frac{d}{dx}\left[F_x \frac{dy}{dx}\right] - \frac{dV}{dx} = 0 \tag{2-94}$$

Assuming Euler bending, $M = EI\frac{d^2y}{dx^2}$, the differential equation of bending for long pipes with variable internal tension (or compression) and exposed to inside and outside hydrostatic pressure is

$$EI\frac{\partial^4 y}{\partial x^4} - \frac{\partial}{\partial x}\left\{\left[(F_B + wx) + (L-x)(A_o\gamma_o - A_i\gamma_i)\right]\frac{\partial y}{\partial x}\right\} + c\frac{\partial y}{\partial t} + m\frac{\partial^2 y}{\partial t^2} = F(x,t) \tag{2-95}$$

The side force, q, has been replaced by damping, inertia and a distributed side force, $F(x,t)$.

2.4.3.1 *Effective Tension*

Equation (2-95) is the general equation of bending for small lateral in-plane motion. The effective tension term varies with x and this term must be considered in predicting lateral motion of drilling and riser pipe. Equation (2-96) shows the location of effective tension, T_{eff}, in the differential equation,

$$EI\frac{\partial^4 y}{\partial x^4} - \frac{\partial}{\partial x}\left\{T_{\text{eff}}\frac{\partial y}{\partial x}\right\} + c\frac{\partial y}{\partial t} + m\frac{\partial^2 y}{\partial t^2} = F(x,t) \tag{2-96}$$

For short beams, T_{eff} is a constant and the solution is obtained by classic methods. For long beams, Eq. (2-96) is composed of the true local tension plus effects of hydrostatic pressure, which vary with x location. When uniform internal pressure is included, the effective tension becomes

$$T_{eff} = \left[(F_B + wx) + (L-x)(A_0\gamma_0 - A_i\gamma_i) - A_i p\right] \tag{2-97}$$

Effective tension is defined for mathematical convenience. It occupies the space in the differential equation set aside for the actual tension, $F_B + wx$. Effective tension alone accounts for the effects of hydrostatic pressure inside and outside riser pipes as well as uniform pressure throughout the pipe from top to bottom. If, for example, a vertical pipe is suspended in air then

$$T_{eff} = F_B + wx \tag{2-98}$$

This is the true internal tension.

If a vertical pipe is suspended in a given fluid with density γ_m (both inside and outside) effective tension becomes

$$T_{eff} = F_B + wx + (L-x)w_m \tag{2-99}$$

where $w_m = A_c\gamma_m$, which has units of force per length. It represents the weight of the fluid (drilling mud) displaced per unit length of pipe.

When the variable terms are combined

$$T_{eff} = F_B + w_m L + x(w - w_m) \tag{2-100}$$

2.4.3.2 Closed Form Solution

A method for obtaining closed form solutions to Eq. (2-95) will be explained by determining natural frequencies and mode shapes. For this case, Eq. (2-95) reduces to

$$EI\frac{\partial^4 y}{\partial x^4} - \frac{\partial}{\partial x}\left\{\left[(F_B + wx) + (L - x)(A_0\gamma_0 - A_i\gamma_i)\right]\frac{\partial y}{\partial x}\right\} + m\frac{\partial^2 y}{\partial t^2} = 0 \tag{2-101}$$

The solution is assumed to be of the form

$$y(x,t) = Y(x)\sin\omega t \tag{2-102}$$

Combining Eq. (2-101) and Eq. (2-102) gives

$$\frac{d^4Y}{d\zeta^4} - \frac{d}{d\zeta}\left[(\beta + \alpha\zeta)\frac{dY}{d\zeta}\right] - \lambda^4 Y = 0 \tag{2-103}$$

where

$$\alpha = \frac{(w - A_0\gamma_0 + A_i\gamma_i)L^3}{EI}$$

$$\beta = \frac{(F_B + LA_0\gamma_0 - LA_i\gamma_i)L^2}{EI}$$

$$\lambda^4 = \frac{m\omega^2 L^4}{EI}$$

$$\zeta = \frac{x}{L}$$

Each of these terms is dimensionless. The term $(\beta + \alpha\zeta)$ is dimensionless effective tension. The components in Eq. (2-101) have been rearranged to separate the variable term, $\alpha\zeta$.

Expanding Eq. (2-103) gives

$$\frac{d^4Y}{d\zeta^4} - \alpha\zeta\frac{d^2Y}{d\zeta^2} - \beta\frac{d^2Y}{d\zeta^2} - \alpha\frac{dY}{d\zeta} - \lambda^4 Y = 0 \tag{2-104}$$

The solution starts by assuming $Y(\zeta)$ takes the form of a power series.

$$Y(\zeta) = \sum_{n=0}^{\infty} a_n \zeta^n \tag{2-105}$$

This series and its derivatives are absolutely uniformly convergent. Therefore, the derivatives of $Y(\zeta)$ are achieved by term-wise differentiation. Substituting Eq. (2-105) and its derivatives into Eq. (2-104) gives

$$\sum_{n=4}^{\infty}\left[a_n n(n-1)(n-2)(n-3)\varsigma^{n-4}\right]-\sum_{n=1}^{\infty}\alpha a_n n(n-1)\varsigma^{n-1}-\sum_{n=2}^{\infty}\beta a_n n(n-1)\zeta^{n-2}$$
$$-\sum_{n=1}^{\infty}\alpha a_n n\zeta^{n-1}-\sum_{n=0}^{\infty}\lambda^4 a_n\zeta^n=0 \quad (2\text{-}106)$$

The coefficients in front of each ζ term in Eq. (2-106) can be expressed in terms of a_0, a_1, a_2, a_3 as follows.

Replace n in the second term by $n - 3$, n in the third term by $n - 3$, n in the fourth term by $n - 3$ and n in the fifth term by $n - 4$. This brings together coefficients of like terms in the overall series. Each coefficient must be equal to zero to satisfy Eq. (2-104), giving the following recurrence equation.

$$\left[a_n n(n-1)(n-2)(n-3)\right]-\left[a_{n-3}\alpha(n-3)(n-4)\right]-\left[a_{n-2}\beta(n-2)(n-3)\right]$$
$$-\left[a_{n-3}\alpha(n-3)\right]-\lambda^4 a_{n-4}=0 \quad (2\text{-}107)$$

Reorganizing Eq. (2-107) produces

$$a_n=\frac{\beta}{n(n-1)}a_{n-2}+\frac{\alpha(n-3)}{n(n-1)(n-2)}a_{n-3}+\frac{\lambda^4}{n(n-1)(n-2)(n-3)}a_{n-4} \quad (2\text{-}108)$$

where $n \geq 4$. By repeated applications of this recurrence equation, starting with $n = 4$, a_n can be expressed as linear combinations of a_0, a_1, a_2, and a_3. Thus, in general

$$a_n = F_n a_0 + G_n a_1 + H_n a_2 + I_n a_3 \quad (2\text{-}109)$$

It follows that for $n = 0, 1, 2, 3$

$$\begin{array}{llll} F_0=1, & F_1=0, & F_2=0, & F_3=0 \\ G_0=0, & G_1=1, & G_2=0, & G_3=0 \\ H_0=0, & H_1=0, & H_2=1, & H_3=0 \\ I_0=0, & I_1=0, & I_2=0, & I_3=1 \end{array} \quad (2\text{-}110)$$

For higher values of n, i.e. $n \geq 4$, $F_n G_n H_n I_n$ are obtained from

$$F_n=\frac{\beta}{n(n-1)}F_{n-2}+\frac{(n-3)\alpha}{n(n-1)(n-2)}F_{n-3}+\frac{\lambda^4}{n(n-1)(n-2)(n-3)}F_{n-4} \quad (2\text{-}111)$$

which can be grouped as

$$\begin{Bmatrix} F_n \\ G_n \\ H_n \\ I_n \end{Bmatrix} = \frac{\beta}{n(n-1)} \begin{Bmatrix} F_{n-2} \\ G_{n-2} \\ H_{n-2} \\ I_{n-2} \end{Bmatrix} + \frac{(n-3)\alpha}{n(n-1)(n-2)} \begin{Bmatrix} F_{n-3} \\ G_{n-3} \\ H_{n-3} \\ I_{n-3} \end{Bmatrix} + \frac{\lambda^4}{n(n-1)(n-2)(n-3)} \begin{Bmatrix} F_{n-4} \\ G_{n-4} \\ H_{n-4} \\ I_{n-4} \end{Bmatrix} \tag{2-112}$$

These equations are established by substituting Eq. (2-109) into Eq. (2-108) for a_n, a_{n-2}, a_{n-3}, a_{n-4}.

Equation (2-112) provides a means of determining all values of F_n, G_n, H_n, I_n which are needed in Eq. (2-109).

The coefficients in the power series can now be expressed in terms of a_0, a_1, a_2, and a_3, as shown by Eq. (2-109). Equation (2-110) give the starting numbers for the calculation of $F_nG_nH_nI_n$ for $n \geq 4$.

The power series solution, Eq. (2-105), is now expressed as

$$Y(\zeta) = \sum_{n=0}^{\infty} (F_n a_0 + G_n a_1 + H_n a_2 + I_n a_3)\zeta^n \tag{2-113}$$

Equation (2-113) can be further expanded into

$$Y(\zeta) = a_0 + \zeta a_1 + \zeta^2 a_2 + \zeta^3 a_3 + \left(\sum_{n=4}^{\infty} F_n \zeta^n\right) a_0 + \left(\sum_{n=4}^{\infty} G_n \zeta^n\right) a_1 + \left(\sum_{n=4}^{\infty} H_n \zeta^n\right) a_2 + \left(\sum_{n=4}^{\infty} I_n \zeta^n\right) a_3 \tag{2-114}$$

This equation contains four constants, a_0, a_1, a_2, a_3 which are determined from four boundary conditions that are compatible with structural supports.

This method of solution will be used to solve various drillstring and marine riser problems in Chapters 3 and 4. The next section uses this method to determine natural frequencies of vertical pipe under different boundary conditions.

2.5 BUCKLING AND FREQUENCIES OF LONG VERTICAL PIPE

Huang and Dareing [15] considered the lateral behavior of long beams (or pipes) under a matrix of 16 different end constraints (see Fig. 2-13). The analysis gives natural frequencies of lateral modes when completely supported at the top. The analysis also considers buckling and natural frequencies when the vertical loads are completely supported at the lower end. Lateral buckling is considered in the later case. Some of the cases (32 in all) address certain drillstring and marine riser scenarios.

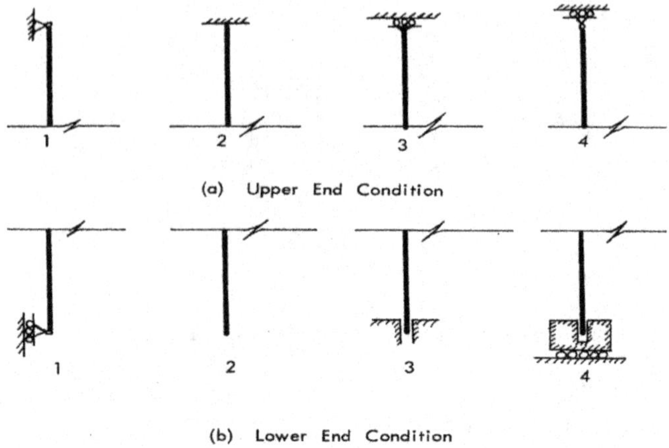

Figure 2-13. Matrix of boundary conditions (with permission from the American Society of Civil Engineers).

The differential equation is derived from the differential element shown in Fig. 2-12. In this case, the free lower end is exposed to hydrostatic pressure only and the top end supports the beam.

Starting with Eq. (2-95) and noting that the actual internal force is

$$F_x(x) = wx - w_m L \tag{2-115}$$

gives

$$EI\frac{\partial^4 y}{\partial x^4} - \frac{\partial}{\partial x}\left[(w - w_m)x\frac{\partial y}{\partial x}\right] + m\frac{\partial^2 y}{\partial t^2} = 0 \tag{2-116}$$

Inertia effects of the surrounding fluid can be included in the mass term, m. Also, the equation applies to free vibration without damping. Assuming a solution of the form

$$y(x,t) = Y(x)\sin\omega t \tag{2-117}$$

Equation (2-116) is now expressed in terms of dimensionless parameters as

$$\frac{d^4Y}{d\zeta^4} - \alpha\zeta\frac{d^2Y}{d\zeta^2} - \alpha\frac{dY}{d\zeta} - \lambda^4 Y = 0 \tag{2-118}$$

where

$$\alpha = \frac{(w - w_m)L^3}{EI}$$

$$\lambda^4 = \frac{m\omega^2 L^4}{EI}$$

$$\zeta = \frac{x}{L}$$

The boundary conditions for each of the 16 cases are:

Lower end condition 1, $$Y(0)=0, \frac{d^2Y}{d\zeta^2}(0)=0 \tag{2-119a}$$

Lower end condition 2, $$\frac{d^2Y}{d\zeta^2}(0)=0, \frac{d^3Y}{d\zeta^3}(0)=0 \tag{2-119b}$$

Lower end condition 3, $$Y(0)=0, \frac{dY}{d\zeta}(0)=0 \tag{2-119c}$$

Lower end condition 4, $$\frac{dY}{d\zeta}(0)=0, \frac{d^3Y}{d\zeta^3}(0)=0 \tag{2-119d}$$

Upper end condition 1, $$Y(1)=0, \frac{d^2Y}{d\zeta^2}(1)=0 \tag{2-119e}$$

Upper end condition 2, $$Y(1)=0, \frac{dY}{d\zeta}(1)=0 \tag{2-119f}$$

Upper end condition 3, $$\frac{dY}{d\zeta}(1)=0, \frac{d^3Y}{d\zeta^3}(1)=0 \tag{2-119g}$$

Upper end condition 4, $$\frac{d^2Y}{d\zeta^2}(1)=0, \frac{d^3Y}{d\zeta^3}(1)-\alpha\frac{dY}{d\zeta}(1)=0 \tag{2-119h}$$

β is not involved because the force at the lower end is assumed to hydrostatic. Clearly, the differential equation contains a term with a variable coefficient. The solution is assumed to be of the form

$$Y(\zeta)=\sum_{n=0}^{\infty} a_n \zeta^n \tag{2-120}$$

Substituting Eq. (2-120) and its derivatives into Eq. (2-118) gives a power series in ζ whose sum is equal to zero. As explained earlier, coefficients can be grouped and each coefficient must be equal to zero giving the following recurrence formula.

$$a_n = \frac{\alpha(n-3)}{n(n-1)(n-2)} a_{n-3} + \frac{\lambda^4}{n(n-1)(n-2)(n-3)} a_{n-4} \tag{2-121}$$

where $n \geq 4$. By repeated applications of this recurrence equation, starting with $n = 4$, we can express a_n as linear combinations of a_0, a_1, a_2, and a_3. Thus in general

$$a_n = F_n a_0 + G_n a_1 + H_n a_2 + I_n a_3 \tag{2-122}$$

for $n \geq 4$, and noting that

$$\begin{matrix} F_0 = 1, & F_1 = 0, & F_2 = 0, & F_3 = 0 \\ G_0 = 0, & G_1 = 1, & G_2 = 0, & G_3 = 0 \\ H_0 = 0, & H_1 = 0, & H_2 = 1, & H_3 = 0 \\ I_0 = 0, & I_1 = 0, & I_2 = 0, & I_3 = 1 \end{matrix} \tag{2-123}$$

when Eq. (2-122) is substituted into Eq. (2-121) for a_n, a_{n-3}, a_{n-4},

$$F_n = \frac{(n-3)\alpha}{n(n-1)(n-2)} F_{n-3} + \frac{\lambda^4}{n(n-1)(n-2)(n-3)} F_{n-4} \tag{2-124}$$

which can be grouped as follows

$$\begin{Bmatrix} F_n \\ G_n \\ H_n \\ I_n \end{Bmatrix} = \frac{(n-3)\alpha}{n(n-1)(n-2)} \begin{Bmatrix} F_{n-3} \\ G_{n-3} \\ H_{n-3} \\ I_{n-3} \end{Bmatrix} + \frac{\lambda^4}{n(n-1)(n-2)(n-3)} \begin{Bmatrix} F_{n-4} \\ G_{n-4} \\ H_{n-4} \\ I_{n-4} \end{Bmatrix} \tag{2-125}$$

for $n \geq 4$

Equation (2-125) provides a means of determining all values of F_n, G_n, H_n, I_n which are needed in Eq. (2-122).

The coefficients in the power series can now be expressed in terms of a_0, a_1, a_2, and a_3, as shown by Eq. (2-122). These four constants are determined from boundary conditions.

Considering Case 11, for example,

$$Y(0) = 0 \tag{2-126a}$$

$$\frac{d^2Y}{d\zeta^2}(0) = 0 \tag{2-126b}$$

$$Y(1) = 0 \tag{2-126c}$$

$$\frac{d^2Y}{d\zeta^2}(1) = 0 \tag{2-126d}$$

Using Eq. (2-120) and the first two boundary conditions, we find that $a_0 = 0$ and $a_2 = 0$. The third and fourth boundary conditions give (refer to Eq. (2-126c, d))

$$a_1 + \sum_{n=3}^{\infty} a_n = 0 \tag{2-127a}$$

$$\sum_{n=3}^{\infty} n(n-1)a_n = 0 \tag{2-127b}$$

Substituting for a_n (Eq. (2-122)), gives

$$\left[1 + \sum_{n=3}^{\infty} G_n\right] a_1 + \left[\sum_{n=3}^{\infty} I_n\right] a_3 = 0 \tag{2-128a}$$

$$\left[\sum_{n=3} n(n-1)G_n\right] a_1 + \left[\sum_{n=3}^{\infty} n(n-1)I_n\right] a_3 = 0 \tag{2-128b}$$

For non-trivial solutions, the determinant, D, of the coefficient matrix must be zero or

$$D\ (\alpha, \lambda) = 0$$

For a given value of α, the solution to this equation yields values for λ. The eigenvalues, λ, are determined by an iterative process [15]. Mode shapes are determined by letting, say, $a_1 = 1$, and then solving for a_3 using Eq. 2-128. The remaining coefficients can be determined by Eq. 2-121 with the desired eigenvalue, λ_i. The mode shape follows with Eq. 2-120.

The relation between α and λ depend on boundary conditions. Solutions were obtained for sixteen sets of boundary conditions given earlier. Natural frequencies for plus (+) values of α are given in Table 2-3. When $\alpha = 0$ the tubular is neutrally buoyant; the fluid force at the bottom end equals the weight of the string of pipe.

2.5.1 Pipes Completely Supported at the Top

When the pipe is completely supported at the top, the tension at the top is essentially the buoyed weight of the pipe. The compression at bottom end is hydrostatic. Table 2-3 gives the values of λ for different values of α for the 16 cases. These numbers apply to positive values of α meaning that the density of pipe material is greater than the density of the surrounding fluid.

With new materials, such as composites, it may be possible for the weight of the pipe to be less than the weight of the fluid. In this case, α would be minus. This situation is addressed later. At certain values of $-\alpha$, the λ or the natural frequency becomes zero. Under this condition, the pipe buckles as a result of buoyancy.

Consider Case 11 with the following conditions for a 4 ½ in. (16.6 lb/ft) drill pipe.

L = 475 ft
$E = 30 \times 10^6$ psi
I = 9.6 in.^{4}
w = 16.6 lb/ft
w_m = 1.91 lb/ft

For this case, $\alpha = 800$ and $\lambda_1 = 7.3$, $\lambda_2 = 10.9$, $\lambda_3 = 14.0$. Corresponding frequencies are $f_1 = 0.073$ cps, $f_2 = 0.17$ cps, $f_3 = 0.46$ cps. The mass of the fluid has not been included in the λ expression.

2.5.2 Synchronous Whirl of Long Pipe

Critical whirling speeds of short pipe are determined from Eq. 2-78 assuming simply supported boundaries [12].

Table 2-3. Values of λ. 0 ≤ α ≤ 1000 (with permission from ASCE).

α		0	200	400	600	800	1000
Case 11	λ_1	3.1416	5.4652	6.3076	6.8804	7.3261	7.6957
	λ_2	6.2832	8.4625	9.5456	10.3113	10.9171	11.4243
	λ_3	9.4248	11.3031	12.4565	13.3145	14.0089	14.5980
Case 12	λ_1	0	4.1293	4.9077	5.4300	5.8342	6.1684
	λ_2	3.9266	6.6210	7.6994	8.4435	9.0269	9.5131
	λ_3	7.0686	9.2779	10.4467	11.2868	11.9574	12.5222
Case 13	λ_1	3.9266	6.0973	6.9535	7.5387	7.9947	8.3732
	λ_2	7.0686	9.1389	10.2306	11.0098	11.6285	12.1474
	λ_3	10.2102	11.9999	13.1454	14.0087	14.7115	15.3097
Case 14	λ_1	1.5708	4.2112	4.9728	5.4867	5.8856	6.2160
	λ_2	4.7124	7.0946	8.1293	8.8488	9.4147	9.8873
	λ_3	7.8540	9.9110	11.0588	11.8913	12.5578	131197
Case 21	λ_1	3.9266	5.6429	6.4353	6.9875	7.4212	7.7828
	λ_2	7.0686	8.7817	9.7720	10.4969	11.0790	11.5706
	λ_3	10.2102	11.7443	12.7812	13.5811	14.2404	14.8056
Case 22	λ_1	1.8751	4.2155	4.9762	5.4905	5.8898	6.2205
	λ_2	4.6941	6.8154	7.8375	8.5586	9.1288	9.6062
	λ_3	7.8548	9.6222	10.6904	11.4853	12.1295	12.6768
Case 23	λ_1	4.7300	6.3202	7.1121	7.6704	8.1109	8.4790
	λ_2	7.8532	9.7720	10.4851	11.2186	11.8105	12.3115
	λ_3	10.9956	12.7812	13.4947	14.2966	14.9617	15.3141
Case 24	λ_1	2.3650	4.3116	5.0489	5.5526	5.9453	6.2714
	λ_2	5.4975	7.3379	8.3004	8.9896	9.5382	9.9992
	λ_3	8.6394	10.2929	11.3331	12.1155	12.7524	13.2945
Case 31	λ_1	1.5708	3.2308	3.7310	4.0653	4.3232	4.5328
	λ_2	4.7124	6.9463	7.9229	8.5999	9.1310	9.5735
	λ_3	7.8540	9.8601	10.9855	11.8019	12.4551	13.0056
Case 32	λ_1	0	0	0	0	0	0
	λ_2	2.3650	5.2935	6.2494	6.8970	7.4004	7.8177
	λ_3	5.4978	7.8838	9.0111	9.8060	10.4352	10.9626
	λ_4	8.6394	10.6371	11.8126	12.6788	13.3776	13.9698
Case 33	λ_1	2.3650	3.8660	4.3873	4.7384	5.0099	5.2341
	λ_2	5.4978	7.6211	8.6123	9.3041	9.8480	10.3017
	λ_3	8.6394	10.5508	11.6768	12.5034	13.1681	13.7295
Case 34	λ_1	0	0	0	0	0	0
	λ_2	3.1416	5.5701	6.4817	7.1063	7.5943	8.0002
	λ_3	6.2832	8.4692	9.5660	10.3445	10.9622	11.4804
	λ_4	9.4248	11.3027	12.4585	13.3198	14.0177	14.6102
Case 41	λ_1	0	3.2304	3.7309	4.0653	4.3231	4.5357
	λ_2	3.9266	6.9312	7.9183	8.5975	9.1295	9.5725
	λ_3	7.0686	9.7908	10.9620	11.7899	12.4477	13.0005

Table 2-3. (*continued*)

α		0	200	400	600	800	1000
Case 42	λ_1	0	0	0	0	0	0
	λ_2	0	5.2906	6.2483	6.8965	7.400	7.8174
	λ_3	4.7300	7.8589	9.0031	9.8019	10.4326	10.9608
	λ_4	7.8532	10.5470	11.7808	12.6623	13.3674	13.9627
Case 43	λ_1	1.8751	3.8650	4.3870	4.7382	5.0098	5.2341
	λ_2	4.6941	7.5970	8.6049	9.3004	9.8457	10.3001
	λ_3	7.8532	10.4609	11.6454	12.4873	13.1581	13.7226
Case 44	λ_1	0	0	0	0	0	0
	λ_2	2.3650	5.5656	6.4803	7.1055	7.5938	7.9998
	λ_3	5.4978	8.4327	9.5542	10.3386	10.9585	11.4778
	λ_4	8.6394	11.1899	12.4173	13.2983	14.0043	14.6010

Table 2-4. Values of α_{cr} (with permission from ASCE).

Case	α_{cr1}	α_{cr2}	α_{cr3}	α_{cr4}
11	–18.569	–86.431	–196.29	
12	0	–25.638	–95.95	–210.68
13	–30.009	–112.09	–234.99	
14	–3.4766	–44.138	–129.26	
21	–52.501	–129.05	–276.24	
22	–7.8373	–55.977	–148.51	
23	–74.629	–157.03	–325.51	
24	–18.956	–81.887	–189.22	
31	–7.8373	–55.977	–148.51	
32	0	–7.8373	–55.977	–148.51
33	–18.956	–81.887	–189.22	
34	0	–18.956	–81.887	–189.22
41	0	–25.638	–95.950	–210.68
42	0	0	–25.638	–95.95
43	–3.4766	–44.138	–129.26	
44	0	–3.4766	–44.138	–129.26

$$\Omega_{cr} = \left(\frac{n\pi}{L}\right)^2 \sqrt{\frac{EI}{m}} \qquad n = 1,2,3 \;.....$$

Critical whirling speeds of long pipe follows a similar procedure except variable tension is considered. The differential equation of motion becomes

$$EI\frac{d^4r}{dx^4} - \frac{d}{dx}\left[(w - w_m)x\frac{dr}{dx}\right] - mr\Omega^2 = 0 \tag{2-129}$$

assuming there is no end thrust and the lower end is open to hydrostatic pressure as before. The dependent variable, r, is the radius of the whirling motion. Equation 2-129 is put in dimensionless form

$$\frac{d^4r}{d\zeta^4} - \alpha\zeta\frac{d^2r}{d\zeta^2} - \alpha\frac{dr}{d\zeta} - \lambda^4 r = 0 \tag{2-130}$$

where

$$\alpha = \frac{(w - w_m)L^3}{EI}$$

$$\lambda^4 = \frac{m\Omega^2 L^4}{EI}$$

$$\zeta = \frac{x}{L}$$

Note the similarity of Eq. 2-130 and 2-118. The only difference is the interpretation of ω (natural frequency) and Ω (critical rotary speed). This means that the mathematical solutions are the same for both problems and the eigenvalues given in Table 2-3 apply to both. For case 11, i.e., pinned-pinned boundaries, $\lambda_1 = \pi$ for $\alpha = 0$, which agrees with the short beam solution. However, the variable tension becomes a factor for nonzero values of α. For example, when $\alpha = 1000$, and $\lambda_1 = 7.6957$

$$\Omega_{cr} = \left(\frac{7.6957}{L}\right)^2 \sqrt{\frac{EI}{m}}$$

In general

$$\Omega_{cr} = \left(\frac{\lambda_i}{L}\right)^2 \sqrt{\frac{EI}{m}}$$

The eigenvalues given in Table 2-3 for the (16) boundary conditions (Fig. 2-13) can therefore be used to predict synchronous whirl. The shape of pipe whirl for long pipe follows the mode deflections characterized by a large wave length at the top and large deflections at the lower end. The addition of the distributed weight along the pipe increases critical rotary speed as expected.

2.5.3 Fluid Density Greater Than Pipe Density

Mathematically, it is possible for α to take on negative (–) values. In this case, the density of the fluid is greater than the density of the pipe. Values of λ for negative values of α are given in ref. [15]. Negative values of α create more and more compression in the pipe and can buckle the pipe when α reaches α_{cr}. Even higher modes of buckling can be established in this manner (see Table 2-4).

2.5.4 Pipes Completely Supported at the Lower End

The data in Table 2-4 can be viewed another way. By supporting the pipes at the bottom instead at the top (or reversing the order of the 16 boundary conditions), the differential equation of bending becomes

$$EI\frac{\partial^4 y}{\partial x^4} + \frac{\partial}{\partial x}\left[(w - w_m)(L - x)\frac{\partial y}{\partial x}\right] + m\frac{\partial^2 y}{\partial t^2} = 0 \tag{2-131}$$

In this development,

$$F_x = -(w - w_m)(L - x) \tag{2-132}$$

If $(L - x)$ is replaced by, say z, then Eq. (2-131) has the exact same form as Eq. (2-116) except the sign after the first term is plus instead of minus. This gives a new interpretation of α, i.e. it applies to sixteen (16) new cases where the top now is located at the bottom end and the bottom end now becomes an unloaded end at the top.

Mathematically, the boundary conditions stay the same but x is taken positive downward from the top instead of up from the bottom. This gives a different but useful application of the results, i.e., the buckling of vertical pipes supported at the bottom under different boundary conditions. In this case, the numbers in Table 2-4 give the critical lengths of bottom-supported pipes.

Considering again Case 11 and the numbers given above. The critical values of α for the first three buckling modes are

$$\alpha_{cr1} = -18.569$$
$$\alpha_{cr2} = -86.431$$
$$\alpha_{cr3} = -196.29$$

Critical lengths are

$$L_{cr1} = 136 \text{ ft}$$
$$L_{cr2} = 227 \text{ ft}$$
$$L_{cr3} = 298 \text{ ft}$$

2.5.5 Stability of Pipe Stands (Case 11)

The potential of buckling of pipe stacked in derricks is considered here. Lubinski [16] shows that buckling will occur when pipe length is 2.65 times the dimensionless unit, m.

$$L_{cr1} = 2.65\sqrt[3]{\frac{EI}{w}} \tag{2-133}$$

Consider α_{cr1} = 18.569 given in Table 2-4 for Case 11 (pinned-pinned boundary conditions). In this case, w_m = 0, so

$$\alpha = \frac{wL^3}{EI} \tag{2-134}$$

$$L_{cr1} = \sqrt[3]{\frac{\alpha_{cr1}EI}{w}} \tag{2-135}$$

or

$$L_{cr1} = 2.648\sqrt[3]{\frac{EI}{w}} \tag{2-136}$$

which is the same as Eq. (2-133).

Critical stacked lengths for various drill pipe and collar sizes are given below [16].

- 107 ft for 3 ½ in. (13.3 lb/ft) drill pipe
- 127 ft for 4 ½ in. (16.6 lb/ft) drill pipe
- 122 ft for 4 ¾ in. (1 ⅞ in ID) drill collars
- 144 ft for 6 ¼ in. (2 ¼ in ID) drill collars
- 159 ft for 7 in. (3 in ID) drill collars

These stands of pipe will not buckle when stacked in the derrick. Longer stands must be supported near the middle to prevent buckling.

2.5.6 Comparison of Pipe and Chain Solutions (Case 12)

Comparison of natural frequency data for a vertical pipe and a vertical chain shows that the lateral vibration frequencies for an elastically flexible pipe and a chain approach the same frequency at high values of α (see Fig. 2-14). For the sake of comparing the two solutions, the effects of buoyancy are not included. For large values of α, the elastic pipe solution approaches that of chains. Also, as α increases, the lower-order frequencies approach those of a chain faster than those of high order.

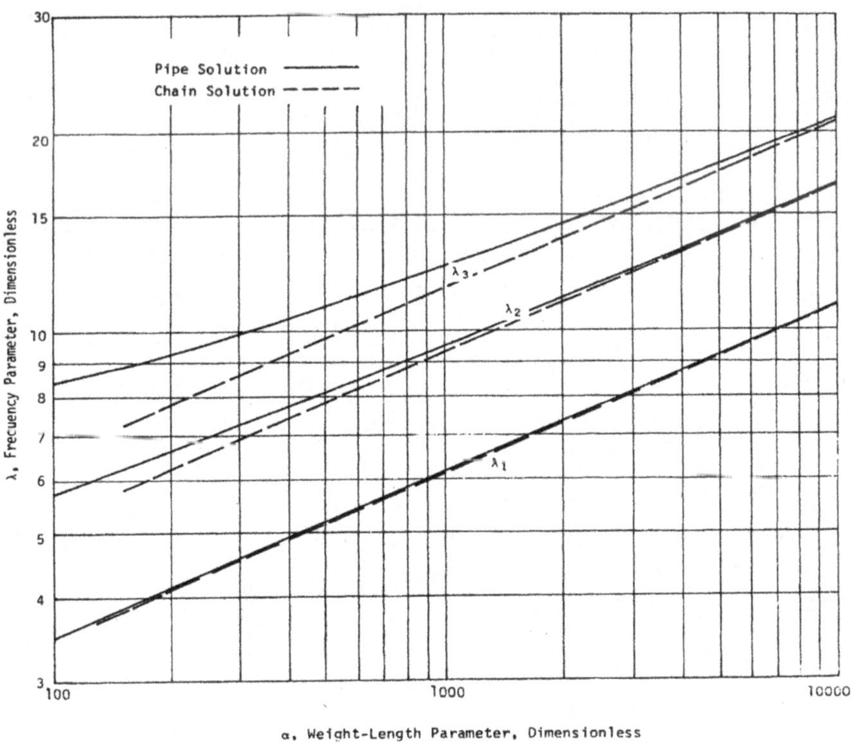

Figure 2-14. Comparison of frequency parameters for flexible pipe and chain (Case 12) (with permission from ASCE).

Table 2-5. Eigenvalues ψ for a Steel Catenary (based on Equation (2-148)), b = 50,000 (force units).

a	ψ_1	ψ_2	ψ_3	ψ_4
10,000	736	1472	2208	2944
20,000	767	1535	2301	3067
30,000	795	1591	2387	3182
40,000	821	1645	2467	3290

2.5.7 Natural Frequencies of Steel Catenaries with Buoyancy

A steel catenary is considered to be suspended pipe that carries the effects of gravity and fluid pressures but without the support of bending stiffness (EI = 0). The differential equation of deflection, based on this model, is greatly simplified because of the elimination of the fourth order term. Predictions give reasonable engineer results except in areas of high localized curvature such as at boundaries and within high vibrations modes. Its application, though attractive, should be used with caution.

Eliminating the stiffness term in Eq. (2-101) and substituting Eq. (2-102) for $y(x,t)$ gives

$$\frac{d}{d\zeta}\left[(b+a\zeta)\frac{dY}{d\zeta}\right]+\psi^2 Y=0 \tag{2-137}$$

$$a\zeta\frac{d^2Y}{d\zeta^2}+a\frac{dY}{d\zeta}+b\frac{d^2Y}{d\zeta}+\psi^2 Y=0 \tag{2-138}$$

where

$$a = (w - A_o\gamma_o + A_i\gamma_i)L$$
$$b = F_B + LA_o\gamma_o - LA_i\gamma_i$$
$$\psi^2 = m\omega^2 L^2$$
$$\zeta = \frac{x}{L}$$

Equation (2-138) could be put into a dimensionless form but for the sake of clarity, a and b are left in units of force.

The solution to Eq. (2-138) is assumed to be of the form

$$Y(\zeta)=\sum_{n=0}^{\infty}a_n\zeta^n \tag{2-139}$$

$$\frac{dY}{dx}=\sum_{n=0}^{\infty}a_n n\zeta^{n-1} \tag{2-140}$$

$$\frac{d^2Y}{d\zeta^2} = \sum_{n=0}^{\infty} a_n n(n-1)\zeta^{n-2} \tag{2-141}$$

Boundary conditions for this case are

$$\mathrm{Y}(0) = 0$$

$$\mathrm{Y}(1) = 0$$

which give $a_0 = 0$ and

$$\sum_{n=1}^{\infty} a_n = 0 \tag{2-142}$$

All of the a_n s can be related to a_1 as follows. Substituting Eq. (2-139) through (2-141) into Eq. (2-138) gives

$$\sum_{2}^{\infty} a_n n(n-1)\zeta^{n-2} = -\frac{a}{b}\sum_{2}^{\infty} a_n n^2 \zeta^{n-1} - \frac{\psi^2}{b}\sum_{2}^{\infty} a_n \zeta^n \tag{2-143}$$

Adjusting the exponents on ζ as explained earlier gives

$$\sum_{2}^{\infty} a_n n(n-1)\zeta^{n-2} = -\frac{a}{b}\sum_{2}^{\infty} a_{n-1}(n-1)^2\zeta^{n-2} - \frac{\psi^2}{b}\sum_{2}^{\infty} a_{n-2}\zeta^{n-2} \tag{2-144}$$

which shows that the coefficient to each ζ with exponent greater than 2 can be expressed in terms of previous coefficients by the recursion formula

$$a_n = -\frac{a}{b}\frac{n-1}{n}a_{n-1} - \frac{\psi^2}{bn(n-1)}a_{n-2} \quad \text{for } n \geq 2 \tag{2-145}$$

By Eq. (2-145) all a_n for $n \geq 2$ can be expressed in terms of a_1 so we let $a_n = G_n a_1$. It is apparent that $G_1 = 1$. Substituting this equality into Eq. (2-145) gives

$$G_n = -\frac{a}{b}\frac{n-1}{n}G_{n-1} - \frac{\psi^2}{bn(n-1)}G_{n-2}, \quad n \geq 2 \tag{2-146}$$

By the second boundary condition

$$a_1\sum_{1}^{\infty} G_n = 0 \tag{2-147}$$

or

$$\sum_{1}^{\infty} G_n = 0 \tag{2-148}$$

This becomes the characteristic equation from which the eigenvalues, ψ, are extracted. For each set of a and b, the summation produces multiple roots of ψ^2 which define the natural frequencies of each mode. Corresponding mode shapes are obtained by substitution of each value of a_n into Eq. (2-139). Table 2-5 gives eigenvalues, ψ, for an assumed value of $b = 50{,}000$ and $a = 10{,}000$ to $40{,}000$. The four eigenvalues are obtained from Eq. (2-148).

Consider the following example.

$D_0 = 20$ in
$D_i = 19$ in
$L = 1000$ ft
$\gamma_0 = 65$ lb/ft^3
$\gamma_i = 80$ lb/ft^3
$F_B = 100{,}000$ lb
$w = 104.23$ lb/ft
$m = 3.237$ slugs/ft

$a = 119{,}936$ lb
$b = 84{,}292$ lb

The eigenvalues for the first four modes are given in Table 2-6.

Equation (2-148) is easily programmed for various combinations of a and b.

Table 2-6. Natural frequencies of first four modes.

Mode	Ψ	Frequency (cps)	Period (sec)
1	496.7	0.044	22.76
2	1140	0.101	9.92
3	1787.5	0.158	6.32
4	2436	0.215	4.64

2.5.8 Comparison of Steel Catenary with Elastic Pipe Solutions

The effect of hydrostatic pressure on the chain model can be added by simplifying Eq. (2-116). Since internal bending resistance is zero, Eq. (2-116) becomes

$$-\frac{\partial}{\partial x}\left[(w - w_m)x\frac{\partial y}{\partial x}\right] + m\frac{\partial^2 y}{\partial t^2} = 0 \tag{2-149}$$

assuming

$$y(x,t) = Y(x)\cos\omega t \tag{2-150}$$

and letting

$$\varsigma = \frac{x}{L} \tag{2-151}$$

gives

$$\zeta \frac{d^2Y}{d\zeta^2} + \frac{dY}{d\zeta} + \lambda^2 Y = 0 \tag{2-152}$$

where

$$\lambda_c^2 = \frac{m\omega^2 L}{(w - w_{\mathrm{m}})} \text{ (steel catenary)} \tag{2-153}$$

Recall that for elastic pipe with the same fluid inside and out

$$\alpha = \frac{(w - w_m)L^3}{EI} \tag{2-154}$$

and

$$\lambda_p^4 = \frac{m\omega^2 L^4}{EI} \text{ (elastic flexible pipe)} \tag{2-155}$$

Note that α does not occur in λ_p^4. As a means of comparing natural frequencies of soft pipe with flexible pipe, the last three expressions combine to give.

$$\lambda_p^4 = \lambda_c^2 \alpha \tag{2-156}$$

This equation gives a way to find an equivalent elastically flexible pipe frequency from cable frequency data assuming zero tension at the lower end.

For example, consider $\lambda_c = 1.2024$, the eigenvalue (see Eq. (2-21)) for the first mode of a hanging cable. The comparable eigenvalue for a pipe (case 12) at $\alpha = 1000$ would be (using Eq. (2-156)) $\lambda_p = 6.166$. Using this equivalency, frequencies for the two cases (chain and pipe) could be compared (see Fig. 2-14). Based on the pipe solution, $\lambda_1 = 6.168$ (see Table 2-3).

For large values of α, the solution for flexible pipe approaches those for soft pipe (or cables). Solutions for cables are given in an earlier section. It can be concluded that as α increases, the lower-order frequencies approach to those of a chain faster than those of high order.

2.6 TORSION BUCKLING OF VERTICAL PIPE

Drill pipe can undergo high levels of torque during directional drilling operations. This is mainly due to wall friction along the entire length of the drill-string. Friction torque develops over curved portions of a planned well path, local dog legs and drill bits. If drill pipe buckles under applied torque, the post buckling configuration will add further wall contact and friction. It will also increase bending stress. An analytical solution to the torsion stability of long vertical pipe does not appear in the literature. This unique problem is addressed below.

2.6.1 Torsion Buckling of Short Pipe

In 1895, Greenhill [17,18] developed and solved the differential equation of bending defining the lateral stability of shafts under end torque. The coordinate system is shown in Fig. 2-15 and the differential equations of bending are

$$M_t \frac{dy}{dx} = -EI \frac{d^2z}{dx^2} \tag{2-157a}$$

$$M_t \frac{dz}{dx} = +EI \frac{d^2y}{dx^2} \tag{2-157b}$$

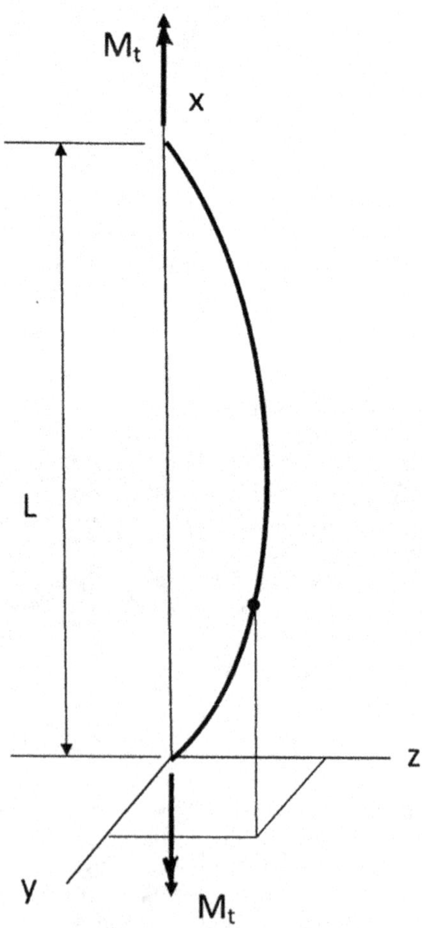

Figure 2-15. Coordinate systems for Greenhill's formulation.

Later, Webb [19] obtained a solution to these equations using a complex function. His solution starts by defining

$$u = y + iz \tag{2-158}$$

Eqs. (2-157a) and (2-157b) are combined by multiplying the second equation by the complex number, i, and adding it to the first giving

$$\frac{d^2u}{dx^2} + i\frac{M_t}{EI}\frac{du}{dx} = 0 \tag{2-159}$$

This is a first-order equation in terms of $\frac{du}{dx}$ having a general solution of

$$\frac{du}{dx} = Ce^{-i\frac{M_t}{EI}x} \tag{2-160}$$

$$\frac{du}{dx} = C_1 \cos\left(\frac{M_t}{EI}x\right) - iC_1 \sin\left(\frac{M_t}{EI}x\right) \tag{2-161}$$

Integration gives

$$u(x) = C_1 \frac{EI}{M_y}\left[\sin\left(\frac{M_t}{EI}x\right) + i\cos\left(\frac{M_t}{EI}x\right) + C_2\right] \tag{2-162}$$

Using the boundary condition at the left end, $y = z = 0$ at $x = 0$ gives $C_2 = -i$ and

$$u = C^*\left\{\sin\left(\frac{M_t}{EI}x\right) - i\left[1 - \cos\left(\frac{M_t}{EI}x\right)\right]\right\} \tag{2-163}$$

Applying the boundary condition at the right end, $y = z = 0$ at $x = L$ gives

$$\sin\left(\frac{M_t}{EI}L\right) - i\left[1 - \cos\left(\frac{M_t}{EI}L\right)\right] = 0 \tag{2-164}$$

Both real and imaginary parts have to be zero simultaneously. This condition is satisfied when

$$\frac{M_t}{EI}L = 2\pi, 4\pi, 6\pi, \text{etc.} \tag{2-165}$$

The lowest critical torque for the first buckle mode is

$$(M_t)_{crit} = \frac{2\pi EI}{L} \tag{2-166}$$

At that time, the feeling was: --- only extremely long and thin shafts run in danger of buckling; the question hardly is a practical one, except possibly for long, thin wires transmitting torque.

When the same problem is solved for both torque and compressive force, Greenhill's equations become

$$EI\frac{d^2z}{dx^2} = -Pz - M_t\frac{dy}{dx}$$

$$EI\frac{d^2y}{dx^2} = -Py + M_t\frac{dz}{dx}$$

The solution follows Webb's approach and is summarized by den Hartog [13, p295]. The result is presented in a compact form as

$$\left(\frac{M_t}{M_{t,Greenhill}}\right)^2 + \frac{P}{P_{Euler}} = 1 \tag{2-167}$$

where

$$M_{t,Greenhill} = \frac{2\pi EI}{L}$$

$$P_{Euler} = \frac{\pi^2 EI}{L^2}$$

When rotation is included in a stability analysis with end compression but in the absence of torque, the bending equation becomes

$$EI\frac{d^4y}{dx^4} + P\frac{d^2y}{dx^2} - m\omega^2 y = 0$$

Synchronous is assumed in this formulation. Its solution is presented as [13]

$$\frac{P}{P_{Euler}} + \left(\frac{\omega}{\omega_{crit}}\right)^2 = 1 \tag{2-168}$$

where

$$\omega_{crit}^2 = \frac{EI}{m}\left(\frac{\pi}{L}\right)^4$$

Southwell and Dough [20] developed the solution for synchronous whirl of shafts subjected to combinations of torque, rotation, and end thrust. Lateral displacement coordinates are x and y with s coordinate being along the length of the shaft.

Differential equations for out-of-plane bending are

$$EI\frac{d^4x}{ds^4} + T\frac{d^3y}{ds^3} + P\frac{d^2x}{ds^2} - m\omega^2 x = 0 \tag{2-169a}$$

$$EI\frac{d^4y}{ds^4} - T\frac{d^3x}{ds^3} + P\frac{d^2y}{ds^2} - m\omega^2 y = 0 \tag{2-169b}$$

The compressive force, P, is taken as a plus number. Note the similarity between Eqs. (2-157) and (2-169). The differences between the two sets of equations are: the compressive force and the acceleration term, which in this case is based on normal acceleration cause by steady whirl. Steps to a solution require a determination of four roots to an algebraic equation, which are determined numerically by trial.

Two sets of boundary conditions are considered: pinned-pinned and fixed-fixed. Results in both cases are presented in graph form, which relate three dimensionless parameters

$$A = \frac{TL}{2EI}$$

$$B = \frac{PL^2}{4EI}$$

$$C = \frac{m\omega^2 L^4}{16EI}$$

The authors show that A, B, and C can be related by a simple formula to give a reasonable approximation.

$$\left(\frac{A}{\pi}\right)^2 + \frac{4B}{\pi^2} + \frac{16C}{\pi^4} = 1 \tag{2-170}$$

The equation is more accurate at small values of C and exact if either A or C is zero. A similar express is given for the fixed-fixed case with each term being multiplied by 0.49, 0.25 and 0.195, respectively.

Based on Eq. (2-167) and Eq. (2-168), it would appear that Eq. (2-170) could be casted in the following form.

$$\frac{P}{P_{Euler}} + \left(\frac{M_t}{M_{t,Greenhill}}\right)^2 + \left(\frac{\omega}{\omega_{crit}}\right)^2 = 1 \tag{2-171}$$

This equation is not exactly correct as pointed out by the authors, but does give a decent approximations [20].

2.6.2 Torsion Buckling of Long Vertical Pipe

The previous solution applies to short pipe where tension is assumed to be constant throughout. This solution might be adaptable to short section of drill pipe and drill collars by using average tension or compression over a given length.

However, in general, variable effective tension has to be considered. The solution given here was developed specifically for this manuscript. The governing differential equations of bending stem from Huang and Dareing [21]. They are similar to Greenhill's equations [13,17], except for the effective tension term.

Figure 2-16 illustrates the model and shows the coordinates used in the following discussion. The resulting differential equations of bending in *xz* and *yz* planes as a result of torque applied at each end are

$$EI\frac{d^4x}{dz^4} - T\frac{d^3y}{dz^3} - \frac{d}{dz}\left[T_{eff}\frac{dx}{dz}\right] = 0 \tag{2-172a}$$

$$EI\frac{d^4y}{dz^4} + T\frac{d^3x}{dz^3} - \frac{d}{dz}\left[T_{eff}\frac{dy}{dz}\right] = 0 \tag{2-172b}$$

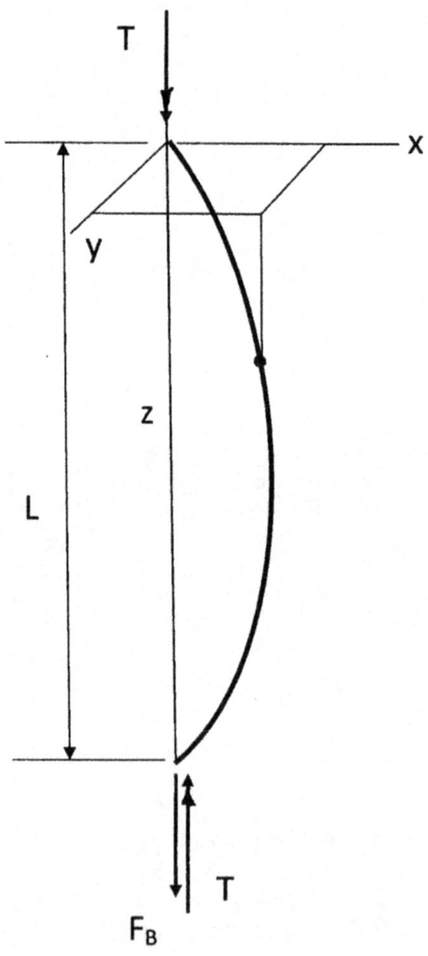

Figure 2-16. Torsion stability model.

The symbol, T, without the subscript, represents applied torque, which is assumed to be constant over the entire length. Effective tension, T_{eff}, however, is assumed to vary along the pipe.

Force, F_B represents the actual internal force in the pipe at $z = L$. The effective tension is expressed in terms of the independent variable, z, is

$$T_{eff} = \left[F_B + w(L - z)\right] + zw_m \qquad (2\text{-}173)$$

The first term is the actual tension in the pipe at location z, while the last term represents the effects of hydrostatic fluid pressure. Here, we assume the fluid density is the same both inside and outside the pipe. If the lower end of the pipe is open to the fluid, $F_B = -Lw_m$ and

$$T_{eff} = (L - z)(w - w_m) \qquad (2\text{-}174)$$

Going back to the more general case and rearranging Eq. (2-173)

$$T_{eff} = (F_B + w_m L) + (w - w_m)(L - z) \qquad (2\text{-}175)$$

Boundary conditions for pinned conditions at both top and bottom are

$$x(0) = y(0) = 0 \qquad (2\text{-}176a)$$

$$x(L) = y(L) = 0 \qquad (2\text{-}176b)$$

$$EI\frac{d^2x}{dz^2}(0) = T\frac{dy}{dz}(0) \qquad (2\text{-}176c)$$

$$EI\frac{d^2y}{dz^2}(0) = -T\frac{dx}{dz}(0) \qquad (2\text{-}176d)$$

$$EI\frac{d^2x}{dz^2}(L) = T\frac{dy}{dz}(L) \qquad (2\text{-}176e)$$

$$EI\frac{d^2y}{dz^2}(L) = -T\frac{dx}{dz}(L) \qquad (2\text{-}176f)$$

Equations (2-172a) and (2-172b) are now combined using the complex variable

$$u = x(z) + iy(z) \qquad (2\text{-}177)$$

Multiplying Eq. (2-172b) by the imaginary number i and adding the resulting equation to Eq. (2-172a) gives

$$EI\frac{d^4u}{dz^4} - iT\frac{d^3u}{dz^3} - \frac{d}{dz}\left\{\left[\left(F_B + w_mL\right) + \left(w - w_m\right)(L - z)\right]\frac{du}{dz}\right\} = 0 \quad (2\text{-}178)$$

Now replace z with the dimensionless number

$$\varsigma = \frac{L - z}{L} \quad (2\text{-}179)$$

and noting that $dz = -Ld\zeta$. Dividing through by EI gives

$$\frac{d^4u}{d\zeta^4} - i\Theta\frac{d^3u}{d\zeta^3} - \frac{d}{d\zeta}\left[(\beta + \alpha\zeta)\frac{du}{d\zeta}\right] = 0 \quad (2\text{-}180)$$

which expands into

$$\frac{d^4u}{d\zeta^4} - i\Theta\frac{d^3u}{d\zeta^3} - \beta\frac{d^2u}{d\zeta^2} - \alpha\zeta\frac{d^2u}{d\zeta^2} - \alpha\frac{du}{d\zeta} = 0 \quad (2\text{-}181)$$

where

$$\alpha = \frac{\left(w - w_m\right)L^3}{EI}$$

$$\beta = \frac{\left(F_B + w_m L\right)L^2}{EI}$$

$$\Theta = \frac{TL}{EI}$$

The parameter, Θ, is introduced, which replaces β in ref. [21]. This was done so the symbol, β, is consistent throughout.

2.6.2.1 Both Top and Bottom Ends Pinned

The boundary conditions (Eq. (2-176)) for this case transform into

$$u(0) = 0 \quad (2\text{-}182a)$$

$$u(1) = 0 \quad (2\text{-}182b)$$

$$\frac{d^2u}{d\zeta^2}(0) = i\Theta\frac{du}{d\zeta}(0) \quad (2\text{-}182c)$$

$$\frac{d^2u}{d\zeta^2}(1) = i\Theta\frac{du}{d\zeta}(1) \quad (2\text{-}182d)$$

Assuming the solution to the differential equation (Eq. (2-181)) of the form

$$u(\zeta) = a_0 + a_1\zeta + a_2\zeta^2 + \ldots\ldots + a_n\zeta^n + \ldots \qquad (2\text{-}183)$$

where the a's are complex constants for a given set of boundary conditions, and ζ is the independent real variable as defined by Eq. (2-179). $u(\zeta)$ is a complex function as defined earlier.

Substituting Eq. (2-183) into Eq. (2-181) gives the recurrence equation

$$a_n = i\frac{\Theta}{n}a_{n-1} + \frac{\beta}{n(n-1)}a_{n-2} + \frac{(n-3)\alpha}{n(n-1)(n-2)}a_{n-3},\ n \geq 4 \qquad (2\text{-}184)$$

The complex constants are dependent on α and β, which are real numbers. Using Eq. (2-183) and its derivatives, the boundary conditions yield

$$a_0 = 0 \qquad (2\text{-}185a)$$

$$\sum_{n=1}^{\infty} a_n = 0 \qquad (2\text{-}185b)$$

$$a_2 = i\frac{\Theta}{2}a_1 \qquad (2\text{-}185c)$$

$$\sum_{n=1}^{\infty} n(n-1)a_n = i\Theta\sum_{n=1}^{\infty} na_n \qquad (2\text{-}185d)$$

These conditions show that $a_0 = 0$ and a_2 depends on a_1. From these conditions and by successive application of the recurrence Eq. (2-184), the coefficient a_n can be expressed as a linear combination of a_1 and a_3 or

$$a_n = (F_n + iJ_n)a_1 + (H_n + iL_n)a_3 \qquad (2\text{-}186)$$

where F_n, H_n, J_n, L_n are real. It is apparent that for $n = 0, 1, 2, 3$ and using boundary conditions Eqs. (2-185a) and (2-185c)

$$\begin{array}{llll} F_0 = 0 & J_0 = 0 & H_0 = 0 & L_0 = 0 \\ F_1 = 1 & J_1 = 0 & H_1 = 0 & L_1 = 0 \\ F_2 = 0 & J_2 = T/2 & H_2 = 0 & L_2 = 0 \\ F_3 = 0 & J_3 = 0 & H_3 = 1 & L_3 = 0 \end{array} \qquad (2\text{-}187)$$

The remaining values for F_n, H_n, J_n, L_n for $(n \geq 4)$ are determined from the recurrence equations derived from Eqs. (2-184) and (2-186). The starting values are used to obtain higher-order values of F_n, H_n, J_n, L_n by repeated application of Eq. (2-188a, b, c, d), which are each functions of Θ, α, and β.

Table 2-7. Critical buckling torque for 5 ½ in. (19.2 lb/ft) drill pipe.

α	*L*, ft	Θ	*T*, lb-ft	*T*, kip-ft
0		6.28318		
1	56.04602	4.99873	3.02E+05	301.8
2	70.61192	4.60423	2.21E+05	220.7
3	80.82938	4.39605	1.84E+05	184.1
4	88.96339	4.27309	1.63E+05	162.6
5	95.8322	4.19655	1.48E+05	148.2
6	101.8363	4.14753	1.38E+05	137.8
7	107.2052	4.11579	1.30E+05	129.9
8	112.0843	4.09544	1.24E+05	123.7
9	116.5719	4.08285	1.19E+05	118.5
10	120.7382	4.07574	1.14E+05	114.2
15	138.2089	4.08336	1.00E+05	100.0
20	152.1171	4.12117	9.17E+04	91.7
25	163.862	4.16903	8.61E+04	86.1
30	174.1283	4.22072	8.20E+04	82.0
35	183.3085	4.27381	7.89E+04	78.9
40	191.6511	4.32723	7.64E+04	76.4
45	199.3244	4.38048	7.44E+04	74.4
50	206.4484	4.43331	7.27E+04	72.7
100	260.1026	4.92777	6.41E+04	64.1
150	297.7392	5.36535	6.10E+04	61.0
200	327.7012	5.75751	5.95E+04	59.5
250	353.0028	6.11295	5.86E+04	58.6
300	375.1192	6.43826	5.81E+04	58.1
350	394.8959	6.73847	5.77E+04	57.7
400	412.8681	7.01756	5.75E+04	57.5
450	429.3985	7.27919	5.74E+04	57.4
500	444.7454	7.52138	5.72E+04	57.2
750	509.0995	8.57198	5.70E+04	57.0
1000	560.3311	9.4216	5.69E+04	56.9
1500	641.4104	10.77679	5.69E+04	56.9

$$F_n = -\frac{\Theta}{n} J_{n-1} + \frac{\beta}{n(n-1)} F_{n-2} + \frac{(n-3)\alpha}{n(n-1)(n-2)} F_{n-3} \tag{2-188a}$$

$$J_n = +\frac{\Theta}{n} F_{n-1} + \frac{\beta}{n(n-1)} J_{n-2} + \frac{(n-3)\alpha}{n(n-1)(n-2)} J_{n-3} \tag{2-188b}$$

$$H_n = -\frac{\Theta}{n} L_{n-1} + \frac{\beta}{n(n-1)} H_{n-2} + \frac{(n-3)\alpha}{n(n-1)(n-2)} H_{n-3} \tag{2-188c}$$

$$L_n = +\frac{\Theta}{n} H_{n-1} + \frac{\beta}{n(n-1)} L_{n-2} + \frac{(n-3)\alpha}{n(n-1)(n-2)} L_{n-3} \tag{2-188d}$$

Next, consider the 2nd and 4th conditions. Substituting Eq. (2-186) into the boundary conditions Eqs. (2-185b) and (2-185d), the following two homogeneous equations are obtained.

$$\left(\sum F_n + i\sum J_n\right)a_1 + \left(\sum H_n + i\sum L_n\right)a_3 = 0 \tag{2-189a}$$

and

$$\begin{aligned}&\left\{\left[\sum n(n-1)F_n + \Theta\sum nJ_n\right] + i\left[\sum n(n-1)J_n - \Theta\sum nF_n\right]\right\}a_1\\&+\left\{\left[\sum n(n-1)H_n + \Theta\sum nL_n\right] + i\left[\sum n(n-1)L_n - \Theta\sum nH_n\right]\right\}a_3 = 0\end{aligned} \tag{2-189b}$$

For nontrivial solutions of a_1 and a_3, the determinant of the coefficients in Eq. (2-190) must be zero. The expansion of the determinate produces a real part and an imaginary part. Both parts have to be zero by each eigenvalue. The real part of the expanded determinate is

$$\begin{aligned}&\left(\sum F_n\right)\left[\sum n(n-1)H_n + \Theta\sum nL_n\right] - \left(\sum J_n\right)\left[\sum n(n-1)L_n - \Theta\sum nH_n\right]\\&-\left(\sum H_n\right)\left[\sum n(n-1)F_n + \Theta\sum nJ_n\right] + \left(\sum L_n\right)\left[\sum n(n-1)J_n - \Theta\sum nF_n\right] = 0\end{aligned} \tag{2-190a}$$

and the imaginary part is

$$\begin{aligned}&\left(\sum J_n\right)\left[\sum n(n-1)H_n + \Theta\sum nL_n\right] + \left(\sum F_n\right)\left[\sum n(n-1)L_n - \Theta\sum nH_n\right]\\&-\left(\sum L_n\right)\left[\sum n(n-1)F_n + \Theta\sum nJ_n\right] - \left(\sum H_n\right)\left[\sum n(n-1)J_n - \Theta\sum nF_n\right] = 0\end{aligned} \tag{2-190b}$$

Let the left sides of Eqs. (2-190a) and (2-190b) be denoted by P and Q, respectively, then Eq. (2-190a, b) can be written as

$$P(\Theta,\alpha,\beta)=0 \tag{2-191a}$$

$$Q(\Theta,\alpha,\beta)=0 \tag{2-191b}$$

Both are real value functions. The solution to these equations gives the critical buckling torque parameter, Θ_{cr} and the critical buckling torque, T_{cr} corresponding to given values of α. Note that for a given value of α, Θ_{cr}must satisfy both P and Q functions simultaneously. There may be some values of Θ that satisfy $P = 0$ but may not satisfy $Q = 0$. In this case, this particular Θ is not the solution or eigenvalue.

2.6.2.2 *Simply Supported at Both Ends with no End Thrust*

In this case, the end pull, F_B, is assumed to be equal to zero and the pipe exposed to hydrostatic pressure, making $\beta = 0$. Solutions to Eqs. (2-190a) and (2-190b) were found by trial. The results are given in Table 2-7.

By assuming the pipe is open ended, $F_B = -w_m L$, so $\beta = 0$. Using the numbers in Table 2-7 and

$$\alpha = \frac{(w - w_m)L^3}{EI}$$

$$\Theta = \frac{TL}{EI}$$

an interesting plot is generated to show the effects of pipe length on the critical buckling torque. The data applies to 5 ½ in. (4.892 in. ID, 19.2 lb/ft) drill pipe. Drilling mud weight can be given various values.

A special case was considered by assuming $w_m = 0$. Critical buckling torque is plotted against pipe length (Fig. 2-17) and is indicated by "Linear Tension"; results apply to open pipe in air for comparison with Greenhill results. The data show that critical buckling reduces to a level of ~55,000 ft/lbs and stays constant at this level. This is expected because nearly all of the buckling takes place at the lower end. Pipe length has little or no effect on critical buckling force, past $L = 400$ ft.

Critical buckling predicted by Greenhill's formula (Eq. 2-166) is hyperbolic with pipe length. Tension is assumed to be equal to zero in this case.

These two plots are based on dry conditions, i.e., $w_m = 0$. Separate calculations show that drilling mud density has a negligible effect on torsion buckling.

Consider the case where 5 ½ in. (19.2 lb/ft) drill pipe is simply supported at both ends, but the bottom is opened to hydrostatic pressure. In this case, $F_B = 0$ as well as β. If length of drill pipe is 10,000 ft, critical buckling torque is essentially the same as for 641.4 ft of pipe or 56,900 ft/lbs (see Table 2-7). Mud weight which factors into α does not affect the outcome. Pipe rotation is not considered in this example.

Including a force, F_B, at the lower end adds a useful extension to the problem. The entire analysis in the previous section applies except that $\beta \neq 0$.

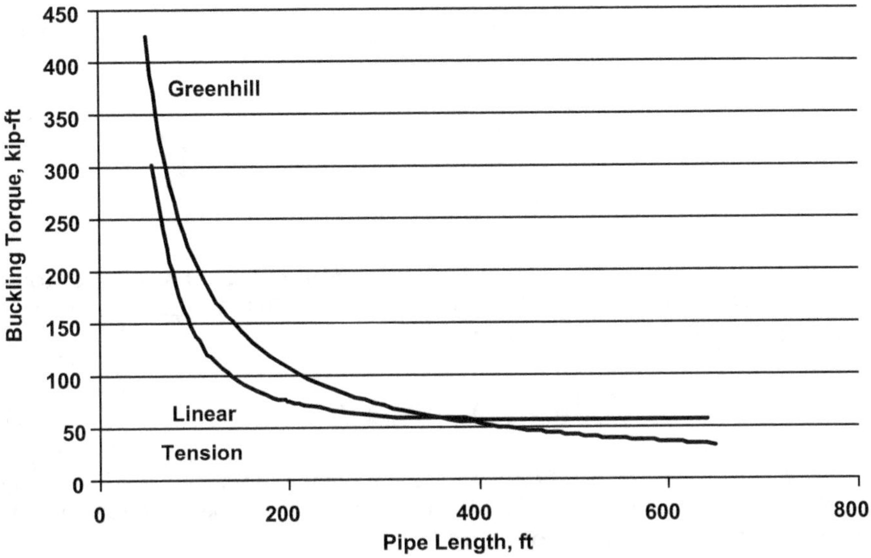

Figure 2-17. Critical buckling torque for first mode (5 ½ in. drill pipe, 19.2 lb/ft).

2.6.2.3 Top End Pinned and Bottom End Fixed

In this case, the bottom end is fixed, and allowance is made for an applied load at the bottom end. The approach to the solution is the same as above with a minor adjustment to the boundary conditions.

Boundary conditions for this case are

$$u(0) = 0 \tag{2-192a}$$

$$u(1) = 0 \tag{2-192b}$$

$$\frac{du}{d\zeta}(0) = 0 \quad \text{(lower end)} \tag{2-192c}$$

$$\frac{d^2u}{d\zeta^2}(1) = i\Theta\frac{du}{d\zeta}(1) \quad \text{(upper end)} \tag{2-192d}$$

Substitution of Eq. (2-183) and its derivatives into these equations gives

$$a_0 = 0 \tag{2-193a}$$

$$\sum_{n=1}^{\infty} a_n = 0 \tag{2-193b}$$

$$a_1 = 0 \tag{2-193c}$$

$$\sum_{n=1}^{\infty} n(n-1)a_n = i\Theta \sum_{n=1}^{\infty} na_n \tag{2-193d}$$

The 1st and 3rd conditions show a_0 and a_1 are both zero. From these conditions and by successive application of the recurrence in Eq. (2-184), the coefficient a_n can be expressed as a linear combination of a_2 and a_3 or

$$a_n = (F_n + iJ_n)a_2 + (H_n + iL_n)a_3 \tag{2-194}$$

where F_n, H_n, J_n, L_n are real. While the mathematical expressions for F_n, H_n, J_n, L_n are the same as before, their initial values are different because of the new set of boundary conditions. Accordingly

$$\begin{array}{llll} F_0 = 0 & J_0 = 0 & H_0 = 0 & L_0 = 0 \\ F_1 = 1 & J_1 = 0 & H_1 = 0 & L_1 = 0 \\ F_2 = 0 & J_2 = 0 & H_2 = 0 & L_2 = 0 \\ F_3 = 0 & J_3 = 0 & H_3 = 1 & L_3 = 0 \end{array} \tag{2-195}$$

From the 2nd and 4th boundary condition

$$\left(\sum F_n + i\sum J_n\right)a_2 + \left(\sum H_n + i\sum L_n\right)a_3 = 0 \tag{2-196a}$$

and

$$\begin{aligned} &\left\{\left[\sum n(n-1)F_n + \Theta\sum nJ_n\right] + i\left[\sum n(n-1)J_n - \Theta\sum nF_n\right]\right\}a_2 \\ &+\left\{\left[\sum n(n-1)H_n + \Theta\sum nL_n\right] + i\left[\sum n(n-1)L_n - \Theta\sum nH_n\right]\right\}a_3 = 0 \end{aligned} \tag{2-196b}$$

In this case, the constants are a_2 and a_3, not a_1 and a_3 as before. For nontrivial solutions of a_2 and a_3, the determinant of the coefficients in Eq. (2-196) must be zero. The real part of the expanded determinate is

$$\begin{aligned} &\left(\sum F_n\right)\left[\sum n(n-1)H_n + \Theta\sum nL_n\right] - \left(\sum J_n\right)\left[\sum n(n-1)L_n - \Theta\sum nH_n\right] \\ &-\left(\sum H_n\right)\left[\sum n(n-1)F_n + \Theta\sum nJ_n\right] + \left(\sum L_n\right)\left[\sum n(n-1)J_n - \Theta\sum nF_n\right] = 0 \end{aligned} \tag{2-197a}$$

and the imaginary part is

$$\left(\sum J_n\right)\left[\sum n(n-1)H_n + \Theta\sum nL_n\right] + \left(\sum F_n\right)\left[\sum n(n-1)\,L_n - \Theta\sum nH_n\right]$$
$$-\left(\sum L_n\right)\left[\sum n(n-1)F_n + \Theta\sum nJ_n\right] - \left(\sum H_n\right)\left[\sum n(n-1)J_n - \Theta\sum nF_n\right] = 0 \tag{2-197b}$$

Let the left sides of Eqs. (2-197a) and (2-197b) be denoted by P and Q, respectively, then Eq. (2-197) can be written as

$$P(\Theta,\alpha,\beta) = 0 \tag{2-198a}$$

$$Q(\Theta,\alpha,\beta) = 0 \tag{2-198b}$$

The steps required to obtain the critical Θ is the same as for the first set of boundary conditions. Only the starting values of F_n, H_n, J_n, L_n are different.

2.7 ROTATIONAL STABILITY AND WHIRL OF VERTICAL PIPE

Drillstring whirl has been an area of interest in recent years. Drill bit performance, pipe wear, and tool damage have been attributed to pipe whirl [22]. Sections of drillstrings can experience, forward, backward, or synchronous whirl. Prediction and control of whirl are complicated by multiple factors that must be considered in any attempt to model this problem. This section predicts conditions under which drillstrings experience these types of behavior.

Synchronous whirl refers to pipe (or a disc) rotation in which one side continuously faces outward [11]. Lateral movement maybe due to dynamic instability or imbalance. Lalanne and Ferraris [23] explain rotor and shaft dynamics for a discs that are non-symmetrical on a shaft. Results predict conditions of forward or backward whirl. As applied to drilling, synchronous whirl has the potential for wearing down one side of drill pipe or drill collars.

2.7.1 Stability of Long Vertical Pipe due to Torque, Rotation, and Damping

The prediction of whirling motion in drillstrings is complicated by multiple drilling parameters and hole confinement. Factors that can affect pipe whirl are torque, axial force, rotary speed, and damping caused by fluid viscosity. This section addresses the special case of vertical pipes rotating and transmitting torque in a viscous media. The results predict conditions of forward or backward whirl, stable or unstable whirl, and synchronous whirl.

It is difficult to account for every constraint but much insight can be obtained by considering tubular behavior in an open space. It is apparent that when portions of drillstrings make contact with a well bore, contact forces and resulting kinematics greatly affect whirling behavior. The analysis that follows is based on conditions of non-confinement with an initial concentric position in a vertical well bore.

The differential equations of out of plane motion of vertical pipe are developed by Huang and Dareing [21].

2.7.2 Lower End Subjected to a Direct Pull Force

The differential equations of motion for this case are expressed in terms of the dependent variables $x(z)$ and $y(z)$. The model is illustrated in Fig. 2-18.

$$EI\frac{\partial^4 x}{\partial z^4} - T\frac{\partial^3 y}{\partial z^3} - \frac{\partial}{\partial z}\left(\left[(F_B + w_m L) + (w - w_m)(L - z)\right]\frac{\partial x}{\partial z}\right) + m\left(\frac{\partial^2 x}{\partial t^2} - 2\omega\frac{\partial y}{\partial t} - \omega^2 x\right) + c\left(\frac{\partial x}{\partial t} - \omega y\right) = 0 \quad \text{(2-199a)}$$

$$EI\frac{\partial^4 y}{\partial z^4} + T\frac{\partial^3 x}{\partial z^3} - \frac{\partial}{\partial z}\left(\left[(F_B + w_m L) + (w - w_m)(L - z)\right]\frac{\partial y}{\partial z}\right) + m\left(\frac{\partial^2 y}{\partial t^2} + 2\omega\frac{\partial x}{\partial t} - \omega^2 y\right) + c\left(\frac{\partial y}{\partial t} + \omega x\right) = 0 \quad \text{(2-199b)}$$

The x and y displacements are measured relative to a rotating reference frame (xyz) which rotates about the Z axis as shown in Fig. 2-18.

These equations convert to

$$\frac{\partial^4 x}{\partial \zeta^4} + \Theta\frac{\partial^3 y}{\partial \zeta^3} - \frac{\partial}{\partial \zeta}\left([\beta + \alpha\zeta]\frac{\partial x}{\partial \zeta}\right) + \frac{mL^4}{EI}\left(\frac{\partial^2 x}{\partial t^2} - 2\omega\frac{\partial y}{\partial t} - \omega^2 x\right) + \frac{cL^4}{EI}\left(\frac{\partial x}{\partial t} - \omega y\right) = 0 \quad \text{(2-200a)}$$

$$\frac{\partial^4 y}{\partial \zeta^4} - \Theta\frac{\partial^3 x}{\partial \zeta^3} - \frac{\partial}{\partial \zeta}\left([\beta + \alpha\zeta]\frac{\partial y}{\partial \zeta}\right) + \frac{mL^4}{EI}\left(\frac{\partial^2 y}{\partial t^2} + 2\omega\frac{\partial x}{\partial t} - \omega^2 y\right) + \frac{cL^4}{EI}\left(\frac{\partial y}{\partial t} + \omega x\right) = 0 \quad \text{(2-200b)}$$

where

$$\zeta = \frac{L - z}{L}$$

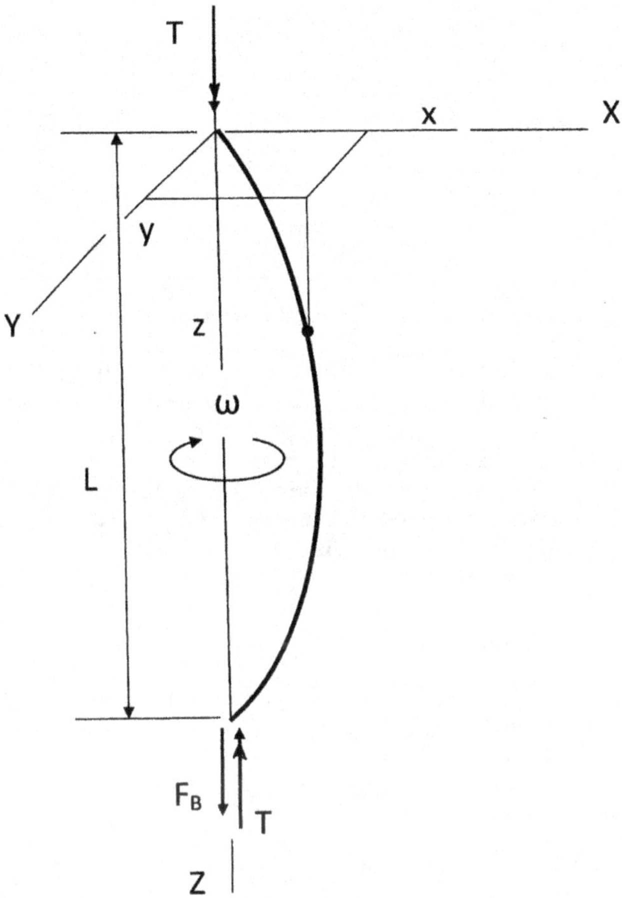

Figure 2-18. Coordinate systems for rotational stability analysis.

$$\Theta = \frac{TL}{EI}$$

$$\alpha = \frac{(w - w_m)L^3}{EI}$$

$$\beta = \frac{(F_B + Lw_m)L^2}{EI}$$

Equation (2-200) are combined into one equation in terms of complex variables

$$\frac{\partial^4 U}{\partial \zeta^4} - i\Theta\frac{\partial^3 U}{\partial \zeta^3} - \beta\frac{\partial^2 U}{\partial \zeta^2} - \alpha\varsigma\frac{\partial^2 U}{\partial \zeta^2} - \alpha\frac{\partial U}{\partial \zeta} + \frac{mL^4}{EI}\left[\frac{\partial^2 U}{\partial t^2} + i\left(2\omega\frac{\partial U}{\partial t}\right) - \omega^2 U\right]$$
$$+ \frac{cL^4}{EI}\left(\frac{\partial U}{\partial t} + i\omega U\right) = 0 \qquad (2\text{-}201)$$

where

$$U = x\,(z,t) + iy\,(z,t)$$

Stability is considered here in the dynamic sense, i.e., the pipe is unstable if the lateral displacement of pipe from its initial centerline position increases with time after an initial lateral disturbance. On the other hand, the pipe is considered to be stable if the displacement decreases with time after some initial disturbance. If the displacement of the pipe neither increases nor decreases with time, the pipe is in stationary equilibrium.

A unique feature of this problem is the variable internal tension from top to bottom and the effects of hydrostatic pressure, which increases linearly with depth. The coordinate system is shown in Fig. 2-18 with *xyz* is a rotating reference frame and *XYZ* is a fixed reference frame.

2.7.3 Lower End Open to Hydrostatic Pressure

Now consider the special case of a simply supported pipe (Fig. 2-18) whose lower end is open to hydrostatic pressure. This assumption ($F_B = -w_m L$) alters the effective tension term as shown below.

In this case, the differential equations of bending (Eq. (2-200a, b)) reduce to

$$EI\frac{\partial^4 x}{\partial z^4} - T\frac{\partial^3 y}{\partial z^3} - \frac{\partial}{\partial z}\left[(w - w_m)(L - z)\frac{\partial x}{\partial z}\right] + m\left(\frac{\partial^2 x}{\partial t^2} - 2\omega\frac{\partial y}{\partial t} - \omega^2 x\right) + c\left(\frac{\partial x}{\partial t} - \omega y\right) = 0 \qquad (2\text{-}202a)$$

$$EI\frac{\partial^4 y}{\partial z^4} + T\frac{\partial^3 x}{\partial z^3} - \frac{\partial}{\partial z}\left[(w - w_m)(L - z)\frac{\partial y}{\partial z}\right] + m\left(\frac{\partial^2 y}{\partial t^2} + 2\omega\frac{\partial x}{\partial t} - \omega^2 y\right) + c\left(\frac{\partial y}{\partial t} + \omega x\right) = 0 \qquad (2\text{-}202b)$$

The nomenclature in these equations is defined below:

EI = cross-sectional flexural rigidity
T = applied torque
ω = angular velocity of drill pipe and *xyz* reference frame attached to the pipe at $z = 0$
w = weight per unit length of pipe
w_m = weight of drilling mud per unit length of pipe
m = mass of pipe per unit length
c = damping coefficient (force per unit length per unit velocity)

The following boundary conditions will apply.

$$x(0) = y(0) = 0 \tag{2-203a}$$

$$x(L) = y(L) = 0 \tag{2-203b}$$

$$EI\frac{d^2x}{dz^2}(0) = T\frac{dy}{dz}(0) \tag{2-203c}$$

$$EI\frac{d^2y}{dz^2}(0) = -T\frac{dx}{dz}(0) \tag{2-203d}$$

$$EI\frac{d^2x}{dz^2}(L) = T\frac{dy}{dz}(L) \tag{2-203e}$$

$$EI\frac{d^2y}{dz^2}(L) = -T\frac{dx}{dz}(L) \tag{2-203f}$$

2.7.3.1 Mathematical Solution

It is convenient to combine Eqs. (2-202a) and (2-202b) into a single equation using the complex function

$$U(z,t) = x(z,t) + iy(z,t) \tag{2-204}$$

The resulting differential equation is

$$\frac{\partial^4 U}{\partial \zeta^4} - i\Theta\frac{\partial^3 U}{\partial \zeta^3} - \alpha\varsigma\frac{\partial^2 U}{\partial \zeta^2} - \alpha\frac{\partial U}{\partial \zeta} + \frac{mL^4}{EI}\left[\frac{\partial^2 U}{\partial t^2} + i\left(2\omega\frac{\partial U}{\partial t}\right) - \omega^2 U\right] + \frac{cL^4}{EI}\left(\frac{\partial U}{\partial t} + i\omega U\right) = 0 \tag{2-205}$$

where

$$\zeta = \frac{L-z}{L}$$

$$\alpha = \frac{(w - w_m)L^3}{EI}$$

$$\Theta = \frac{TL}{EI}$$

β does not appear in Eq. (2-205). The boundary conditions become

Table 2-8. Eigenvalues γ and critical damping factor η_{cr}.

Weight parameter	Torque parameter	Eigenvalue $\gamma = R_\gamma + iI_\gamma$		Damping factor, η_{cr}
α	Θ	R_γ	I_γ	$\frac{I_\gamma}{\sqrt{R_\gamma}}$
	1	100.36	0.247	0.025
	2	94.31	0.508	0.052
1	3	83.87	0.777	0.085
	4	68.39	1.069	0.156
	5	46.48	1.419	0.208
	1	143.70	2.291	0.19
	2	138.55	4.641	0.40
10	3	128.57	7.187	0.64
	4	114.47	9.815	0.92
	5	94.97	12.956	1.33
	1	520.07	13.57	0.59
	2	516.51	27.64	1.22
100	3	510.37	41.89	1.85
	4	501.57	56.66	2.53
	5	489.91	72.17	3.26
	1	3507.3	47.3	0.80
	2	3505.2	95.2	1.61
1000	3	3503.1	142.3	2.41
	4	3498.4	191.6	3.24
	5	3493.2	240.9	4.08
	6	3486.6	290.1	4.91

$$U(0,t) = 0 \tag{2-206a}$$

$$U(1,t) = 0 \tag{2-206b}$$

$$\frac{d^2U}{d\zeta^2}(0,t) = i\Theta\frac{dU}{d\zeta}(0,t) \tag{2-206c}$$

$$\frac{d^2U}{d\zeta^2}(1,t) = i\Theta\frac{dU}{d\zeta}(1,t) \tag{2-206d}$$

Using the method of separation of variables, Eq. (2-205) is expressed in the form

$$U(\zeta,t) = u(\zeta)\tau(t) \tag{2-207}$$

The two components u(ζ) and τ(t) are separated and equated to a complex constant represented by

$$\gamma = R_\gamma + iI_\gamma \tag{2-208}$$

The two resulting differential equations are

$$\frac{d^4u}{d\zeta^4} - i\Theta\frac{d^3u}{d\zeta^3} - \alpha\varsigma\frac{d^2u}{d\zeta^2} - \alpha\frac{du}{d\zeta} - (R_\gamma + iI_\gamma)u = 0 \tag{2-209}$$

$$\frac{d^2\tau}{dt^2} + 2\left(\frac{c}{2m} + i\omega\right)\frac{d\tau}{dt} + \frac{EI}{mL^4}\left[R_\gamma - \Omega^2 + i\left(I_\gamma + \frac{c\omega L^4}{EI}\right)\right]\tau = 0 \tag{2-210}$$

where

$$\Omega^2 = \frac{m\omega^2L^4}{EI}$$

2.7.3.2 *Stability Analysis*

The solution to Eq. (2-209) describes the shape of the deflected pipe with respect to the rotating reference frame while the solution to Eq. (2-210) describes the time history of displacements along the pipe. The time history solution establishes whether or not the pipe is dynamically stable or unstable.

The boundary conditions that apply to Eq. (2-209) are

$$u(0) = 0 \tag{2-211a}$$

$$u(1) = 0 \tag{2-211b}$$

$$\frac{d^2u}{d\zeta^2}(0) = i\Theta\frac{du}{d\zeta}(0) \tag{2-211c}$$

$$\frac{d^2u}{d\zeta^2}(1) = i\Theta\frac{du}{d\zeta}(1) \tag{2-211d}$$

First consider Eq. (2-209) along with these boundary conditions. The goal here is to determine the complex eigenvalue, γ. Mode shape and the time-dependent behavior of the total pipe depend on both the real and imaginary parts of this number.

The solution to Eq. (2-209) solution is assumed to be of the form of a power series.

$$u(\varsigma) = a_0 + a_1\varsigma + a_2\varsigma^2 + \ldots\ldots + a_n\varsigma^n + \ldots \tag{2-212}$$

where the a's are complex constants, to be determined from boundary conditions. ζ is the independent real variable as defined earlier.

Substituting Eq. (2-212) and its derivatives into Eq. (2-209), give the recurrence equation

$$a_n = i\frac{\Theta}{n} a_{n-1} + \frac{(n-3)\alpha}{n(n-1)(n-2)} a_{n-3} + \frac{R_\gamma + iI_\gamma}{n(n-1)(n-2)(n-3)} a_{n-4},\ n = 4, 5, 6\ldots \quad (2\text{-}213)$$

The constants, a_n, are complex and dependent on α, Θ, R_γ, I_γ.

Using Eq. (2-212) and its derivatives, the boundary conditions yield

$$a_0 = 0 \quad (2\text{-}214a)$$

$$\sum_{n=1}^{\infty} a_n = 0 \quad (2\text{-}214b)$$

$$a_2 = i\frac{\Theta}{2} a_1 \quad (2\text{-}214c)$$

$$\sum_{n=1}^{\infty} n(n-1)a_n = i\Theta \sum_{n=1}^{\infty} na_n \quad (2\text{-}214d)$$

The constants, a_0 and a_2, are established by Eq. (2-214a, c). By successive application of the recurrence Eq. (2-213), coefficients a_n can be expressed as a linear combination of a_1 and a_3.

$$a_n = (F_n + iJ_n)a_1 + (H_n + iL_n)a_3 \quad (2\text{-}215)$$

Following the steps explained earlier, F_n, J_n, H_n, L_n are expressed in terms of their previous values (see ref. [21]). The eigenvalues are determined from the characteristic equation. The real and imaginary parts lead to two functions

$$P(\alpha, \Theta, R_\gamma, I_\gamma) = 0 \quad (2\text{-}216a)$$

$$Q(\alpha, \Theta, R_\gamma, I_\gamma) = 0 \quad (2\text{-}216b)$$

which are solved numerically by trail, to obtain R_γ and I_γ. Table 2-8 gives these eigenvalues for different combinations of α and Θ.

Next consider the time-dependent differential equation, Eq. (2-210). Its solution is of the form

$$\tau(t) = A \exp\left[\lambda_1 \left(\frac{EI}{mL^4}\right)^{\frac{1}{2}} t\right] + B \exp\left[\lambda_2 \left(\frac{EI}{mL^4}\right)^{\frac{1}{2}} t\right] \quad (2\text{-}217)$$

where A and B are arbitrary complex constants, which depend on initial conditions. The complex constants, λ_1 and λ_2 are defined as follows.

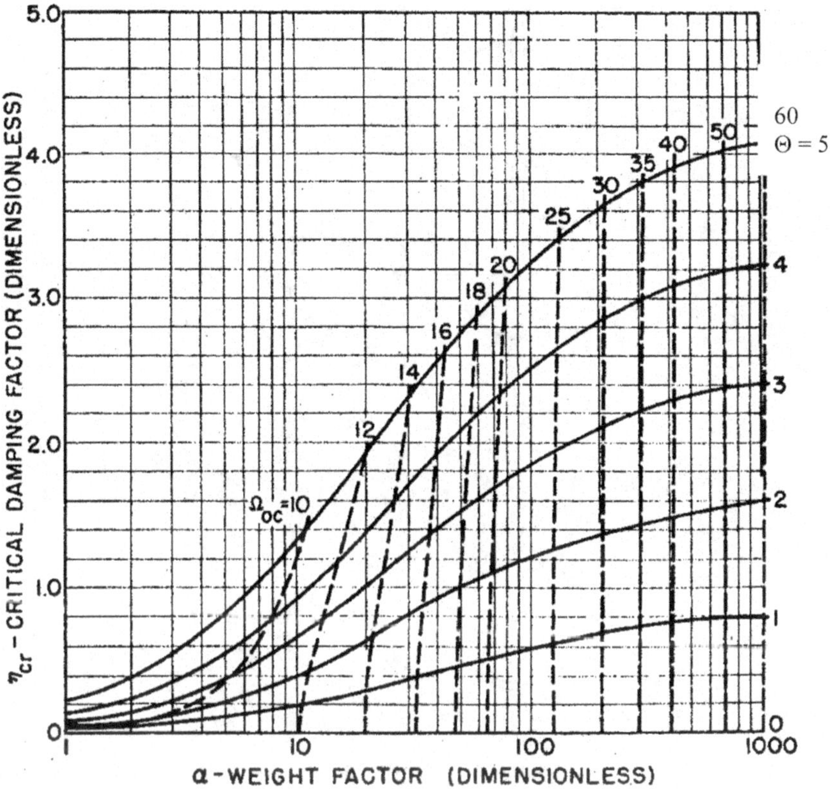

Figure 2-19. Stability data for first mode.

$$\lambda_{1,2} = -\frac{\eta}{2} \pm \frac{1}{\sqrt{2}}\left\{\left[\left(R_\gamma - \frac{\eta^2}{4}\right)^2 + I_\gamma^2\right]^{\frac{1}{2}} - \left(R_\gamma - \frac{\eta^2}{4}\right)\right\}^{\frac{1}{2}}$$

$$+i\left\langle \Omega \mp \frac{1}{\sqrt{2}}\left\{\left[\left(R_\gamma - \frac{\eta^2}{4}\right)^2 + I_\gamma^2\right]^{\frac{1}{2}} + \left(R_\gamma - \frac{\eta^2}{4}\right)\right\}^{\frac{1}{2}}\right\rangle \tag{2-218}$$

where

$$R_\gamma > \frac{\eta^2}{4}$$

$$I_\gamma \geq 0$$

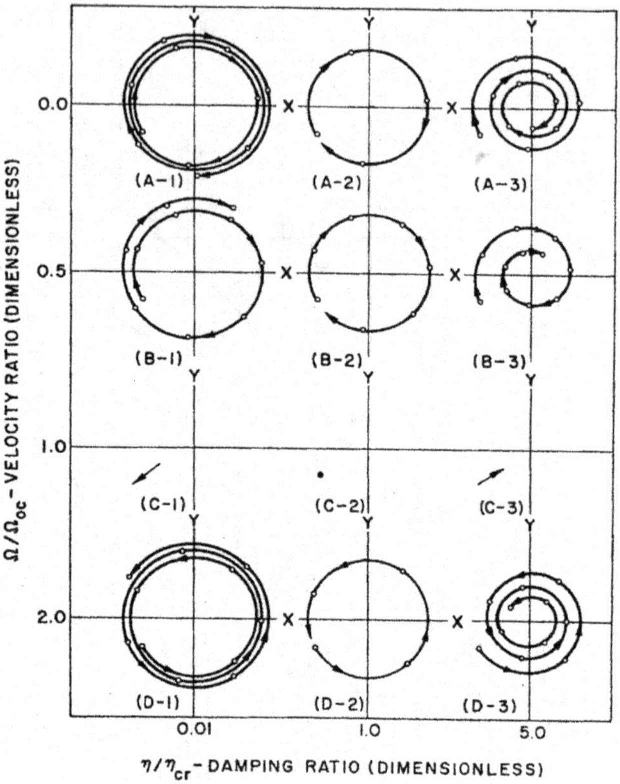

Figure 2-20. Centerline motion of pipe relative to rotating reference frame.

$$\eta = c\left(\frac{L^4}{EIm}\right)^{\frac{1}{2}}$$

$$\Omega = \omega\left(\frac{mL^4}{EI}\right)^{\frac{1}{2}}$$

The imaginary parts of $\lambda_{1,2}$ represent the harmonic motion, while the real parts represent the decay or growth of the amplitudes. Since the real part of λ_2 is negative, the second term in Eq. (2-218) approaches zero with time. However, the real part of λ_1 may be positive or negative, and it is this term that dictates stability or instability of motion.

The amplitude of pipe motion will not be bounded if the real part is positive. Therefore, for stability,

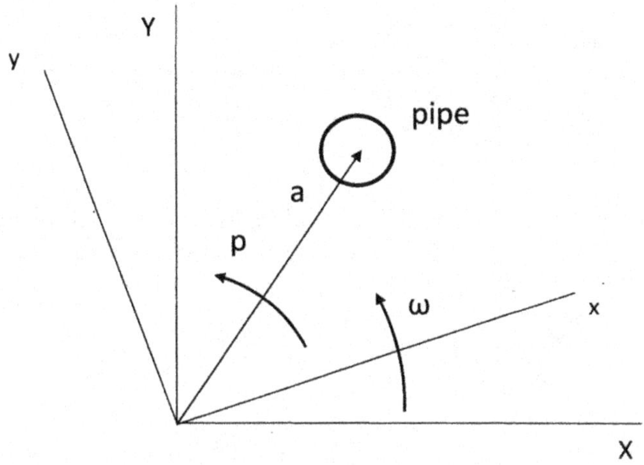

Figure 2-21. Angular velocity *p* with respect to *xy* rotating reference frame.

$$-\frac{\eta}{2}+\frac{1}{\sqrt{2}}\left\{\left[\left(R_\gamma-\frac{\eta^2}{4}\right)^2+I_\gamma^2\right]^{\frac{1}{2}}-\left(R_\gamma-\frac{\eta^2}{4}\right)\right\}^{\frac{1}{2}}\leq 0 \qquad (2\text{-}219)$$

By expanding the left side and simplifying, this inequality is satisfied when

$$\eta^2 R_\gamma \geq I_\gamma^2 \qquad (2\text{-}220)$$

For stability, R_γ must be positive. A critical damping factor is defined from this inequality as

$$\eta_{cr} = \frac{|I_\gamma|}{\sqrt{R_\gamma}} \qquad (2\text{-}221)$$

Values of η_{cr} corresponding to various values of α and Θ are also given in Table 2-8. If $\eta < \eta_{cr}$ the amplitude of pipe motion grows with time.

The angular velocity parameter, Ω, does not appear in this stability criteria. It does, however, affect the vibration frequency as observed from the rotating reference frame.

The circular frequency, p, of the position vector, measured from the well bore center to the pipe center is determined from the imaginary part of Eq. (2-218).

$$p = \left(\frac{EI}{mL^4}\right)^{\frac{1}{2}} \left\langle \Omega - \frac{1}{\sqrt{2}} \left\{ \left[\left(R_\gamma - \frac{\eta^2}{4} \right)^2 + I_\gamma^2 \right]^{\frac{1}{2}} + \left(R_\gamma - \frac{\eta^2}{4} \right) \right\}^{\frac{1}{2}} \right\rangle \tag{2-222}$$

The whirling (buckled) pipe will not exhibit lateral vibration relative to the rotating reference from if $p = 0$. The particular angular velocity parameter that corresponds to zero vibration is

$$\Omega_0^2 = \frac{1}{2} \left\{ \left[\left(R_\gamma - \frac{\eta^2}{4} \right)^2 + I_\gamma^2 \right]^{\frac{1}{2}} + \left(R_\gamma - \frac{\eta^2}{4} \right) \right\} \tag{2-223}$$

In this case, the center of the pipe either grows, is stationary, or decays relative to the rotating reference frame. If the damping factor is critical, then the point is stationary and under this condition, the pipe is in a state of synchronous whirl. This condition is defined by

$$\Omega_{oc} = R_\gamma^{\frac{1}{2}} \tag{2-224}$$

These critical parameters are related graphically (Fig. 2-19) for different values of α and Θ. For each combination of α and Θ, there is a critical damping factor, η_{cr} and Ω_{oc}. Under this condition, the center of the pipe is stationary relative to the rotating reference frame. Note that in the absence of torque and damping $R_\gamma = \lambda_1^4$ (see Table 2-3).

Various possible motions relative to the rotating reference are given in a phase plane diagram (Fig. 2-20) where motions are illustrated for different combinations $\frac{\eta}{\eta_{cr}}$ and $\frac{\Omega}{\Omega_{oc}}$. These diagrams show that if $\frac{\eta}{\eta_{cr}} < 1$, the displacement increases with time, and the phase-plane diagram is a diverging spiral; hence, the system is unstable. If $\frac{\eta}{\eta_{cr}} = 1$, the displacement remains constant, and the phase-plane diagram is a circle; therefore, the system is in the critical condition. Finally, if $\frac{\eta}{\eta_{cr}} > 1$, the displacement approaches zero, and the phase-plane diagram is a converging spiral so the system is stable.

The influence of the angular velocity of the pipe on its whirling behavior is also shown in Fig. 2-20. First, refer to the phase-plane diagrams in the center column corresponding to $\frac{\eta}{\eta_{cr}} = 1$. For $\frac{\Omega}{\Omega_{oc}} < 1$, the motion relative to the rotating frame is clockwise; for $\frac{\Omega}{\Omega_{cr}} > 1$, the motion is counterclockwise; and for $\frac{\Omega}{\Omega_{oc}} = 1$, point A is stationary.

2.7.4 Relative Whirl versus Absolute Whirl

The centerline motions shown in Fig. 2-20 illustrate the various types of whirling motion in pipes relative to the rotating *xyz* reference frame. These motions can be forward, backward, or synchronous whirl when viewed from the fixed *XZY* reference frame. For example, consider the whirling condition corresponding to (C-2). In this case, the velocity ratio is 1.0, meaning there is no vibration motion of the pipe relative to the rotating *xyz* reference frame. Also, the damping ratio is also 1.0, meaning the motion is stable. This is synchronous whirl when viewed from the fixed *XYZ* reference frame. The pipe whirls about the wellbore centerline with the same side outward.

Now, consider (D-2). Here, the damping ratio is 1.0 meaning the center of the pipe stays in the same circle about the wellbore. The motion is stable, and since the velocity ratio is less than 1.0, the centerline of the pipe moves counterclockwise about the wellbore centerline while the pipe rotates. The center of the pipe moves counterclockwise or in the same direction as the rotating reference frame. This motion is clearly forward whirl when viewed from the fixed *XYZ* reference frame.

Next consider (B-2). The damping ratio is still 1.0, meaning the motion is stable. The velocity ratio is 0.5, meaning there is pipe motion relative to the rotating reference frame. However, in this case, the center of the pipe moves in the clockwise direction while the pipe is rotating counterclockwise. There is a possibility of backward whirl relative to the fixed *XYZ* reference frame. Whether or not backward whirl occurs depending on the magnitude of p in the clockwise direction. If $p > \omega$ there is backward whirl relative to the well bore. All other motions in this figure indicate stable whirl, unstable whirl, or radial movement toward or away from the wellbore centerline.

Figure 2-21 shows the relation between p and ω. The circular frequency, p, is relative to the xy rotating reference frame, which rotates with ω angular velocity. The radial position of the pipe with respect to the center of the well bore is designated by the symbol **a**. The angular velocity of line **a** is defined by frequency, p, in radians per second. The absolute angular velocity of line **a** is $\omega + p$, keeping in mind that p can be plus or minus.

2.8 PARAMETRIC RESONANCE

The problem here is one in which an axial time varying force is applied the end of a beam as shown in Fig. 2-22. The magnitude of the applied axial force varies between $P + S$ and $P - S$. The frequency of the pulsating force is Ω rad/sec. Ω is used again for convenience and does not relate to the previous discussion of pipe whirl. By assuming that the period of the pulsating force is large in comparison with the period of the natural frequencies of the longitudinal modes, the internal axial force can be assumed constant along the beam. The differential equation of motion in the case is [24,25]

$$EI\frac{\partial^4 y}{\partial x^4} + (P + S\cos\Omega t)\frac{\partial^2 y}{\partial x^2} + m\frac{\partial^2 y}{\partial t^2} = 0 \qquad (2\text{-}225)$$

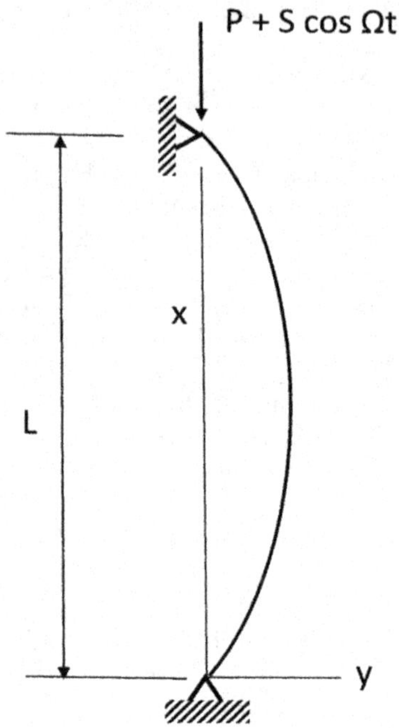

Figure 2-22. Vertical beam under time varying end loads.

It is similar to the short beam equation (Eq. (2-35)) given earlier except the axial force varies with time. The solution has potential application to short beam columns or drill collars between stabilizers.

Note that *P* and *S* are compressive forces, which account for the plus sign in front of the second term.

Another key assumption is that the beam is given a small lateral displacement or the beam is initially curved. This assumption allows the beam to respond directly to the pulsating end forces.

If *S* is equal to zero, the equation represents the free vibration of a beam and its solution gives the eigenvalues and eigenvectors for a constant applied load, *P*. For this case, Euler's critical buckling force can be determined from a dynamic approach by noting that buckling occurs when the natural frequency is zero.

The basic problem is to find conditions under which this system is dynamically unstable. When this occurs, the applied end force would be doing positive work on the beam, and the beam would respond with higher kinetic energy and higher amplitudes of vibration.

The solution proposed by Belajev [24] starts by assuming the

$$y(x,t) = Af(t)\sin\frac{\pi x}{L} \tag{2-226}$$

This equation satisfies the pinned boundary conditions shown in Fig. 2-22. The results of this classical problem would, at best, apply to section of drill collars between stabilizers in a vertical hole. Substituting Eq. (2-226) into Eq. (2-225) and using the following substitutions

$$p = \frac{P}{P_e} \tag{2-227}$$

$$s = \frac{S}{P_e} \tag{2-228}$$

$$\tau = \Omega t \tag{2-229}$$

gives the Mathieu equation,

$$\frac{d^2 f(\tau)}{d\tau^2} + (a + b\cos\tau) f(\tau) = 0 \tag{2-230}$$

where

$$a = \frac{\omega_o^2}{\Omega^2}(1 - p)$$

$$b = -\frac{\omega_o^2}{\Omega^2} s$$

$$\omega_o^2 = \frac{\pi^4 EI}{mL^4} \quad \text{(fundamental natural frequency)} \tag{2-231}$$

$$P_e = \frac{\pi^2 EI}{L^2} \quad \text{(Euler buckling)} \tag{2-232}$$

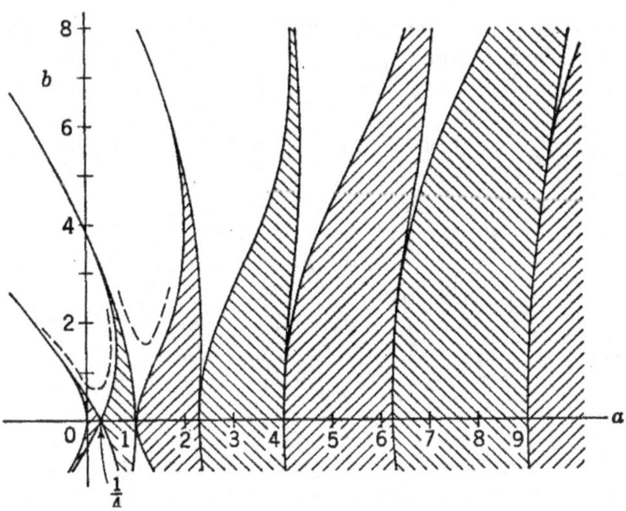

Figure 2-23. Stability data [26].

The real time solution to Eq. (2-230) is not necessary to determine dynamic stability. A stability analysis developed by Strutt [25–27] shows that dynamic stability depends on the relation between a and b (see Fig. 2-23). The shaded area in Fig. 2-23 indicates areas of stability while the clear areas indicate instability. Other references are given by Lubkin and Stoker [28].

This stability diagram shows the system is unstable for low values of *s* when $a = 1/4, 1, 2\ 1/4$, ---- . When $a = 1/4$, and $p = 0$, for example, the critical frequency, $\Omega = 2\omega_0$. In this case, there is no average compressive force. Only the time varying time component exists. But if $p = 1/2$, and b is small, the critical speed becomes $\Omega = \sqrt{2\omega_0}$. In either case, lateral vibrations would become severe.

Figure 2-23 shows that dynamic stability conditions exits for values of $P + S$ that are greater than Euler's static buckling force. For example, when $p = 0$ and $\frac{1}{4} < a < 1$, there are values of b that are greater than two (2) in the region of stability. This means that the total dynamic force $(P + S)$ can be more than twice Euler's static buckling force without causing dynamic instability.

REFERENCES

1. Wai-Fah Chen and Toshio Atsuta, *Theory of Beam-Columns, Vol. 1: In-Plane Behavior and Design*, McGraw-Hill, 1976.
2. William Riley, Leroy Sturges, and Don Morris; *Mechanics of Materials*, 6th ed. John Wiley & Sons, Inc, 2007.
3. A. G. Greenhill, *Proc. Cambridge Phil. Soc.*, vol. 4, 1881.
4. S. Timoshenko; *Strength of Materials; Part II Advanced Theory and Problems*, 3rd ed., D. Van Nostrand Company, Inc. 1956, pp. 167–168.
5. Willers, Fr. A.; "Das Knicken Schwerer Gestangeg," Z. Angew., Math, Mech., Vol. 21, No. 1, 1941, pp. 43–51.
6. H. W. Reddick and F. H. Miller, *Advanced Mathematics for Engineers*, John Wiley & Sons, 3rd ed. 1956.
7. Louis A. Pipes, *Applied Mathematics for Engineers and Physicists*, 2nd McGraw-Hill Book Company, NY, 1958, pp. 437–439.
8. Tseng Huang and D. W. Dareing; "Frequencies of a Hanging Chain", *J. of Acoustical Society of America*, Vol. 45, No. 4, April 1969, pp. 1046–1049.
9. Lubinski, A., Althouse, W. S., and Logan, J. L. ,"Helical Buckling of Tubing Sealed in Packers", J. Pet. Tech. (June 1962) 655–70.
10. Herman Betz, Paul B. Burcham and George M. Ewing, *Differential Equations with Applications,* Harper & Brothers Publishers, 1954.
11. William T. Thomson and Marie Dillon Dahleh, *Theory of Vibrations with Applications*, 5th Ed. Prentice Hall, 1998.
12. S. Timoshenko, D. W. Young and W. Weaver, Jr.; *Vibration Problems in Engineering,* 4th Ed., John Wiley & Sons, 1974.
13. J. P. Den Hartog, Advanced Strength of Materials, McGraw-Hill Book Co., 1952 (see p. 297).
14. Roy R. Craig, Jr., *Structural Dynamics*, John Wiley & Sons, 1981.
15. Tseng Huang and D. W. Dareing; "Buckling and Frequencies of Long Vertical Pipes", ASCE Journal of the Engineering Mechanics Division, Vol. 95, No. EM1, proc. paper 6400, Feb. 1969, 167–181.

16. A. Lubinski, "A study of the buckling of rotary drilling string", *API Drilling and Production Practice*, **17** (1950), pp. 178–214.
17. Stephen P. Timoshenko and James M. Gere; *Theory of Elastic Stability*, McGraw-Hill Book Company, 1961, p. 157.
18. A. G. Greenhill, "On the Strength of Shafting When Exposed both to Torsion and End Thrust," Proceedings of the Institution of Mechanical Engineers, London, England, Vol. 6, 1883, pp. 182–209.
19. H. A. Webb solution is taken from the English text by John Case, *Strength of Materials*, Edward Arnold & Co. London 1925 (from ref. 13).
20. R. V. Southwell and B. S. Dough; "On the Stability of a Rotating Shaft Subjected Simultaneously to End Thrust and Twist", British Association for the Advancement of Science, Vol. 8 (1921) pp. 345–355.
21. Huang, T. and Dareing, D. W., "Predicting the Stability of Long Vertical Pipe Transmitting Torque in a Viscous Medium," Trans., ASME, *Journal of Engineering for Industry,* Vol. 88, Series B, No. 2, May 1966, pp. 191–200.
22. J. Kim Vandiver, James W. Nicholson, Rong-Juin Shyu, "Case Studies of the Bending Vibration and whirling Motion of Drill Collars", SPE Drilling Engineering, December 1990 pp. 282–290.
23. Michel Lalanne and Guy Ferraris, *Rotordynamics Prediction in Engineering*, 2nd Edition, (1998), John Wiley & Sons Inc.
24. N.M. Belajev, "Collection of Papers", Engineering Structures and Structural Mechanics, pp149-167, Leningrad, 1924.
25. Timoshenko, Stephen P. and James M. Gere, Theory of Elastic Stability, (p. 158), McGraw-Hill Book Company, 1961.
26. Strutt, M. J. O. (1931) Diffraction of a plane wave at a slit of finite width. Zeitschrift für Physik, 69(9), 597–617.
27. Strutt, M. J. O. (1932) Lamésche, Mathieusche und Verwandte Funktionen in Physik und Technik. Ergebnisse der Mathematik und ihrer Grenzgebiete Herausgegeben von der Schriftleitung des "Zentralblatt for Mathematik," 1(3). Springer, Berlin.
28. S. Lubkin and J. J. Stoker, Quart. Appl. Math., Vol. 1, No. 3, p 215, 1943.

SELECTED PAPERS BY DR. TSENG HUANG

1. Huang, Tseng and Kang, Qing Liang, "Three Dimensional Analysis of A Marine Riser with Large Displacements", International Society of Offshore and Polar Engineers, Vol 1, No. 4 (1991), pp. 300–306.
2. Huang, Tseng and Chucheepsakul, Somchai, "Large Displacement Analysis of a Marine Riser", Proceedings of ASME Third International Offshore Mechanics and Arctic Engineering Symposium, New Orleans, La, (1984) Vol. 1, pp. 471–476.
3. Chuckeepsakul, Somchai and Huang, Tseng, "Influence of Transported Mass on the Equilibrium Configuration of Risers", International Society of Offshore and Polar Engineers (1994), Vol. II, pp. 246–250.
4. Huang, Tseng, "Kinematics of Transported Mass Inside Risers and Pipes" International Society of Offshore and Polar Engineers, (1993), Vol. II, pp. 331–336.

5. Chucheepsakul, Somchai and Huang, Tseng, "Natural Frequencies of a Marine Riser in Three Dimensions", (ASME) Proceedings of 8th International Conference on Offshore Mechanics and Arctic Engineering (1989), pp. 651–658.
6. Huang, Tseng and Saha, K. G., "Polar Coordinates and Riser Analysis", (ASME) 8th International Conference on Offshore Mechanics and Arctic Engineering (1989), pp. 467–475.
7. Chucheepsakul, Somchai, Huang, Tseng and Monprapussorn, Tinnakorn, "Stability Analysis of Extensible Flexible Marine Pipes Transporting Fluid," Proceeding of the Eleventh International Offshore and Polar Engineering Conf., Stavanger, Norway, June 17–22, 2001.
8. Chucheepsakul, Somchai, Huang, Tseng and Monprapussorn, Tinnakorn, "Buckling of Marine Elastica Pipes with Large Strain Transporting Fluid," Proceedings of the First International Conference on Structural Stability and Dynamics, Taipei, Taiwan, Dec 7–9, 2000.

Chapter 3

DRILLSTRING MECHANICS

The overall behavior of drillstrings is complex. Dynamic behavior is the combination of the effects of axial, tortional and lateral modes of vibration. Well bore confinement offers the greatest challenge in predicting drillstring behavior. Solutions to the overall dynamic behavior of drillstrings are possible through finite element techniques. However, much can be learned through mathematical models of the various modes (axial, tortion, lateral) of motion. Coupling these analyses with measurements of downhole parameters while drilling provides much information for the design and control of drillstrings.

3.1 BUCKLING AND LATERAL VIBRATIONS OF DRILL PIPE

Buckling of structures can be predicted from a static analysis or a dynamic analysis. The static analysis approach is based on seeking eigenvalues from static equations of bending. The principle of minimum potential energy is useful in this regard. The dynamic analysis approach predicts buckling from natural frequency predictions; when natural frequency is zero, buckling occurs. The following discussion approaches drill pipe buckling from the dynamic perspective following Huang and Dareing [1]. The physical model is illustrated in Fig. 3-1.

3.1.1 Differential Equation of Bending

Because of the relative bending stiffness of the drill collars, the lower end of the drill pipe is modeled as a fixed end with zero slope. The top end is pinned. The axial force applied to the pipe at $x = 0$ is the internal force between the pipe and collars. This force is designated as F_B in the expression for effective tension (see Eq. (2-97)). Plus F_B means tension and minus means compression.

The analysis is based on vertical pipe under static conditions and unconstrained laterally. Vibration and buckling modes occur in a vertical plane. The basic equation of motion is developed from Eq. (2-95) (Ch 2), noting that fluid density is the same both inside and out. In this case, effective tension is

$$T_{eff} = (F_B + wx) + (L - x)w_m \tag{3-1}$$

giving

$$EI\frac{\partial^4 y}{\partial x^4} - \frac{\partial}{\partial x}\left\{\left[(F_B + w_m L) + (w - w_m)x\right]\frac{\partial y}{\partial x}\right\} + m\frac{\partial^2 y}{\partial t^2} = 0 \tag{3-2}$$

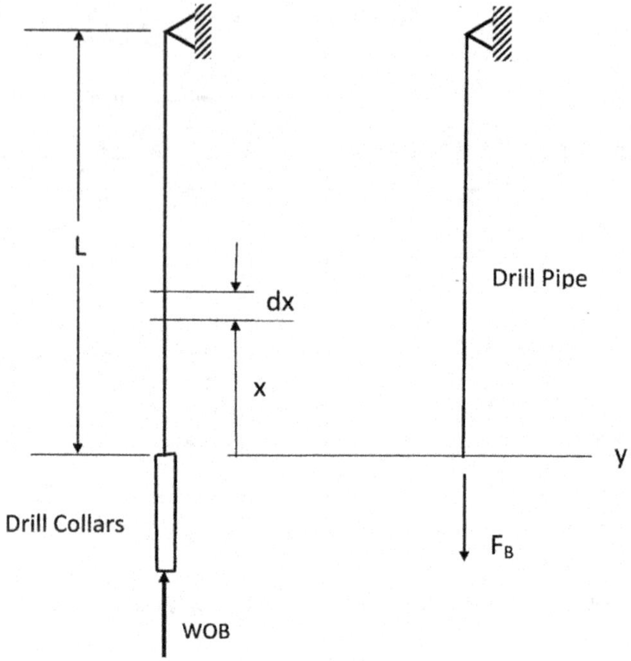

Figure 3-1. Drill pipe loading.

Consider the case where weight on bit (WOB) puts the neutral point at the top of the drill collars. This means that the internal stress at the lower end of the drill pipe is equal to the local hydrostatic pressure, and the force in the drill pipe at this interface is,

$$F_{B} = -Lw_{m} \tag{3-3}$$

Under this condition, the pipe will not buckle as explained in Ch 1. However, if WOB is increased further by F, then the force at the bottom of the drill pipe becomes

$$F_{B} = -Lw_{m} - F \tag{3-4}$$

This additional force can cause drill pipe buckling. The critical buckling force, F_{cr}, applied to the bottom of the drill pipe will be discussed next.

Substituting Eq. (3-4) into Eq. (3-2) gives

$$EI\frac{\partial^4 y}{\partial x^4} - \frac{\partial}{\partial x}\left\{\left[-F + (w - w_m)x\right]\frac{\partial y}{\partial x}\right\} + m\frac{\partial^2 y}{\partial t^2} = 0 \tag{3-5}$$

Here, F is a plus number and represents added compressive force beyond hydrostatic compression. The minus sign accounts for compression. The boundary conditions at the lower and upper ends are

$$y(0,t) = 0 \tag{3-6a}$$

$$\frac{\partial y}{\partial x}(0,t) = 0 \tag{3-6b}$$

$$y(L,t) = 0 \tag{3-6c}$$

$$\frac{\partial^2 y}{\partial x^2}(L,t) = 0 \tag{3-6d}$$

Assuming the solution to Eq. (3-5) is of the form

$$y(x,t) = Y(x)\sin\omega t \tag{3-7}$$

then Eq. (3-5) converts to

$$\frac{d^4Y}{d\zeta^4} - \alpha\zeta\frac{d^2Y}{d\zeta^2} + \beta\frac{d^2Y}{d\zeta^2} - \alpha\frac{dY}{d\zeta} - \lambda^4 Y = 0 \tag{3-8}$$

where

$$\varsigma = \frac{x}{L}$$

$$\alpha = \frac{(w - w_m)L^3}{EI}$$

$$\beta = \frac{FL^2}{EI}$$

$$\lambda^4 = \frac{m\omega^2 L^4}{EI}$$

Boundary conditions become

$$Y(0) = 0 \tag{3-9a}$$

$$\frac{dY}{dx}(0) = 0 \tag{3-9b}$$

$$Y(L) = 0 \tag{3-9c}$$

$$\frac{d^2Y}{dx^2}(L) = 0 \tag{3-9d}$$

3.1.2 Method of Solution

The solution to Eq. (3-8) is expressed by

$$Y(\zeta) = \sum_{n=0}^{\infty} a_n \zeta^n \tag{3-10}$$

The series and its derivatives are uniformly convergent; therefore, the derivatives of Y are achieved by term wise differentiation. Substituting Eq. (3-10) and its derivatives into Eq. (3-8) produces an equation involving a power series in ζ, whose sum is equal to zero. Each coefficient in the series must be zero, giving

$$a_n = -\frac{\beta}{n(n-1)} a_{n-2} + \frac{(n-3)\alpha}{n(n-1)(n-2)} a_{n-3} + \frac{\lambda^4}{n(n-1)(n-2)(n-3)} a_{n-4} \tag{3-11}$$

where $n \geq 4$.

By repeated application of this recurrence equation, starting with $n = 4$, each a_n can be expressed as a linear combination of a_0, a_1, a_2, a_3. In general,

$$a_n = F_n a_0 + G_n a_1 + H_n a_2 + I_n a_3 \tag{3-12}$$

Values of $F_n G_n H_n I_n$ for $n = 0, 1, 2, 3$ are given by Eq. (2-110)

The remaining values are determined by combining Eq. (3-11) and Eq. (3-12) giving

$$\begin{bmatrix} F_n \\ G_n \\ H_n \\ I_n \end{bmatrix} = -\frac{\beta}{n(n-1)} \begin{bmatrix} F_{n-2} \\ G_{n-2} \\ H_{n-2} \\ I_{n-2} \end{bmatrix} + \frac{(n-3)\alpha}{n(n-1)(n-2)} \begin{bmatrix} F_{n-3} \\ G_{n-3} \\ H_{n-3} \\ I_{n-3} \end{bmatrix} + \frac{\lambda^4}{n(n-1)(n-2)(n-3)} \begin{bmatrix} F_{n-4} \\ G_{n-4} \\ H_{n-4} \\ I_{n-4} \end{bmatrix} \tag{3-13}$$

With the starting values for $n = 0, 1, 2, 3$, F_n, G_n, H_n, I_n for $n \geq 4$ are expressed in terms of α, β, λ from Eq. (3-13). λ is determined from a characteristic equation containing known values of α and β.

The first two boundary conditions (Eq. (3-9)) give $a_0 = a_1 = 0$. The remaining two boundary conditions give

$$\left[\sum_{2}^{\infty} H_n\right] a_2 + \left[\sum_{2}^{\infty} I_n\right] a_3 = 0 \tag{3-14}$$

$$\left[\sum_{2}^{\infty} n(n-1)H_n\right] a_2 + \left[\sum_{2}^{\infty} n(n-1)I_n\right] a_3 = 0 \tag{3-15}$$

For non-trivial solutions, the determinate of the coefficients must be zero. The values of λ that satisfy this determinate for given values of α and β are the eigenvalues. These eigenvalues give: 1) critical buckling forces applied to the bottom end of the drill pipe and 2) natural lateral vibration mode shapes and frequencies.

Table 3-1. Buckling parameters.

α	β_{cr1}	β_{cr2}	β_{cr3}	β_{cr4}
0	20.19	59.68	118.90	197.86
1	20.84	60.21	119.42	198.37
2	21.50	60.73	119.94	198.88
5	23.44	62.31	121.50	200.40
10	26.66	64.94	124.12	202.95
20	32.99	70.22	129.36	208.07
50	51.06	86.18	145.31	223.51
100	77.92	113.25	172.67	249.60
200	121.61	166.59	230.50	303.21
500	218.06	297.87	403.85	474.41
1000	339.59	465.36	634.66	740.10
2000	529.83	729.55	997.20	1162.37
4000	827.87	1146.48	1569.43	1830.50

3.1.3 Use of Buckling Solution

Critical values of β, which contain buckling force, F, are given in Table 3-1 for the first four modes of buckling. The critical buckling force for the first mode defines the on-set of buckling.

Values of α are limited to 4000 because of the high degree of numerical precision required to execute the calculations. The relation between α and β_{cr} for higher values of α can be determined as follows.

When the values of β_{cr} versus α (Table 3-2) are plotted on log-log scale, they become straight and parallel with a slope of about 2/3. A straight line on log-log graph can be expressed as

$$\ell n\beta_{cr} = \frac{2}{3}\ell n\alpha + \ell n\phi \quad (3\text{-}16)$$

Table 3-2. Relation between ϕ values and α.

α	ϕ_1	ϕ_2	ϕ_3	ϕ_4
1	20.89	60.21	119.42	198.37
2	13.54	38.26	75.56	125.82
5	8.02	21.31	41.55	68.54
10	5.74	13.99	26.74	43.73
20	4.48	9.53	17.56	28.24
50	3.76	6.35	10.71	16.47
100	3.62	5.26	8.02	11.59
200	3.56	4.87	6.74	8.87
500	3.46	4.73	6.41	7.53
1000	3.40	4.65	6.35	7.40
2000	3.34	4.60	6.28	7.32
4000	3.29	4.55	6.23	7.26

which simplifies to

$$\beta_{cr} = \phi\alpha^{\frac{2}{3}} \tag{3-17}$$

giving

$$\phi_{cr} = \frac{\beta_{cr}}{\alpha^{\frac{2}{3}}} = \frac{F_{cr}}{(EI)^{\frac{1}{3}}(w - w_m)^{\frac{2}{3}}} \tag{3-18}$$

Pipe length, L, does not appear in this equation. The buckling force becomes independent of pipe length as would be expected. The same conclusion applies to other boundary conditions at $x = 0$. According to Table 3-2, ϕ_1 is 3.29 for the first mode; ϕ_n for higher buckling modes are also given in this table.

The critical buckling force equation for long pipe can now be approximated by

$$F_{cr} = \phi_{cr}\,(EI)^{\frac{1}{3}}(w - w_m)^{\frac{2}{3}} \tag{3-19}$$

where F_{cr} is a force over and above hydrostatic compression. The critical buckling forces for the first four modes can be determined from

$$F_{cr1} = 3.29(EI)^{\frac{1}{3}}(w - w_m)^{\frac{2}{3}} \tag{3-20a}$$

$$F_{cr2} = 4.55(EI)^{\frac{1}{3}}(w - w_m)^{\frac{2}{3}} \tag{3-20b}$$

$$F_{cr3} = 6.23(EI)^{\frac{1}{3}}(w - w_m)^{\frac{2}{3}} \tag{3-20c}$$

$$F_{cr4} = 7.26(EI)^{\frac{1}{3}}(w - w_m)^{\frac{2}{3}} \tag{3-20d}$$

For example, consider 5 ½ in. (19.2 lb/ft) drill pipe along with the following parameters.

$E = 30(10)^6$ psi
$I = 16.8045$ in.^{4}
$\gamma_m = 10$ ppg
$\gamma_{stl} = 490$ lb/ft^3
$L = 1000$ ft

Applying these numbers to

$$\alpha = \frac{(w - w_m)L^3}{EI} = \frac{BFwL^3}{EI} \tag{3-21}$$

gives

$$\alpha = \frac{0.847(19.2)1000^3}{30(10)^6 16.8045}\left|\frac{144\,in^2}{1\,ft^2}\right| = 4645.2$$

Since α is greater than 4000, critical buckling parameter β_{cr} is determined by Eq. (3-17), where ϕ_1, ϕ_2, ϕ_3, ϕ_4 are 3.29, 4.55, 6.23, and 7.26, respectively. The critical buckling parameter for the first mode is

$$(\beta_{cr})_1 = 3.29(4645.2)^{\frac{2}{3}} = 916$$

Applying

$$\beta_{cr} = \frac{F_{cr}L^2}{EI}$$

gives

$$(F_{cr})_1 = 916\frac{30(10)^6 16.8045}{1000^2}\left|\frac{1ft^2}{144in^2}\right| = 3,207 \text{ lb}$$

The application of the higher modes of drill pipe buckling can only be viewed as an approximation because in reality the pipe makes contact with the well bore, but even this could be helpful in horizontal drilling where the vertical portion of drill pipe is sometimes put in compression.

For another example, consider the following set of parameters, which incorporates drill collars and WOB.

Drill depth — 8000 ft
Drill Pipe — 4.5 in. OD × 3.826 in. ID
Drill Collars — 6.5 in. OD × 2.5 in. ID
Drill Collar Length — 720 ft
Mud weight — 10 lb/gal
Weight on Bit — 50,000 lb

Drill collar length is typically calculated using

$$L_c = \frac{\text{WOB}}{0.85BFw} \tag{3-22}$$

which includes a 1.15 safety factor against drill pipe buckling. The buoyancy factor, corresponding to 10 ppg mud is 0.846. In this case, collar length is 720 ft and the neutral point is 610 ft from the bit (see Chapter 1). Using Eq. (3-20a), the critical buckling force required to buckle the drill pipe into the first mode is F = 2200 lb. This force is over and above the hydrostatic force (16,900 lbs) at the bottom of the drill pipe. Actually, the total internal force at the bottom of the drill pipe at the onset of buckling is 19,100 lbs ($p_{7280}A_p + F_{cr}$ = 16,900 + 2200 lbs). Since, under the drilling conditions stated above, the compressive force at the bottom of the pipe is 8425 lb (based on a free body diagram of the 720-ft drill collar section), an added 10,675-lb bit weight would cause the 4 ½ in drill pipe to buckle.

The shapes of the first two buckling modes (for α = 4000) are shown in Fig. 3-2. The corresponding modal moment for each mode is also given. The mode shapes show zero slope at the drill pipe/collar interface. The bending moments for the first two buckle modes are

$$\frac{ML^2}{EIY_0} = 200 \qquad \text{(first buckle mode)}$$

and

$$\frac{ML^2}{EIY_0} = 250 \qquad \text{(second buckle mode)}$$

Assuming 5 1/2in. (19.2 lb/ft) drill pipe

I = 16.8 in.^{4}
E = 29(10)^6 psi
L = 5000 ft
Y_0 = 2.5 in.

The bending moment for the first and second modes are:

$$M_1 = 200\frac{EIY_0}{L^2} = 200\frac{29(10)^6 16.8(2.5)}{5000^2}\left|\frac{1ft}{12in}\right|^3 = 5.64 ft - lb$$

$$M_2 = \frac{250}{200}5.64 = 7.05 ft - lb$$

Corresponding bending stresses are

$$\sigma_{B1} = 11.08 \text{ psi}$$

$$\sigma_{B2} = 13.85 \text{ psi}$$

These bending stress levels are quite low and support the practice of sometimes running drill pipe in compression in horizontal drilling. However, bending in this area becomes significant in high curvature situations.

Note that the critical buckling parameters for the first and second modes (see Table 3-1) are

$$\beta_{cr1} = 827.87$$

$$\beta_{cr2} = 1146.48$$

Corresponding buckling forces are determined from

$$F_{cr1} = \beta_{cr1}\frac{EI}{L^2} \tag{3-23a}$$

$$F_{cr2} = \beta_{cr2}\frac{EI}{L^2} \tag{3-23b}$$

3.1.4 Natural Frequencies of Drill Pipe (Lateral Modes)

Eigenvalues for this problem are determined from the characteristic equation derived from the expansion of the coefficient matrix of Eqs. (3-14) and (3-15). Values of λ that satisfy the characteristic equation were determined through trial and error and are given in Table 3-3. Figure 3-3 shows how λ relates to α for the special case of $\beta = 0$.

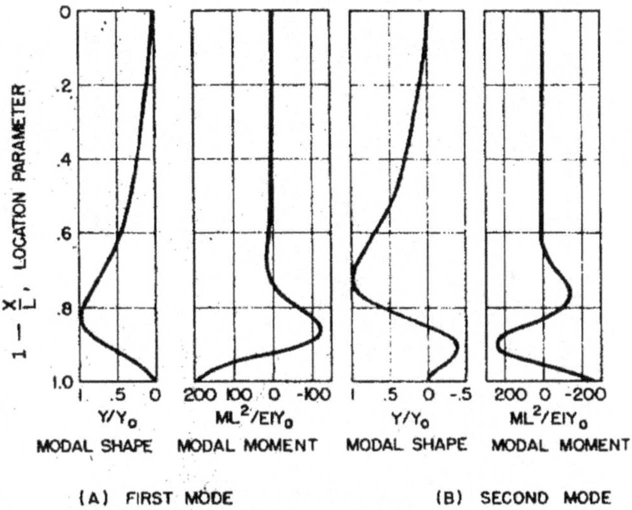

Figure 3-2. Buckling modes for α = 4000.

Table 3-3. Frequency parameters, λ.

α	$\frac{\beta}{\alpha}$	λ_1	λ_2	λ_3
100	0	5.409	8.336	11.232
	0.1	5.230	8.147	11.063
	0.2	5.029	7.944	10.886
	0.3	4.801	7.724	10.700
250	0	6.354	9.458	12.325
	0.1	6.070	9.130	12.001
	0.2	5.733	8.761	11.648
	0.3	5.315	8.336	11.260
600	0	7.534	11.010	14.009
	0.1	7.100	10.488	13.463
	0.2	6.531	9.863	12.838
	0.3	5.667	9.073	12.104
1000	0	8.373	12.147	15.310
	0.1	7.807	11.470	14.589
	0.2	6.999	10.613	13.731
	0.3	5.335	9.416	12.659
2500	0	10.178	14.664	18.277
	0.1	9.262	13.579	17.109
	0.2	7.390	11.952	15.547
	0.3	....	8.585	13.155
6000	0	12.341	17.726	21.963
	0.1	10.809	15.980	20.093
	0.2	....	12.236	16.980
	0.3	....	....	10.859

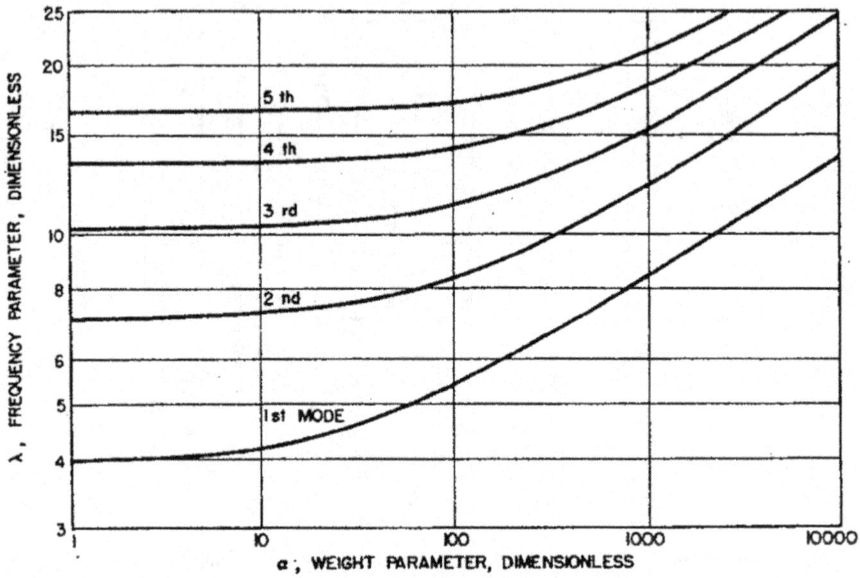

Figure 3-3. Natural frequency data corresponding to $\frac{\beta}{\alpha} = 0$.

Figure 3-4 shows this relationship for the special case of $\frac{\beta}{\alpha} = 0.3$. This graph shows that $\lambda = 0$ at $\alpha = 1413$. Under this condition ($\omega = 0$) the pipe moves outward but does not return, meaning the pipe has buckled. Critical buckling force for the first mode is determined as follows.

$$\beta_{cr} = 0.3(1{,}413)$$

and by Eq. (3-23a)

$$F_{cr} = 0.3(1{,}413)\frac{EI}{L^2}$$

For another scenario, consider a case where $\alpha = 10{,}000$ and $\beta = 0$. This data would apply to

Drill pipe = 4 ½ × 16.6 lb/ft
Drill pipe length = 1090 ft
Mud weight = 10 ppg
$F = 0$ (neutral point is located at the lower end of drill pipe)

In this case, λ for the first three modes are 13.8, 20, and 24.5 (Fig. 3-3). Corresponding natural frequencies of the lateral modes of vibration are: 0.0492, 0.1036, and 0.155 cps. These modes could be excited at rotary speeds of 2.95, 6.21, and 9.33 rpm. In reality, much higher lateral modes would be excited because practical rotary speeds are near 80 to 100 rpm.

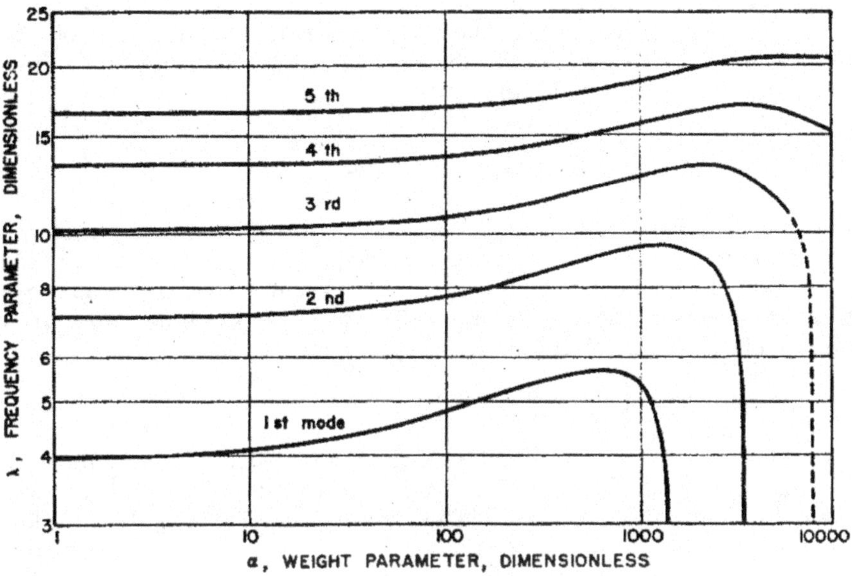

Figure 3-4. Natural frequency data corresponding to $\frac{\beta}{\alpha} = 0.3$.

Mode shapes and their bending moments are of interest because they show the intensity of bending at the interface between drill pipe and drill collars. Mode shapes are indicative of the effects of variable tension along drill pipe. Bending at the top is low because of high tension and high at the collars because of low tension and the structural constraint at the interface. Figure 3-5 gives mode shapes for the first two vibration modes and corresponding modal moments for both modes (corresponding to $\alpha = 4000$ and $\beta = 800$). Bending

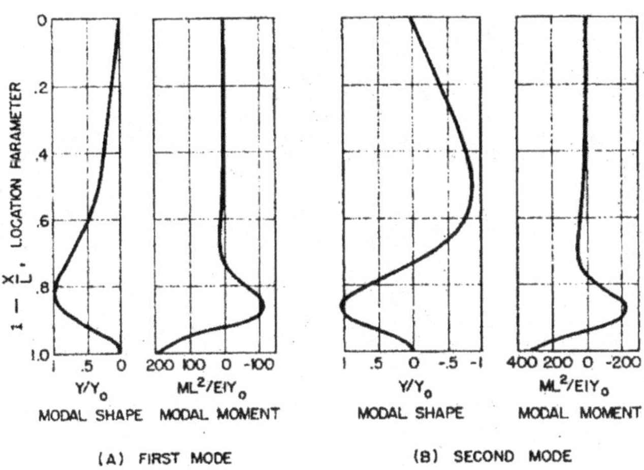

Figure 3-5. Vibration modes for $\alpha = 4000$, $\beta = 800$.

moments are greatest at the interface between the drill pipe and collars even though as shown above they are relatively small. Fatigue damage is most likely caused by drillstring rotation with the drill pipe/drill collar interface being located in a dog leg under conditions of high tension.

3.2 BUCKLING OF DRILL COLLARS

Drill collars affect the rotary drilling process in many ways: rate of penetration, directional control, and the performance of downhole tools to mention a few. Even though the drill collar section may be several hundred feet long, the bottom 150 ft affects everything. Lubinski [2] recognized the importance of drill collar buckling and its effect on well bore direction.

3.2.1 Bending Equation

The development of Lubinski's differential equation is summarized below. His coordinate system is shown in Fig. 3-6 as X,Y where positive X is measured downward from the neutral point. This drawing also includes the coordinate system used in the previous discussion of drill pipe buckling and frequencies. The drill bit is located at the origin, point 0, of the xy coordinate system.

Summing forces in the x and y directions gives

$$F_x + (L - x)w_m + F - (w - w_m)x = 0 \tag{3-24}$$

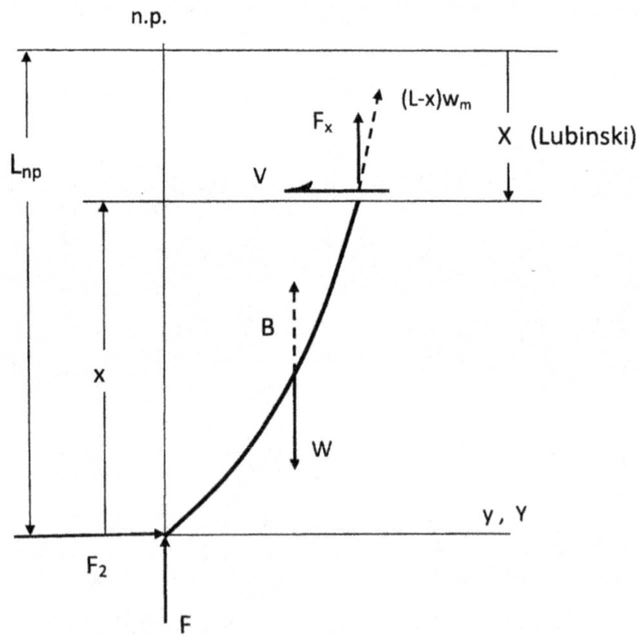

Figure 3-6. Freebody of drill collar section.

or

$$F_x = wx - Lw_m - F \tag{3-25}$$

as shown earlier. Summing forces in the y direction gives

$$V = F_2 + (L - x)w_m \frac{dy}{dx} \tag{3-26}$$

Recall from Chapter 2 that

$$V + F_x \frac{dy}{dx} = \frac{dM}{dx} \tag{3-27}$$

Substitution of Eqs. (3-25) and (3-26) into Eq. (3-27) gives

$$\frac{dM}{dx} = F_2 + \left[(w - w_m)x - F\right]\frac{dy}{dx} \tag{3-28}$$

and

$$EI\frac{d^3y}{dx^3} = F_2 + \left[(w - w_m)x - F\right]\frac{dy}{dx} \tag{3-29}$$

which is a direct spinoff from Eq. (3-5). To convert the coordinate x over to Lubinski's coordinate, X, and eliminate force, F, which is actually WOB or W_2, note that

$$x = L_{np} - X \tag{3-30}$$

and

$$F = L_{np}(w - w_m) \tag{3-31}$$

Substituting these last two equations into Eq. (3-29) and modifying the derivatives according to Eq. (3-30), gives Lubinski's Eq. (18).

$$EI\frac{d^3Y}{dX^3} + pX\frac{dY}{dX} + F_2 = 0 \tag{3-32}$$

where, following Lubinski's notation, $p = (w - w_m)$.

3.2.2 Lubinski's Solution

This is a third-order equation, which directly incorporates side forces on the collars. Lubinski's buckling analysis includes well bore confinement forces, which is a special aspect of his work.

Equation (3-32) naturally contains a variable coefficient to account for the variable compression and thus the solution cannot be expressed in terms of familiar functions.

Equation (3-32) is simplified by converting units.

$$X = mx$$
$$Y = my$$

giving

$$\frac{d^3y}{dx^3} + x\frac{dy}{dx} + c = 0 \tag{3-33}$$

where

$$m^3 = \frac{EI}{p}$$

$$p = (w - w_m)$$

$$c = \frac{F_2}{pm}$$

Equation (3-33) is simplified further by letting $z = \frac{dy}{dx}$, which leads to

$$\frac{d^2z}{dz^2} + xz + c = 0 \tag{3-34}$$

The solution for $z(x)$ is assumed to be of the form

$$z(x) = \sum_{n=0}^{\infty} a_n x^n \tag{3-35}$$

Equation (3-35) and its derivatives are substituted into Eq. (3-34). Since the coefficients of all x terms must be equal to zero, relationships are established such that all coefficients a_n can be expressed as functions of a_0, a_1, c. Thus, Eq. (3-35) is expressed as the sum of three series with a_0, a_1, c being coefficients for each series. Lubinski recognized that the resulting power series and its derivatives can be expressed in terms of Bessel functions. This was important at the time since digital computers were not a common research tool, but Bessel functions were tabulated [3] for easy use. A characteristic equation with m as the eigenvalue was established after substituting Bessel functions into Eq. (3-35) and applying boundary conditions. Eigenvalues for the first two modes are 1.94 and 3.75.

The interpretation of these numbers is as follows. The distance to the neutral point, L_{np}, corresponding to buckling of the first and second modes is given as 1.94 m and 3.75 m Values of m for drill collars and drill pipe are calculated from a characteristic equation resulting from boundary conditions being applied to Eq. (3-35). For 6 ¼ in drill collars (2 ¼ in ID) in 10 ppg mud, m = 58.385 ft.

This particular drill collar size buckles into its first mode when

$$L_{np} = 1.94m = 113.3ft \tag{3-36}$$

or

$$WOB_{cr1} = (w - w_m)L_{np} \tag{3-37}$$

$$WOB_{cr1} = 8{,}855 \text{ lbs}$$

and buckles into its second mode when

$$L_{np} = 3.75m = 218.9ft \tag{3-38}$$

or

$$WOB_{cr2} = (w - w_m)L_{np} \tag{3-39}$$

$$WOB_{cr2} = 17{,}127 \text{ lbs}$$

Critical weight on bit (WOB) for two drill collar sizes is given in Fig. 1-2 (Chapter 1). These curves were developed from

$$WOB_{cr1} = 1.94m(w - w_m) \tag{3-40}$$

and for the second buckling mode

$$WOB_{cr2} = 3.75m(w - w_m) \tag{3-41}$$

These equations predict conditions of instability. Lubinski took the analysis further to predict side force at the drill bit and side force when drill collars made contact with the well bore. This required consideration of collars as a multi-spanned beam column supported laterally at the contact points. Through a post buckling analysis, he located the contact points and quantified the magnitude of contact side forces. Most importantly, he quantified the magnitude of the side force at the drill bit and established the orientation of the resultant bit force. Bit orientation was also predicted. This opened up a whole new discussion of the effects of bit orientation and resultant bit force on well bore direction.

Lubinski pointed out that it is impossible to drill a straight vertical hole with a slick BHA because of drill collar buckling. Furthermore, drill bits tend to move off course in the direction of the resultant bit force. Lubinski's conclusions set the foundation for subsequent and current directional drilling control methods.

The shapes of the first and second buckle modes are shown in Fig. 3-7. The first mode of buckling, indicated by (1), shows the neutral point 1.94

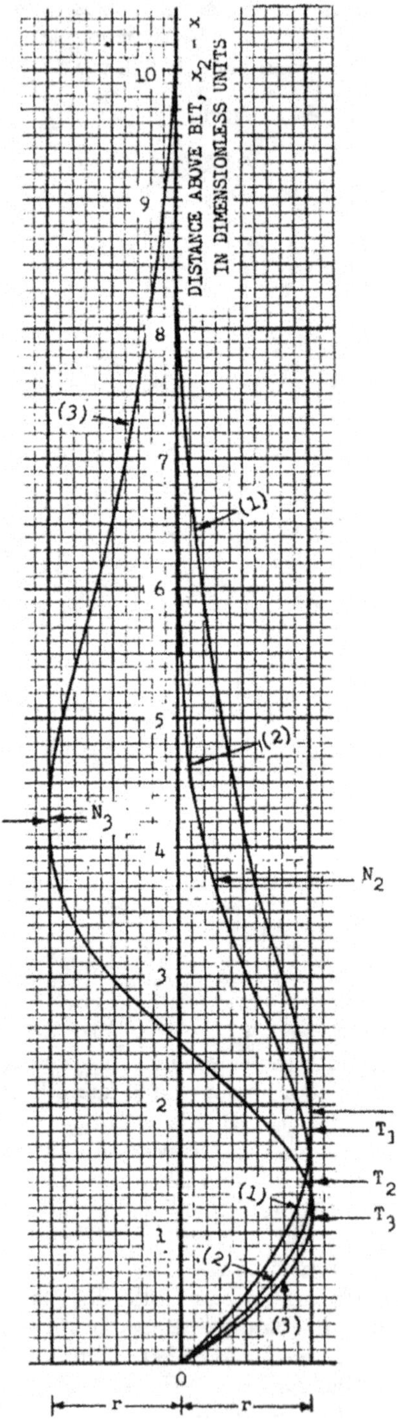

Figure 3-7. Shape of buckled curves [2]. (r is the apparent radius of the hole, i.e., the maximum possible deflection).

**Curve (1) — Critical, First Order;
Curve (2) — Critical Second Order;
Curve (3) — Immediately Prior to Second Buckle contacting the Wall of the Hole.**

Reproduced courtesy of the American Petroleum Institute.

dimensionless units from the bit. This point is marked by $N_1 = 1.94$. The drill collars touch the wall at T_1. The second buckle mode is indicated by (2). The neutral point in this case is indicated by $N_2 = 3.75$. With a slight increase in bit weight indicated by an upward movement of the neutral point (N_3), the second mode touches the well bore at T_3'.

The results show that when the second mode touches the well bore at T_3', the force at the lower tangent point, T_3, is equal to

$$F = 2.7pr = 2.7(w - w_m)r \tag{3-42}$$

where r is radial clearance between drill collars and the well bore. His numbers show this force is quite small, a few hundred pounds at the most.

Lubinski points out that the inclination of the bit, α, and the direction of the bit resultant bit force, β, are different. Both angles are measured from the vertical axis and are determined by

$$\alpha = t\frac{r}{m} \tag{3-43}$$

$$\beta = n\frac{r}{m} \tag{3-44}$$

Constants t and n have been plotted with reference to the neutral point distance. This distance is expressed in terms of the dimensionless unit, m (see Table 3-4). The angles are expressed in units of radians.

These data show that

- The inclination of the force on the bit is smaller than the slope of the bit.
- The inclination of the force on the bit increases with WOB between the first and second modes.
- The inclination of the force on the bit reaches a maximum between critical buckling of the second mode and contact of the second mode shape.

Table 3-4. Bit inclination (*t*) and bit force inclination (*n*) coefficients. (Numbers read from Lubinski's [2] graph).

	X_2 "m"	*t*	*n*
1st buckle mode	1.94	1.0	0.25
	2.6	1.1	0.37
	3.2	1.2	0.46
2nd buckle mode	3.75	1.3	0.52
	4.0	1.36	0.54
2nd buckle contacts wall	4.25	1.52	0.52

3.2.3 Helical Buckling Within Well Bore or Casing

Much work has been done to predict buckling of production tubing inside casing. Initial studies by Lubinski et al [4] give the relation between applied end force and pitch of helical buckling confined within casing.

$$F = \frac{8EI\pi^2}{p^2} \tag{3-45}$$

where p is pitch of the helix.

This equation can be developed by use of the principle of minimum potential energy assuming the buckled tubing is a perfect helix, friction is neglected, and the tubing is weightless. Based on mathematics of space functions, curvature of defected tubing under these assumptions is

$$\frac{1}{R} = \frac{4\pi^2 r}{p^2 + 4\pi^2 r^2} \tag{3-46}$$

where

R — radius of curvature of helix
r — radial clearance between tubing and casing
p — pitch of helix

This equation is useful for determining bending stress in helically curved tubing using the Euler formula, $\sigma = \frac{c}{R}E$. The authors also show the effects of fluid pressure, temperature, and pretension on tubing performance.

Mitchell [5,6] extended this initial work by predicting tubing deflections between packers and the fully developed helix. The results predict bending and shear absorbed by packers under conditions of buckling. These loads are important in packer design. Results also predict contact force between tubing and casing wall.

Cheatham and Pattillo [7] expanded the discussion by considering the post buckling behavior of tubing. Their analysis shows that the buckling formula given by Eq. (3-45) is true when end load is being applied. However, when the externally applied force is unloaded, the relation between load and pitch is defined by

$$F = \frac{4EI\pi^2}{p^2} \tag{3-47}$$

Note that Eqs. (2-45) and (2-47) differ by a factor of 2.

Sorenson and Cheatham [8] give a classic solution to the post buckling behavior of tubing for cases of pinned and fixed end conditions. Their work shows three distinct portions of the buckled shape that have to be considered: 1) an end section, 2) a transition section, and 3) the contact section where the tubing engages the casing throughout. The end section lies between the end point and first contact point. The transition section lies between the initial contact point and the fully engaged section. Boundary conditions are satisfied between each section. This work gives shear and bending loads to the packer. If this work is applied to drill collars directly above the bit, their solution also gives (out of plane) bit orientation.

3.3 STABILIZER PLACEMENT ANALYSIS

It is difficult to predict exact well bore curvature produced by a given stabilized BHA because formation properties play a huge role. However, bit side force predictions are helpful in determining hole direction. Hock [9] appears to be the first to mathematically predict bit side forces with two stabilizers. Analytical predictions along with statistical data and experience with BHA performance are useful guides for stabilizer placement.

Analytical tools and software are designed to predict bit side loads for different BHA configurations. Mathematical models of drill collar bending between stabilizers have been based on multi-spanned beam column theory [10] and the finite element method [11]. Analytical predictions of bit side forces are best made by use of computers because of the complexity of these calculations. These two mathematical models require simultaneous solutions to many algebraic equations (four times the number of spans). The method of transfer matric, described below, requires the solution to only two algebraic equations [12].

3.3.1 Method of Solution

The equation of bending is developed for each span between adjacent stabilizers. The first span is located between the drill bit and the first stabilizer. The last span is located between the last stabilizer and tangent or contact point between drill collar and well bore. Intermediate spans are supported between adjacent stabilizers. Figure 3-8 illustrates the ith span between supports i and $i + 1$. Hole inclination, ϕ, affects the magnitude of the weight component transverse to the collars.

Position constraints are imposed on the continuous beam column in addition to a compressive longitudinal load at the bit. The analysis allows for any position constraints within a vertical plane. Radial clearance between well bore and stabilizer is important and is input to the model. The length of the last or nth span is unknown initially and is determined by an iterative process. A reasonable guess is needed, initially.

A typical drill collar span between two stabilizers is shown in Fig. 3-8, along with a freebody diagram of a differential element, dx. Summing forces and moments on the element gives the differential equation of bending for the span.

$$\frac{dV}{dx} = q \tag{3-48}$$

and

$$\frac{dM}{dx} + Q\frac{dy}{dx} - V = 0 \tag{3-49}$$

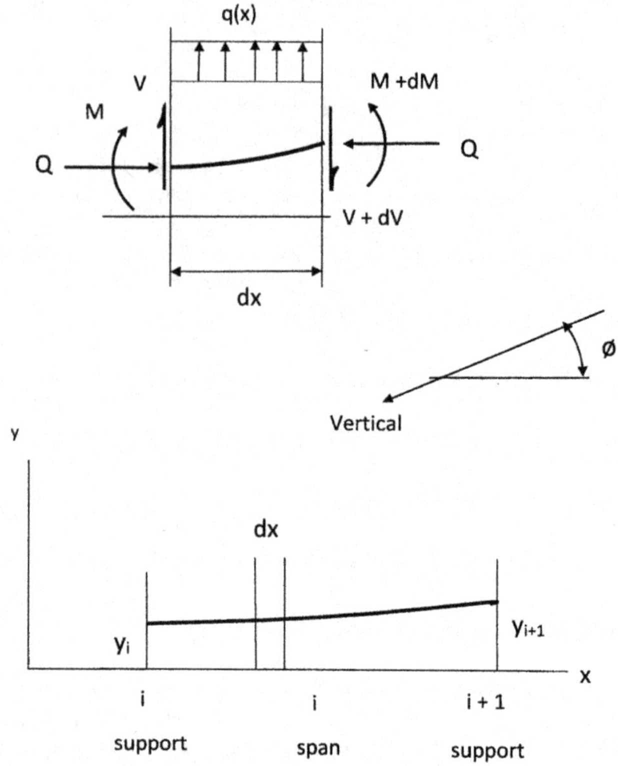

Figure 3-8. Drill collar span between stabilizers with freebody diagram.

Combining the two equations gives

$$\frac{d^2M}{dx^2} + Q\frac{d^2y}{dx^2} = q \tag{3-50}$$

Here, the rate of side loading, q, is plus in the plus y direction. Relating moment to curvature by

$$M = EI\frac{d^2y}{dx^2} \tag{3-51}$$

Combing Eqs. (3-50) and (3-51) gives

$$\frac{d^4y}{dx^4} + \lambda^2\frac{d^2y}{dx^2} = \frac{q}{EI} \tag{3-52}$$

where

$$\lambda^2 = \frac{Q}{EI}$$

The solution (particular plus complimentary) is

$$y(x) = \frac{q}{Q}\frac{x^2}{2} + A_i x + B_i + C_i \cos\lambda x + D_i \sin\lambda x \tag{3-53}$$

At $x = 0$, $y = \psi_i$ giving

$$y_i = B_i + C_i \tag{3-54}$$

At $x = L_i$, $y = y_{i+1}$ giving

$$y_{i+1} = \frac{q}{Q}\frac{L_i^2}{2} + A_i L_i + B_i + C_i \cos\lambda_i L_i + D_i \sin\lambda_i L_i \tag{3-55}$$

Subtracting Eq. (3-54) from Eq. (3-55) and solving for A_i gives

$$A_i = \psi_i - \frac{q}{Q}\frac{L_i}{2} + C_i \frac{1-\cos\lambda_i L_i}{L_i} - D_i \frac{\sin\lambda_i L_i}{L_i} \tag{3-56}$$

where

$$\psi_i = \frac{y_{i+1} - y_i}{L_i} \tag{3-57}$$

Consider now the slope and curvature of the displacement.

$$\frac{dy}{dx} = \frac{q}{Q}x + A_i - C_i\lambda \sin\lambda x + D_i\lambda \cos\lambda x \tag{3-58}$$

$$\frac{d^2y}{dx^2} = \frac{q}{Q} - C_i\lambda^2 \cos\lambda x - D_i\lambda^2 \sin\lambda x \tag{3-59}$$

$$\frac{d^3y}{dx^3} = C_i\lambda^3 \sin\lambda x - D_i\lambda^3 \cos\lambda x \tag{3-60}$$

At $x = 0$

$$\theta_i = A_i + D_i\lambda_i \tag{3-61}$$

$$M_i = \frac{w}{\lambda^2} - QC_i \tag{3-62}$$

Substituting for A_i gives

$$\begin{pmatrix}\theta_i \\ M_i\end{pmatrix} = \begin{bmatrix}\dfrac{1-\cos\lambda_i L_i}{L_i} & \dfrac{\lambda_i L_i - \sin\lambda_i L_i}{L_i} \\ -Q & 0\end{bmatrix}\begin{pmatrix}C_i \\ D_i\end{pmatrix} + \begin{pmatrix}\psi_i - \dfrac{q}{Q}\dfrac{L_i}{2} \\ \dfrac{q}{\lambda^2}\end{pmatrix} \tag{3-63}$$

By inversion

$$\begin{pmatrix} C_i \\ D_i \end{pmatrix} = [\alpha]_i \left(\begin{pmatrix} \theta_i \\ M_i \end{pmatrix} - \begin{pmatrix} \psi_i - \dfrac{q}{Q}\dfrac{L_i}{2} \\ \dfrac{q}{\lambda^2} \end{pmatrix} \right) \tag{3-64}$$

where

$$[\alpha]_i = \frac{1}{-Q(\lambda_i L_i - \sin\lambda_i L_i)} \begin{bmatrix} 0 & \lambda_i L_i - \sin\lambda_i L_i \\ -QL_i & \cos\lambda_i L_i - 1 \end{bmatrix} \tag{3-65}$$

Also at $x = L_i$

$$\theta_{i+1} = \frac{q}{Q} L_i + A_i - C_i \lambda_i \sin\lambda_i L_i + D_i \lambda_i \cos\lambda_i L_i \tag{3-66}$$

$$M_{i+1} = \frac{q}{\lambda^2} - C_i Q \cos\lambda_i L_i - D_i Q \sin\lambda_i L_i) \tag{3-67}$$

Substituting for A_i

$$\begin{pmatrix} \theta_{i+1} \\ M_{i+1} \end{pmatrix} = [\beta]_i \begin{pmatrix} C_i \\ D_i \end{pmatrix} + \begin{pmatrix} \dfrac{q}{Q}\dfrac{L_i}{2} + \psi_i \\ \dfrac{q}{\lambda^2} \end{pmatrix} \tag{3-68}$$

where

$$[\beta]_i = \frac{1}{L_i} \begin{bmatrix} 1 - \cos\lambda_i L_i - \lambda_i L_i \sin\lambda_i L_i & \lambda_i L_i \cos\lambda_i L_i - \sin\lambda_i L_i \\ -QL_i \cos\lambda_i L_i & -QL_i \sin\lambda_i L_i \end{bmatrix} \tag{3-69}$$

substituting for $\begin{pmatrix} C_i \\ D_i \end{pmatrix}$ (Eq. (3-64)), into Eq. (3-68).

$$\begin{pmatrix} \theta_{i+1} \\ M_{i+1} \end{pmatrix} = [\beta]_i [\alpha]_i \left(\begin{pmatrix} \theta_i \\ M_i \end{pmatrix} - \begin{pmatrix} \psi_i - \dfrac{q}{Q}\dfrac{L_i}{2} \\ \dfrac{q}{\lambda^2} \end{pmatrix} \right) + \begin{pmatrix} \dfrac{q}{Q}\dfrac{L_i}{2} + \psi_i \\ \dfrac{q}{\lambda^2} \end{pmatrix} \tag{3-70}$$

According to the above equation, it is possible to express θ_i and M_i in terms of θ_1 and M_1 by successive substitution. This relationship can be expressed in the form

$$\begin{pmatrix} \theta_i \\ M_i \end{pmatrix} = [\gamma]_i \begin{pmatrix} \theta_1 \\ M_1 \end{pmatrix} + (\delta)_i \tag{3-71}$$

From which

$$[\gamma]_1 = \begin{bmatrix} 1 & 0 \\ 0 & 1 \end{bmatrix} \text{and } (\delta)_1 = \begin{pmatrix} 0 \\ 0 \end{pmatrix}$$

By substituting Eq. (3-71) into Eq. (3-70)

$$\begin{pmatrix} \theta_{i+1} \\ M_{i+1} \end{pmatrix} = [\gamma]_{i+1} \begin{pmatrix} \theta_1 \\ M_1 \end{pmatrix} + (\delta)_{i+1} \tag{3-72}$$

where

$$[\gamma]_{i+1} = [\beta]_i [\alpha]_i [\gamma]_i \tag{3-73}$$

and

$$(\delta)_{i+1} = [\beta]_i [\alpha]_i \left((\delta)_i - \begin{pmatrix} \psi_i - \dfrac{q}{Q} \dfrac{L_i}{2} \\ \dfrac{q}{\lambda^2} \end{pmatrix} \right) + \begin{pmatrix} \dfrac{q}{Q} \dfrac{L_i}{2} + \psi_i \\ \dfrac{q}{\lambda^2} \end{pmatrix} \tag{3-74}$$

Above the top stabilizer, where the collars lay on the low side of the hole, the point of contact is identified by $n + 1$. At this point

$$\begin{pmatrix} \theta_{n+1} \\ M_{n+1} \end{pmatrix} = [\gamma]_{n+1} \begin{pmatrix} \theta_1 \\ M_1 \end{pmatrix} + (\delta)_{n+1} \tag{3-75}$$

which is also written

$$\begin{pmatrix} \theta_{n+1} \\ M_{n+1} \end{pmatrix} = \begin{bmatrix} \gamma_{11} & \gamma_{12} \\ \gamma_{21} & \gamma_{22} \end{bmatrix}_{n+1} \begin{pmatrix} \theta_1 \\ M_1 \end{pmatrix} + \begin{pmatrix} \delta_1 \\ \delta_2 \end{pmatrix}_{n+1} \tag{3-76}$$

This pair of algebraic equations relates slope and moment at points 1 and $n + 1$. Using known boundary conditions at these two points, the unknowns can be reduced to two.

$$\theta_{n+1} = (\gamma_{11})_{n+1} \theta_1 + (\gamma_{12})_{n+1} M_1 + (\delta_1)_{n+1} \tag{3-77}$$

$$M_{n+1} = (\gamma_{21})_{n+1} \theta_1 + (\gamma_{22})_{n+1} M_1 + (\delta_2)_{n+1} \tag{3-78}$$

The moment at the drill bit is typically equal to zero, so

$$M_1 = 0$$

At the upper end

$$\theta_{n+1} = 0$$

$$M_{n+1} = 0$$

The distance, L_n, however, is unknown. Through trial and error, L_n is varied until $\theta_{n+1} = 0$ is satisfied. The outcome of the solution are θ_1 and M_1, which are used in Eq. (3-71) to determine all values of θ_i and M_i. Values of $A_i B_i C_i D_i$ are determined subsequently.

The logic for a computer program based on this method is given in Appendix A at the end of this chapter. A numerical example is also given. This method is spelled out in detail because it is used later in this chapter to predict friction in curved portions of directional wells. It is also used in Chapter 6 to predict bending of tubular bundle attached to production risers.

To account for well bore direction and hydrostatic effect, note that effective tension for the differential element in Fig. 3-8 is

$$T_{eff} = (F_B + w_m L \cos\phi) + (w - w_m)x \cos\phi \quad (3\text{-}79)$$

giving

$$Q = WOB - BFwx \cos\phi \quad (3\text{-}80)$$

while

$$q = -(w - w_m)\sin\phi \quad (3\text{-}81)$$

$$q = -wBF \sin\phi \quad (3\text{-}82)$$

The proof is left to the reader.

3.3.2 Example Calculation

An example of the use of this method is given for the following stabilized drill collar:

Hole size = 8 ½ in.
Collar size = 6 ¾ × 2 13/16 in.
$I = 71.83$ in.^{4}
$A = 24.47$ in.^{2}
$w = 83.27$ lb/ft

Mud weight = 12 lb/gal (BF = 0.817)
Stabilizer = 8 15/32 in.
WOB = 40,000 lbs
Hole angle = 30 deg.

Plots of the results are shown in Figs. 3-9 and 3-10. The step-by-step process for computing these plots is given in Appendix A. Side force at the drill bit and stabilizers are given in Table 3-5.

These calculations show that the stabilized assembly is a mild holding assembly, since the side force on the drill bit is only 171 lbs.

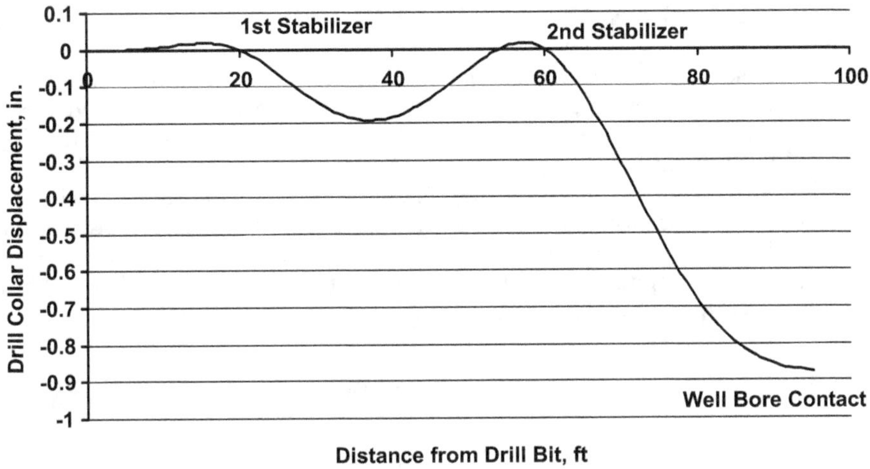

Figure 3-9. Drill collar deflection; two stabilizers (ϕ = 30 deg).

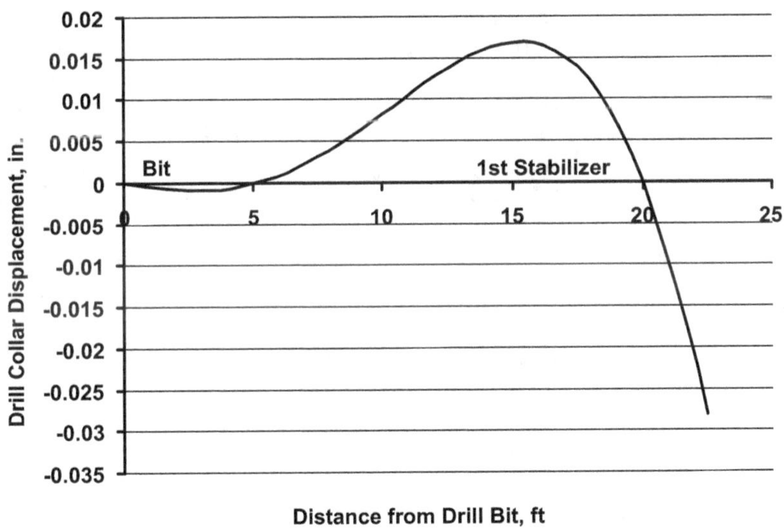

Figure 3-10. Drill collar deflection between drill bit and first stabilizer.

Table 3-5. Transverse support reactions

	Support	Reaction, lbs
(bit)	1	181
	2	1088
	3	1481
(tangent)	4	471*

*Shear is actually minus at the upper tangent location.

3.4 BUCKLING ON AN INCLINED PLANE

Drill pipe is sometimes put in compression under certain conditions of directional drilling in order to generate WOB especially in horizontal well bores. While bit thrust is achieved in this manner, drill pipe fatigue and torsion friction can be significant.

Pasley and Bogy [13,14] give a mathematical basis for predicting drill pipe buckling under conditions of hole size and hole inclination. Later, Dawson and Pasley [15] developed a formula for approximating buckling within an inclined hole. The formula relates critical buckling force to hole size, pipe size and hole angle. It is assumed that the pipe is lying on the low side of the hole.

$$F_{crit} = 2\left(\frac{EIw\sin\theta}{r}\right)^{\frac{1}{2}} \tag{3-83}$$

where

EI = flexure stiffness, lb-in.^{2}
w = weight per unit length
r = radial clearance between pipe and hole
θ = hole angle measured from vertical

This equation is used to generate stability curves of the type shown in Fig. 3-11. These curves (based on pipe size) show regions of stability and instability. They show that pipe stability increases with well bore size and hole inclination. Stability curves can easily be generated for other pipe sizes. This information allows drill

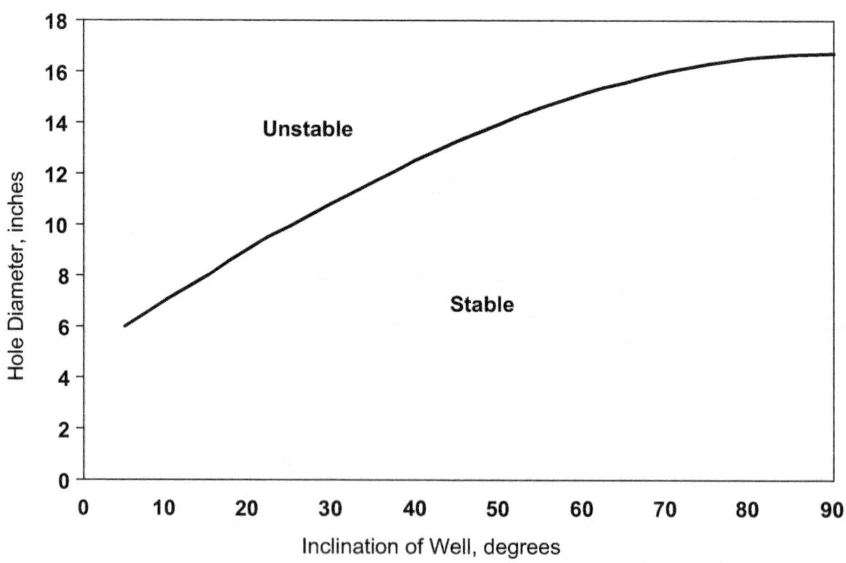

Figure 3-11. Stability data for 5 in. (19.5 lb/ft) drill pipe (compressive force, F_{crit}, is 20,000 lbs).

pipe to carry high levels of compression which eliminate the need for additional drill collars and greatly reducing friction torque and drag.

3.4.1 Buckling of Drillstrings in Directional Wells

Much work has been conducted to model and predict buckling of pipe in curved sections of directional wells [16-19]. These contributions are summarized in API RP 7G [20] (Table 20) relating hole curvature to critical buckling force for a wide range of pipe sizes and hole sizes. Extensive experimental studies [21,22] support these predictions.

3.5 DRILL COLLAR DYNAMICS

The overall dynamic behavior of drill collars is very complex. Axial, torsion, and bending modes of vibrations may occur simultaneously, and well bore contact and geometry are difficult to model mathematically. These modes may be coupled or act independently; research is ongoing. However, much has been learned through downhole measurements and analytical studies over the past 50 years. Because of the complexity of the problem, most studies have focused on specific aspects of the overall dynamics.

Bersalsow and Payne [23] summarize different mechanisms that can excite axial, torsion and bending modes of vibrations in bottom hole assemblies. Conclusions are supported by surface measurements of downhole vibrations taken at a test well and at a drilling site.

By comparing natural frequencies of the different vibration modes with rotary speeds, damaging vibrations can possibly be avoided through proper BHA design and operating practices. The authors conclude that mechanisms of excitation and vibration modes are

> "Mass imbalance can excite lateral modes of vibration. This can happen when rotation speed is equal to a natural frequency of any of the lateral modes. This mechanism can also produce an axial vibration ($2 \times \omega$), but is small by comparison. Torsion vibrations may also be induced.
>
> Three cone drill bits cause axial and torsion vibrations at $3 \times \omega$. A secondary lateral vibration is induced at a frequency of $3/2 \times \omega$ under certain conditions. At low WOB more harmonics of ω may be seen."

3.5.1 Whirling Motion in Drill Collars

Whirling motion of portions of drillstrings, especially within the bottom hole assembly, has received much attention in recent years [23–25]. These gyrations are perceived to be directly responsible for damaging MWD, drilling motors and other equipment located near drill bits.

Pipe whirl is affected by many factors, such as torque, rotary speed, drilling mud properties, axial force, hole inclination, and interaction with the well bore. While a total predictive effort is monumental, much can be learned about the onset of pipe whirl by considering the dynamic stability of vertical pipe.

Greenhill [26] and Southwell and Dough [27] predict instability under conditions of rotation, axial force and torque. Huang and Dareing [28] allowed for non-synchronous whirl and damping by addressing stability of pipe subjected to torque, drill mud, rotational speed and axial force. Variation in tension with length is a unique aspect of their analysis. Results predict conditions of forward, backward, and stationary whirl under various combinations of these parameters. Bottom end of the vertical pipe was assumed to be pinned and open to hydrostatic fluid pressure. Axial force at the lower end was not included in the analysis. The analysis given in ref. [28] is now applied to drill collar between two stabilizers and will include the effects of weight on bit (WOB).

3.5.1.1 *Differential Equations of Motion*

The general equations of motion for long pipe are developed in ref. [28] to include variable tension, torque, rotation, and damping. The equations of motion account for pipe deflection out of a single plane. The mathematical model is shown in Fig. 2-18 along with rotating (*xyz*) and fixed (*XYZ*) coordinate systems. The reference frame *xyz* rotates with angular velocity, ω. The model represented by Fig. 2-18 deals with dynamic stability of vertical pipes under axial load, torque, and rotation and does not represent a forced vibration problem, such as an unbalanced rotation mass.

When applying Eqs. (2-200a) and (2-200b) to a section of drill collars between stabilizers, certain assumptions can be made that greatly simplify the model and still produce reasonable engineering results. Weight on bit (WOB) is large by comparison with collar weight between stabilizers. Also, bit torque is of the order of 1,500 to 2,000 ft-lb. This level of bit torque is small by comparison with torque at the surface. With this in mind, the torque term in Eqs. (2-200a, b) will be dropped, and the effective tension term will be replaced with the average effective tension between stabilizers. These simplifying assumptions reduce the coupled equations of motion to

$$EI\frac{\partial^4 x}{\partial z^4} - T_{eff}\frac{\partial^2 x}{\partial z^2} + m\left(\frac{\partial^2 x}{\partial t^2} - 2\omega\frac{\partial y}{\partial t} - \omega^2 x\right) + c\left(\frac{\partial x}{\partial t} - \omega y\right) = 0 \qquad \text{(3-84a)}$$

$$EI\frac{\partial^4 y}{\partial z^4} - T_{eff}\frac{\partial^2 y}{\partial z^2} + m\left(\frac{\partial^2 y}{\partial t^2} + 2\omega\frac{\partial x}{\partial t} - \omega^2 y\right) + c\left(\frac{\partial y}{\partial t} + \omega x\right) = 0 \qquad \text{(3-84b)}$$

where T_{eff} is the average of the effective tensions between the two stabilizers.

3.5.1.2 *Effective Tension*

Figure 3-12 illustrates a drill collar section between stabilizers 1 and 2. The internal force at the lower (point 1) is WOB plus hydrostatic force Lw_m or

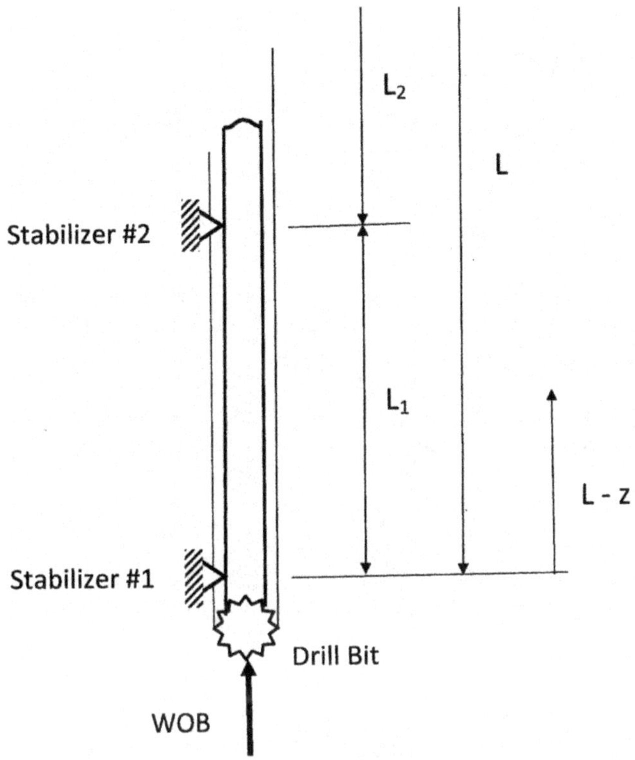

Figure 3-12. Stabilized drill collar section.

$F_B = -\text{WOB} - Lw_m$. L is the distance from the rig floor down to point 1. The average effective tension between points 1 and 2 is defined by

$$T_{eff} = (T_{eff})_{ave} = \frac{(T_{eff})_{top} + (T_{eff})_{bottom}}{2} \tag{3-85}$$

Noting that local effective tension is

$$T_{eff} = F_B + w(L - z) + zw_m \tag{3-86}$$

The average effective tension over Section 1-2 is

$$(T_{eff})_{ave} = (F_B + w_m L) + \frac{1}{2}(W - B) \tag{3-87}$$

W is air weight of drill collar section. Distance, L, is measured from the rig elevation.

Assuming the interval L_1 is the length of drill collars between two stabilizers directly above the drill bit

$$(T_{eff})_{ave} = \left[(-WOB - w_m L) + w_m L\right] + \frac{1}{2}(W - B) \tag{3-88}$$

$$(T_{eff})_{ave} = -WOB + \frac{1}{2} w L_1 BF \tag{3-89}$$

Using the average effective tension eliminates the variable coefficient term and the dimensionless parameter, α. If WOB = 0 the bottom end is open only to hydrostatic pressure so, $(T_{eff})_{ave} = +\frac{1}{2} w L_1 BF$.

Boundary conditions at points 1 and 2 are assumed to be points of structurally pinned connections. Defining the independent variable as

$$\varsigma = \frac{L - z}{L} \tag{3-90}$$

Boundary conditions, using ς as the independent variable, are

$$x(0) = 0 \tag{3-91a}$$

$$y(0) = 0 \tag{3-91b}$$

$$\frac{d^2x}{dz^2}(0) = 0 \tag{3-91c}$$

$$\frac{d^2y}{dz^2}(0) = 0 \tag{3-91d}$$

$$x(\varsigma_1) = 0 \tag{3-91e}$$

$$y(\varsigma_1) = 0 \tag{3-91f}$$

$$\frac{d^2x}{dz^2}(\varsigma_1) = 0 \tag{3-91g}$$

$$\frac{d^2y}{dz^2}(\varsigma_1) = 0 \tag{3-91h}$$

where

$$\varsigma_1 = \frac{L - L_2}{L} = \frac{L_1}{L}$$

3.5.1.3 Mathematical Solution

Equations (3-84a) and (3-84b) are combined into a single equation using the complex function

$$U(z,t) = x(z,t) + iy(z,t) \tag{3-92}$$

The resulting differential equation is

$$\frac{\partial^4 U}{\partial \zeta^4} - \beta \frac{\partial^2 U}{\partial \zeta^2} + \frac{mL^4}{EI}\left[\frac{\partial^2 U}{\partial t^2} + i\left(2\omega \frac{\partial U}{\partial t}\right) - \omega^2 U\right] + \frac{cL^4}{EI}\left(\frac{\partial U}{\partial t} + i\omega U\right) = 0 \tag{3-93}$$

where

$$\zeta = \frac{L - z}{L}$$

$$\beta = \frac{T_{eff} L^2}{EI}$$

Using the method of separation of variables, Eq. (3-92) is expressed in the form

$$U(\varsigma,t) = u(\varsigma)\tau(t) \tag{3-94}$$

The two components $u(\zeta)$ and $\tau(t)$ are separated and equated to a complex constant,

$$\gamma = R_\gamma + iI_\gamma \quad .$$

The separation of variables gives two separate ordinary differential equations.

$$\frac{d^4 u}{d\zeta^4} - \beta \frac{d^2 u}{d\zeta^2} - (R_\gamma + iI_\gamma)u = 0 \tag{3-95}$$

$$\frac{d^2\tau}{dt^2} + 2\left(\frac{c}{2m} + i\omega\right)\frac{d\tau}{dt} + \frac{EI}{mL^4}\left[R_\gamma - \Omega^2 + i\left(I_\gamma + \frac{c\omega L^4}{EI}\right)\right]\tau = 0 \tag{3-96}$$

where

$$\gamma = R_\gamma + iI_\gamma$$

$$\beta = \frac{T_{eff} L^2}{EI}$$

$$\Omega^2 = \frac{m\omega^2 L^4}{EI}$$

The solution to Eq. (3-95) along with boundary conditions gives R_γ and I_γ and the shape of the whirl mode. The solution to Eq. (3-96) gives conditions of stability. Consider the solution to Eq. (3-95) first.

3.5.1.4 Establishing R_γ and I_γ

Values of R_γ and I_γ directly affect dynamic stability of whirl. These values are determined by solving Eq. (3-95) and satisfying boundary conditions.

$$u(0) = 0 \tag{3-97a}$$

$$\frac{d^2u}{d\zeta^2}(0) = 0 \tag{3-97b}$$

$$u(\zeta_1) = 0 \tag{3-97c}$$

$$\frac{d^2u}{d\zeta^2}(\zeta_1) = 0 \tag{3-97d}$$

A solution to Eq. (3-95) is assumed to be of the form

$$u = Ce^{a\zeta} \tag{3-98}$$

where a can be complex. After substitution and simplifying

$$a^4 - \beta a^2 - \gamma = 0 \tag{3-99}$$

The four roots to this equation are

$$a = \pm\left[+\frac{\beta}{2} \pm \sqrt{\left(\frac{\beta}{2}\right)^2 + \gamma}\right]^{\frac{1}{2}} \tag{3-100}$$

Note that $a_3 = -a_1$ and $a_4 = -a_2$.

These four roots (a_1, a_2, a_3, a_4) give four components to the solution.

$$u(\zeta) = C_1e^{a_1\zeta} + C_2e^{a_2\zeta} + C_3e^{a_3\zeta} + C_4e^{a_4\zeta} \tag{3-101}$$

Applying the four boundary conditions of Eq. (3-97) to Eq. (3-101) sets up a matrix, whose expansion gives the characteristic equation. Bringing all of the algebra together leads to

$$\left[-\frac{\beta}{2} + \sqrt{\left(\frac{\beta}{2}\right)^2 + \gamma}\right] = \left(\frac{n\pi}{\zeta_1}\right)^2 \tag{3-102}$$

$$\gamma = \left[\left(\frac{n\pi}{\zeta_1}\right)^2 + \frac{\beta}{2}\right]^2 - \left(\frac{\beta}{2}\right)^2 \tag{3-103}$$

From this we conclude that $I_\gamma = 0$ and $R_\gamma = \gamma$.

$$R_\gamma = \left(\frac{\pi}{\zeta_1}\right)^2 \left[\beta + \left(\frac{\pi}{\zeta_1}\right)^2\right] \tag{3-104}$$

In general, $I_\gamma = 0$ in cases of zero torque.

3.5.2 Criteria for Dynamic Stability

Equation (2-219) is greatly simplified when $I_\gamma = 0$. The second term can be dropping ($B = 0$) by appropriate initial conditions leaving

$$\lambda_1 = -\frac{\eta}{2} + i\left[\Omega - \sqrt{\left(R_\gamma - \left(\frac{\eta}{2}\right)^2\right)}\right] \tag{3-105}$$

The time history of the motion then becomes

$$\tau(t) = e^{-\frac{\eta}{2}\sqrt{\frac{EI}{mL^4}}t} A\left[\cos\sqrt{\frac{EI}{mL^4}}\left(\Omega - \sqrt{R_\gamma - \left(\frac{\eta}{2}\right)^2}\right)t + i\sin\sqrt{\frac{EI}{mL^4}}\left(\Omega - \sqrt{R_\gamma - \left(\frac{\eta}{2}\right)^2}\right)t\right] \tag{3-106}$$

From Eqs. (3-92) and (3-94), it follows that

$$x(z,t) = u\,(\varsigma)e^{-\frac{\eta}{2}\sqrt{\frac{EI}{mL^4}}t} A\cos\sqrt{\frac{EI}{mL^4}}\left(\Omega - \sqrt{R_\gamma - \left(\frac{\eta}{2}\right)^2}\right)t \tag{3-107a}$$

$$y(z,t) = u(\varsigma)e^{-\frac{\eta}{2}\sqrt{\frac{EI}{mL^4}}t} A\sin\sqrt{\frac{EI}{mL^4}}\left(\Omega - \sqrt{R_\gamma - \left(\frac{\eta}{2}\right)^2}t\right) \tag{3-107b}$$

Equations (3-107a, b) define the lateral position of any location, ς, within the stabilized drill collar interval. These equations show that both displacement components reduce with time, so motion with any level of damping is stable. These two equations are represented by position vector, $\bar{a}$, which is illustrated in Fig. 2-21.

The position vector rotates with angular velocity, p, defined by

$$p = \sqrt{\frac{EI}{mL^4}}\left[\Omega - \sqrt{R_\gamma - \left(\frac{\eta}{2}\right)^2}\right] \tag{3-108}$$

This angular rotation frequency, p, is measured relative to the xy rotating reference frame. Recall the angular velocity parameter, Ω_0, relates to R_γ and ς by

$$\Omega_0 = \sqrt{R_\gamma - \left(\frac{\eta}{2}\right)^2},$$

the parameter corresponding to zero rotation of position vector, $\bar{a}$. If damping is critical, the vector, $\bar{a}$, neither grows or reduces. In our case, critical damping is equal to zero so damping of any kind will cause the vector $\bar{a}$ to reduce with time as shown in Fig. 3-13.

The position vector can rotate counterclockwise or clockwise with respect to the rotating xy reference frame. The path that the collars take relative to the rotation xy reference frame will determine forward or backward whirl as it spirals back toward the well bore center. The direction of the spiral depends on the sign (+ or –) of the argument. The frequency of the motion is defined by

$$p = \left(\frac{EI}{mL^4}\right)^{\frac{1}{2}} (\Omega - \Omega_O), \text{ rad/sec.} \tag{3-109}$$

The spiraling motion is clockwise when $\Omega > \Omega_0$. The spiraling motion is counterclockwise when $\Omega < \Omega_0$. If $\frac{\Omega}{\Omega_O} = 1$, the center of the pipe moves straight inward without spiraling.

The tightness of the spiral depends on the frequency parameter, p. If p is small, the spiral is more open, $\frac{\Omega}{\Omega_o} = 0.5$. If p is large, the spiral is tighter, $\frac{\Omega}{\Omega_O} = 0$.

3.5.3 Relative Whirl versus Absolute Whirl

Figure 3-13 shows various types of whirl relative to the xy rotating reference frame, which rotates with ω angular velocity. The angular velocity, ω, is the

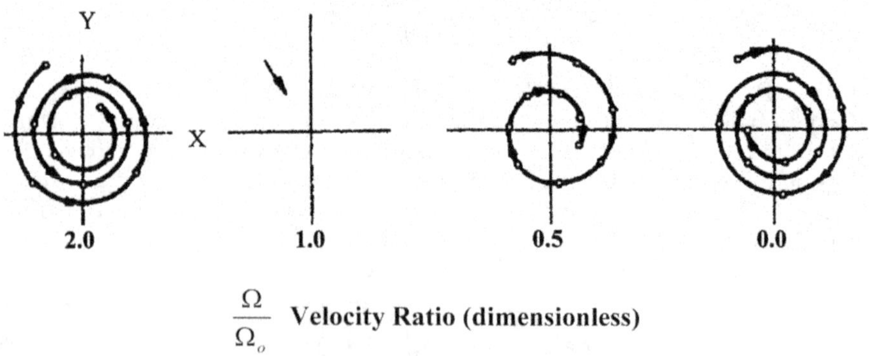

Figure 3-13. Stable whirl of drill collar span.

same as input rotary speed of the drillstring. The absolute angular velocity of position vector $\bar{a}$

$$\Phi = \omega + p \tag{3-110}$$

There are various scenarios for Eq. (3-110), such as $p = 0$ ($\Phi = \omega$). In this case, vector $\bar{a}$ decrease and the outside surface of the pipe is always looking outward.

These results show that sections of drill collars between well bore stabilizers are dynamically stable. If drill collar sections are disturbed, they will spiral back to the centerline of the well bore. The direction of the spiral relative to a rotating reference frame depends on rotary speed and fluid damping. Absolute whirl is the algebraic sum of relative whirl and rotary speed.

This conclusion, however, does not rule out the possibility of other mechanism of whirl caused by gear tracking related to contact interactions between collars and well bore or whirl caused by imbalance of the collars.

3.5.4 Synchronous Whirl Based on Zero Damping

In the absence of damping, Eqs. (3-107a, b) become

$$x(z,t) = u(\varsigma)A\cos\sqrt{\frac{EI}{mL^4}}\left(\Omega - \sqrt{R_\gamma}\right)t \tag{3-111a}$$

$$y(z,t) = u(\varsigma)A\sin\sqrt{\frac{EI}{mL^4}}\left(\Omega - \sqrt{R_\gamma}\right)t \tag{3-111b}$$

When $\Omega = \sqrt{R_\gamma}$, $y(z,t) = 0$ and $x(z,t) = Au(\zeta)$. The collars rotate with synchronous whirl. In this case

$$\frac{m\omega^2 L^4}{EI} = R_\gamma \tag{3-112}$$

Combining Eqs. 3-104 and 3-112 gives

$$\omega_{cr}^2\left(\frac{mL^4}{EI}\right) = \left(\frac{n\pi}{\zeta_1}\right)^2\left[\beta + \left(\frac{n\pi}{\zeta_1}\right)^2\right] \tag{3-113}$$

$$\omega_{cr}^2\left(\frac{mL^4}{EI}\right) = \left(\frac{n\pi L}{L_1}\right)^2\left[\frac{T_{eff}L^2}{EI} + \left(\frac{n\pi L}{L_1}\right)^2\right] \tag{3-114}$$

$$\omega_{cr}^2 = \left(\frac{n\pi}{L_1}\right)^2\frac{EI}{m}\left[\frac{T_{eff}}{EI} + \left(\frac{n\pi}{L_1}\right)^2\right] \tag{3-115}$$

which agrees with Eq. (2-79) (Ch. 2). T_{eff} is assumed to be tension (a positive number) in the earlier derivation. In this scenario, T_{eff} is defined by Eq. (3-89).

The direction of the rotation of position vector, $\bar{a}$ is determined from

$$p = \left(\frac{EI}{mL^4}\right)^{\frac{1}{2}} \left[\Omega - \left(R_\gamma - \frac{\eta^2}{4}\right)^{\frac{1}{2}}\right]$$

where

$$R_\gamma = \left(\frac{\pi L}{L_1}\right)^2 \left[\frac{T_{eff} L^2}{EI} + \left(\frac{\pi L}{L_1}\right)^2\right]$$

Assuming damping is zero

$$p = \omega - \left\{\left(\frac{\pi}{L_1}\right)^2 \left[\frac{T_{eff}}{m} + \left(\frac{\pi}{L_1}\right)^2 \frac{EI}{m}\right]\right\}^{\frac{1}{2}} \tag{3-116}$$

or

$$p = \omega - \omega_{cr} = 0$$

3.5.5 Experimental Measurements of Drill Colar Whirl

The above discussion does not account for other mechanisms of whirl, such as imbalance and gear tracking. Drill collar whirl from whatever the cause has been identified experimentally by Vandiver, Nicholson, and Shyu [24], who discuss the kinematics of different modes of pipe whirl and show that in an extreme case of forward synchronous whirl, sliding velocities may account for flat spots observed on drill collars and drill pipe tool joints.

A significant aspect of this work is the use of strain gauge data to extract time varying bending moment components, which are further used to determine whirl velocity, Ω. The moment vector, $\bar{B}$, varies as the collars rotate within the confines of the well bore. Figure 3-14 illustrates the parameters. B_x and B_y are determined from the strain gauge measurements aligned with the x and y axes.

The radius of curvature of bending is represented by the symbol, ρ. The center of curvature is shown as point "c". Other definitions are:

Ω – whirl of drill collars (absolute whirl)
ω – rotation of drill collars
$\bar{B}$ – resultant bending moment reference from xy rotating frame
α – position of moment vector $\bar{B}$, measured from the x axis

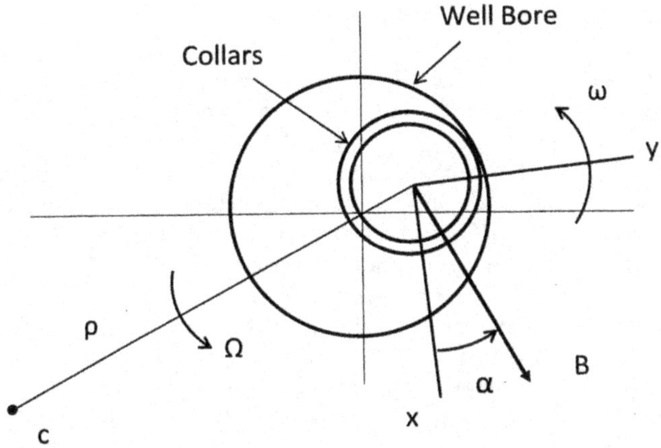

Figure 3-14. Relation of bending vector with drill collar whirl and rotation.

The bending moment vector is defined as

$$\overline{B}(t) = B_x(t)i + B_y(t)j \tag{3-117}$$

$$\overline{B}(t) = B\left[\cos\alpha(t)i + \sin\alpha(t)j\right] \tag{3-118}$$

The phase angle, α, is determined from

$$\tan\alpha = \frac{B_y(t)}{B_x(t)} \tag{3-119}$$

It is related to ω and Ω by

$$\Delta\alpha = \Omega\Delta t - \omega\Delta t \tag{3-120}$$

$$\frac{d\alpha}{dt} = \Omega - \omega \tag{3-121}$$

This equation is comparable to Eq. (3-110) where $\frac{d\alpha}{dt} = p$ and $\Omega = \phi$.

$$\alpha(t) = (\Omega - \omega)t + \alpha_0 \tag{3-122}$$

where α_0 is a constant depending of initial conditions.

Eq. 3-121 is of particular interest because $\frac{d\alpha}{dt}$ can be determined from the slope of the measured $\alpha(t)$ data. Drill collar rotational speed, ω, is known from rig data. Equation (3-121) defines the whirling rate of drill collars. The authors use this approach to determine the whirling motion of drill collars at several drilling sites to predict slip velocities between drill collars and well bore.

3.5.6 Unbalanced Drill Collars

Dykstra, Chen, Warren, and Asar [25] consider the effect of unbalanced drill collars on lateral vibrations in BHA. Drill collars can be unbalanced due to bore misalignment, initial curvature, and gradual wear during service. Severe lateral vibrations were predicted for unbalanced inertia forces under resonant conditions. Natural frequencies and mode shapes were predicted with a finite element analysis. Measurements indicated lateral accelerations reached 50 g and higher. Under resonant conditions, surface torque increased from around 500 ft/lbs to 3000 ft/lbs, indicating backward whirl or gear tracking against the wall.

Spanos, Payne, and Secora [29] give a mathematical model for predicting vibration responses of bottom hole assemblies. BHAs are modeled mathematically by the finite element method and the dynamic responses are predicted using modal summation. Damping is experimentally determined from measured wave reflection data. The authors use this model to predict the lateral vibrations of a BHA containing a number of stabilizers. The source of excitation was a simple harmonic force applied laterally to the drill bit. Response predictions clearly show mode behavior at resonant frequencies.

3.5.7 Excitation by Positive Displacement Motors (PDM)

Another source of lateral excitation comes from positive displacement motors. The rotor (Fig. 3-15) moves around the centerline of the stator as indicated. The stator is made up with the drill collars. PDM rotors move as a rigid body with planetary motion within stators. Assuming planar motion (all forces lie within a transverse plane), the equations of motion give

$$\sum F_r = M_r a_{Gr} \tag{3-123}$$

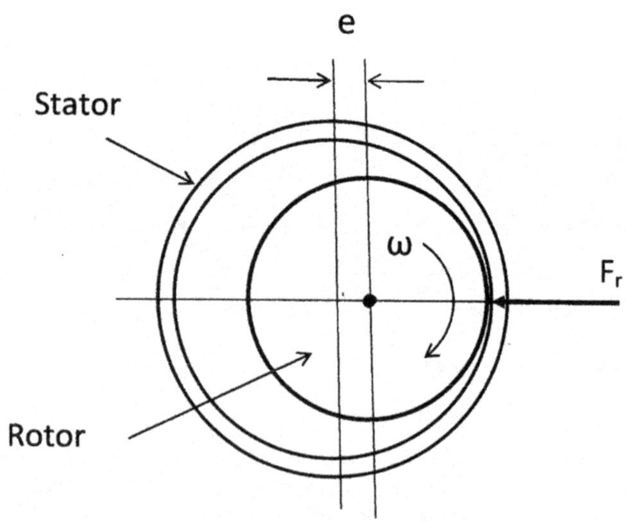

Figure 3-15. PDM rotor dynamics.

where a_{Gr} is the radial component of the acceleration of the center of gravity of the rotor and F_r is the radial force that produces this acceleration component. This radial force also pushes against the stator to excite the drill collars. This force is significant and cyclic.

Consider the following parameters:

$\frac{n_s}{n_r} = 2$ (lobe ratio)

N = 200 rpm (bit speed)

$\omega = \frac{2\pi N}{60}$ rad/sec

$W_r = 200$ lb (rotor weight)

r_r = pitch radius of rotor (= 2 inches)

From the kinematics of the stator/rotor

$$\omega = \left(1 - \frac{n_s}{n_r}\right)\omega_e$$

meaning that vector e rotates counterclockwise with angular velocity of ω. The acceleration of the rotor center is $a_{Gr} = \omega^2 e = \omega^2 r_r$. The force, $F_r = Ma_{Gr}$ or

$$F_r - \frac{200}{32.2}\left(\frac{2\pi 200}{60}\right)^2 \frac{2}{12} = 454 \text{ lbs}$$

The force vector rotates with ω_e = 20.94 rad/sec or a frequency of 3.33 cps. Force magnitude and frequency are both important in determining possible resonance and response of lateral modes between stabilizers. These numbers depend on stator/rotor ratios, output speed and rotor mass.

3.5.8 Coupling of Axial and Lateral Vibrations

Vertical pipes under compression can buckle statically as an Euler column or from their own weight as indicated earlier. If compressive end loads fluctuate with time about a mean force, drill collars may vibrate laterally and become quite severe if the frequency of the dynamic force component is twice the natural frequency of the lateral mode of the collars (see Fig. 2-23). The axial dynamic force is typically due to bit/formation interaction and can become quite large if bit force frequency is equal to the natural frequency of an axial mode of vibration. Dunayevsky et al [30] consider the effects of wall contact and along with axial forces and predict critical rotary speeds causing the onset of severe lateral vibrations.

3.5.9 Stick-Slip-Induced Vibration

Lin and Wang [31] analyzed torsion motions of drill strings with a one degree of freedom model, which included negative damping. Angular motion predicted by this model is used to explain field measurements. They conclude that beating observed in field data is the result of stick-slip vibration induced by the total resistance, i.e. the dry friction, viscous damping and bit torque. Yigit and Christoforou [32] use a similar model to explore coupling between rotational stick-slip and bit bounce.

3.6 THEORIES OF AXIAL AND TORSION VIBRATION

3.6.1 Forced Vibrations

Axial vibrations have been observed at the surface through vertical vibrations of the Kelly for many years, especially while drilling hard formations. Accelerometer measurements of the swivel bounce indicated that the frequency of these vibrations were three cycles per rotation of the drillstring [33]. The tops of core samples show a three lobed pattern, which is consistent with frequencies observed at the surface (see Fig. 1-4). Three lobed patterns in core tops are common.

3.6.2 Roller Cone Drill Bits

The development of three lobed patterns (Fig. 1-4) seems to be independent of rotary speed or drillstring compliance, which implies that resulting vibrations (axial and torsion) are forced as opposed to self-excited vibrations.

How these lobed patterns are formed is unclear. Since they commonly occur in hard formations, it is clear that any slight reorientation of the axis of the drill collars directly the bit would cause the WOB to be concentrated off of well bore center. This would start the process of forming the three lobes. Field measurements of bit force and drill collar bending directly above the bit (Fig. 3-16) seem to support this possibility.

These simultaneous measurements show a clear pattern of bending at the beginning of a build up of axial bit force vibrations. In this case, BHA stabilization

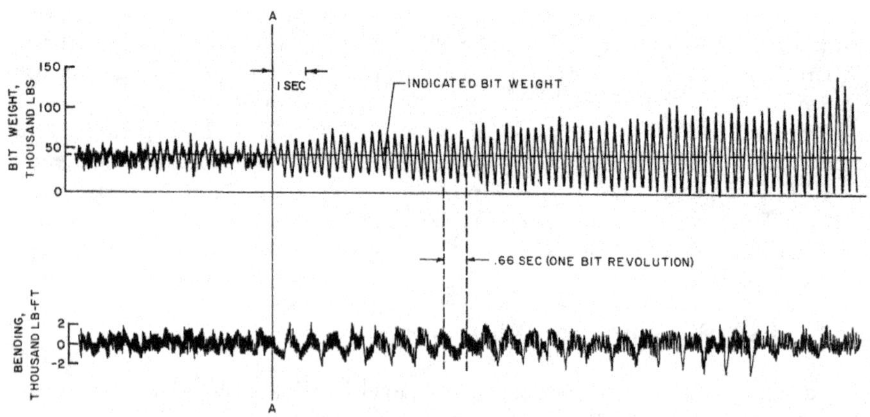

Figure 3-16. Transition into periodic bit force [33].

may help alleviate axial vibrations. Once the lobes in the hole bottom form they can cause boundary movement and corresponding boundary forces, which can drive longitudinal vibrations. The severity of the vibration response would depend on frequency tuning. Non-homogeneous formations may also play a role in the formation of these lobes.

Spanos et al [34] explain analytically how these lobed patterns are modulated during drilling. The lobed pattern is a boundary condition to the overall vibration response of drill strings. The analysis tracks the motion the drill bit as it interacts with the lobed pattern and accounts for zero bit force when the bit disengages the formation. The greatest modulation of bit motion and lobe amplitude comes when bit cones move downward into the low area and disengage the rock over the peaks of the lobes. When the bit disengages the lobes, bit force is zero.

The assumption is made that the lobes exists prior to each calculation. Axial and torsion modes of vibration are considered in conjunction with the three lobes pattern to determine operating conditions which modulate the existing lobe geometry. It is suggested that the initial lobes may be the result of compaction during the initial application of WOB or from non uniformity of the formation. The results from this analysis show responses that are consistent with measured data of bit forces and motion.

The authors point out two types of boundary conditions at the drill bit. In one case, drill bit forces can excite drill strings. In this case, the lower end is considered as a free end. In a second case the drill string is excited by bit displacement much like base motion. In the first case, axial vibration modes are considered free at the bit end. In the second case, the modes are considered fixed at the lower end.

3.6.2.1 Axial and Torsion Vibration Model

Dareing and Livesay [35] modeled drillstrings as a system of two continuous pipes vibrating in a linearly damped environment (see Fig. 3-17). The analysis considers vibration response caused by boundary displacements induced by the three lobes associated with roller cone drill bits (Fig. 1-4). Kreisle and Vance [36] consider a similar model.

Summing forces on a differential element and equating them to the inertia term gives

$$AE\frac{\partial^2 u}{\partial x^2} = \gamma\frac{\partial u}{\partial t} + m\frac{\partial^2 u}{\partial t^2} \tag{3-124}$$

which applies to each tubular section. Variable tension caused by the weight is a static load and does not affect the dynamic analysis. The damping coefficient, γ (having units of lbs per ft/sec per ft of pipe length), is best determined experimentally.

Damping distribution is needed, to make vibration response predictions. The mass and elastic can easily be calculated. Damping normally has to be determined experimentally because of the many factors that affect it.

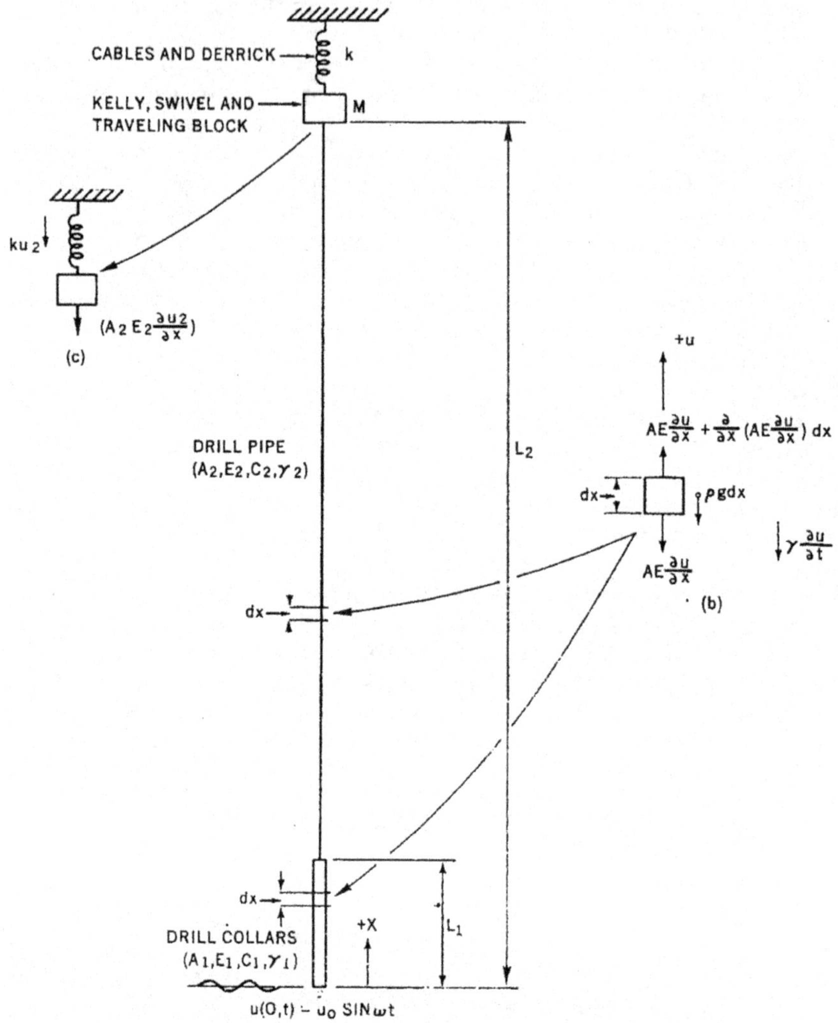

Figure 3-17. Model of a drillstring.

The solution to Eq. (3-124) applies to both the drill pipe and drill collar sections. Axial displacements in each section are identified by

u_1 — drill collar section
u_2 — drill pipe section

These functions are linked together with appropriate boundary conditions.

The boundary condition at the surface considers the mass of swivel and traveling block, M and stiffness of the cables in the draw works, k. From a free body of the top mass, M

$$A_2E_2\frac{\partial u_2}{\partial x}+ku_2+M\frac{\partial^2 u_2}{\partial t^2}=0 \tag{3-125}$$

at $x = L_2$

The connection between the drill pipe and drill collars must satisfy

$$u_1(L_1,t)=u_2(L_1,t) \tag{3-126}$$

$$E_1A_1\frac{\partial u_1}{\partial x}\bigg|_{x=L_1}=E_2A_2\frac{\partial u_2}{\partial x}\bigg|_{x=L_1} \tag{3-127}$$

Finally, bit displacement (caused by the three lobed pattern found in core tops) is defined by

$$u_1(0,t)=U_0\sin\omega t \tag{3-128}$$

The solution to Eq. (3-124) can be expressed in terms of the complex function

$$u(x,t)=\bar{u}(x)e^{i\omega t} \tag{3-129}$$

After substitution, Eq. (3-124) becomes

$$\frac{d^2\bar{u}}{dx^2}+\eta^2\bar{u}=0 \tag{3w-130}$$

where

$$\eta^2=\frac{\omega^2}{c^2}-i\frac{\gamma\omega}{AE}$$

The solution to Eq. (3-130) becomes

$$\bar{u}(x)=\bar{B}\sin(\eta x+\bar{b}) \tag{3-131}$$

B and b depend on the boundary conditions given earlier. The total complex solution becomes

$$\bar{u}(x)=\bar{B}\sin(\eta x+\bar{b})e^{i\omega t} \tag{3-132}$$

The function $u(x,t)$ has complex and imaginary parts. The choice between the real and imaginary solutions depends on the mathematical form of the boundary conditions. Since the boundary condition at the bit can be expresses as

$$u_0\sin\omega t=\mathrm{Re}\left[-iu_0e^{i\omega t}\right] \tag{3-133}$$

The real part of Eq. (3-132) is the desired solution if $[-iu_0e^{i\omega t}]$ is satisfied at $x = 0$. The solution to the damped drillstring then is the real parts of

$$u_1(x,t)=B_1\sin\left(\eta_1x+b_1\right)e^{i\omega t},\ 0\le x\le L_1 \tag{3-134}$$

$$u_2(x,t)=B_2\sin\left(\eta_2x+b_2\right)e^{i\omega t},\ L_1\le x\le L_2 \tag{3-135}$$

where B_1, B_2, b_1, b_2 are complex numbers determined from boundary conditions. The boundary condition expressed by Eq. (3-128) gives

$$B_1 = -\frac{iu_0}{\sin b_1} \tag{3-136}$$

The boundary conditions at the top of the collars, represented by Eqs. (3-126) and (3-127) give

$$B_2 = -\frac{iu_0}{\sin b_1}\frac{\sin(\eta_1 L_1 + b_1)}{\sin(\eta_2 L_1 + b_2)} \tag{3-137}$$

and

$$b_1 = \tan^{-1}\left[\frac{A_1 E_1 \eta_1}{A_2 E_2 \eta_2}\tan(\eta_2 L_1 + b_2)\right] - \eta_1 L_1 \tag{3-138}$$

The boundary condition at the top of the drill pipe section, represented by Eq. (3-125) gives

$$b_2 = \tan^{-1}\left(\frac{A_2 E_2 \eta_2}{M\omega^2 - k}\right) - \eta_2 L_2 \tag{3-139}$$

The solution to both drillstring sections can now be determined. The dynamic force acting on the drill bit as a result of bit cone movement over the three (3) lobes is determined from

$$F_{bit} = EA\left.\frac{\partial u}{\partial x}\right|_{x=0} \tag{3-140}$$

The solution gives the overall shape of the axial amplitudes (Fig. 3-18) and the frequency response (Fig. 3-19) corresponding to various levels of assumed damping for both drill pipe and drill collars. The phase angle diagrams have been omitted but are included in ref [35]. The damping coefficients used in these calculations should be considered reasonable approximations.

Even though higher modes have been observed through measurements in shallow test wells, it will be more difficult to develop the higher modes in typical drillstrings because of the narrow bandwidth of the frequency response as shown in Fig. 3-19. It takes time to put energy into the higher modes and this has to be done at a fairly constant rotary speed. This suggests that drill collars dictate the axial vibrations in the drill pipe.

Note that the frequency response curves (Fig. 3-19) are bounded along the lower edge by the response of the drill collars. The vibration of the drill collar section alone dictates the magnitude of dynamic forces on the drill bit and other tools near the bit. The drill collar response alone is significant to magnitude of dynamic forces applied drill bits.

An application of this study was to predict downhole drillstring vibrations, including dynamic forces at the drill bit, based on surface measurements. One of the most difficult parameters to quantify is damping, which no doubt varies

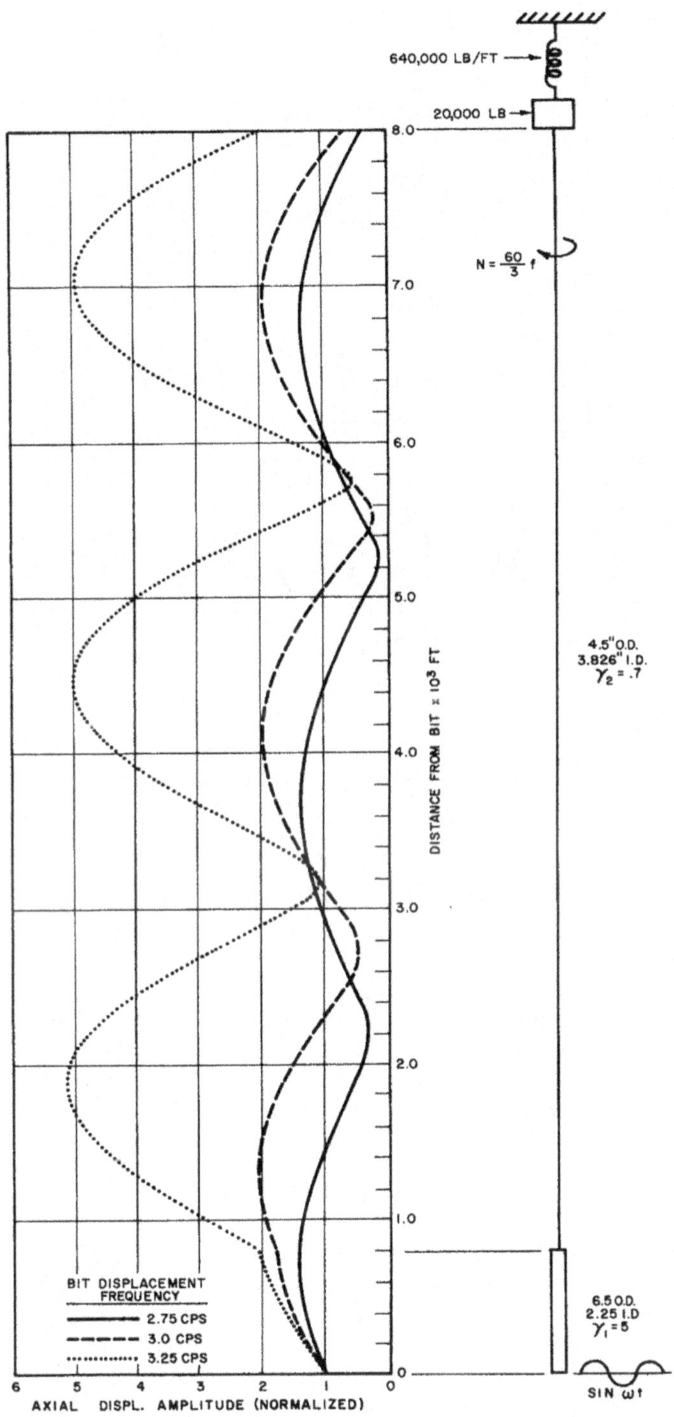

Figure 3-18. Frequency response of a drill string.

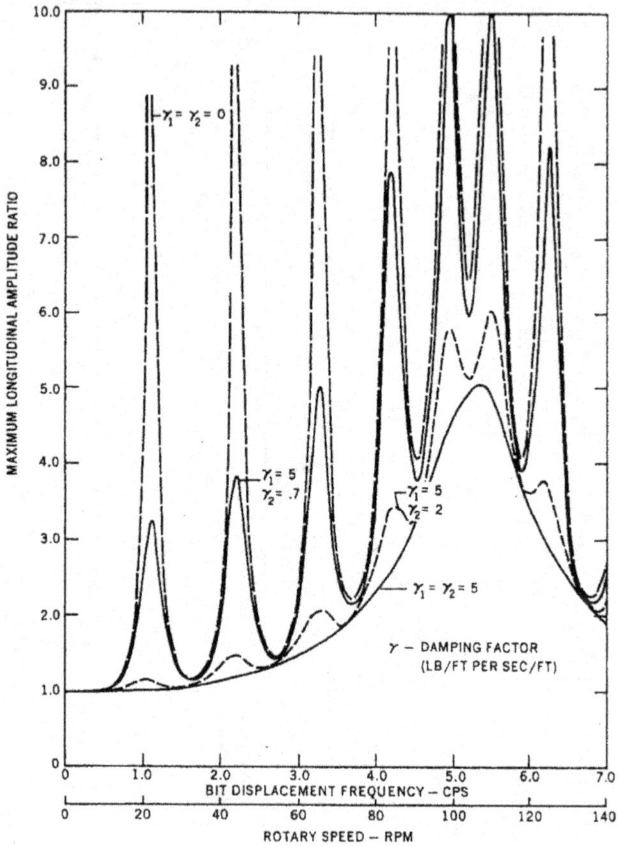

Figure 3-19. Frequency response of a drill string.

along the entire length of the drillstring. Fortunately, downhole measurements of vibration motion and forces of all near bit tools can be measured in real time and transmitted back to the surface through Measurement While Drilling (MWD) systems as drilling progresses.

3.6.2.2 Quantifying Damping for Axial Modes

Sengupta [37] predicts vibration damping of axial and torsion modes assuming the drilling mud to be a Newtonian viscous fluid. Graphs show how damping coefficients for both axial and torsion modes are dependent on vibration frequency. The results provide a good baseline even though they do not include effects of mechanical friction, which could be substantial.

The transient free vibration taken at a drill site (Fig. 3-20) provides some insight into the affects of mechanical friction on the damping coefficient, γ.

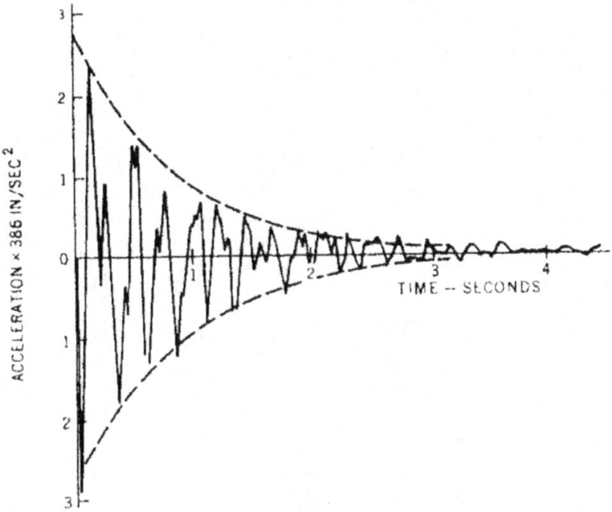

Figure 3-20. Damped free vibration of drillstring taken at the surface [35].

The frequency of this response is about 4 cps. Assuming this is the response of a particular mode in the drillstring, then the damping factor, ζ_n, for this mode can be determined from the log decrement

$$\delta = \frac{1}{n} \ln \frac{X_i}{X_{i+1}} \tag{3-141}$$

From the graph

$$\delta = \frac{1}{4} \ln \frac{1.5}{0.5} = 0.2747 \tag{3-142}$$

Using

$$\delta = \frac{2\pi\zeta}{\sqrt{1-\zeta^2}} \cong 2\pi\zeta \tag{3-143}$$

$$\zeta = 0.0437$$

Recall that the modal damping factor is

$$\zeta = \frac{C}{C_{cr}} \tag{3-144}$$

from which

$$C_{cr} = 2M\omega \tag{3-145}$$

$$C = \zeta(2M\omega) \tag{3-146}$$

where C and M are modal damping and modal mass, which can be related to local damping, γ, and local mass, ρ as follows.

Using modal analysis techniques, the solution to Eq. (3-124) can be represented by

$$u(x,t) = \sum_{i=1}^{\infty} X_i(x)\eta_i(t) \tag{3-147}$$

The substitution of this function and its derivatives into Eq. (3-124), and taking advantage of the orthogonal properties of the modes, gives (assuming proportional damping)

$$M_i\ddot{\eta}_i + C_i\dot{\eta}_i + K_i\eta_i = 0 \tag{3-148}$$

where

$$M_i = m\int_0^L X_i^2 dx$$

and

$$C_i = \gamma\int_0^L X_i^2 dx$$

Substituting into Eq. (3-146) gives

$$\gamma = 2\zeta m\omega \tag{3-149}$$

Assuming 4 ½ in. drill pipe we obtain

$$\gamma = 2(0.0437)\frac{16.6}{32.2}2\pi(4) = 1.13 \text{ lb/fps per ft of pipe length}$$

This number represents a reasonable approximation for the modal damping coefficient. It includes the effects of all friction forces acting on the drill string that existed at the time of this measurement. A more accurate method would be to experimentally determine the transfer function for the axial modes. Modal damping could be taken from the imaginary component of each modal response. This could be done with current MWD technology.

3.6.3 Natural Frequencies of Axial, Torsion, and Lateral Modes

Natural frequencies and mode shapes are important information in the design and operation of equipment. Resonance and severe vibrations can be avoided by simply avoiding frequency tuning.

Drillstrings can be viewed as having two different sets of boundary conditions. In one case the bottom end is fixed as shown in Figs. 3-21a and 3-21b. In the second case, the bottom end of the drill collar section is free-ended.

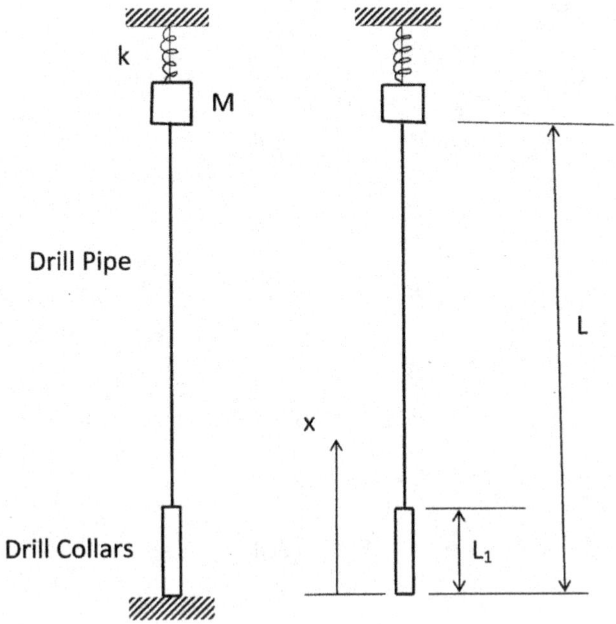

Figure 3-21. Axial vibration models of drillstring.

With reference to Fig. 3-21a, consider a drillstring made up of a drill collar section, a drill pipe section and a top mass representing the swivel and traveling block. The top spring accounts for the flexibility of the draw works cable.

3.6.3.1 Natural Frequencies of Axial Modes

Natural frequencies are determined from undamped free vibrations analysis. The equation of motion for the axial mode becomes

$$AE\frac{\partial^2 u}{\partial x^2} = m\frac{\partial^2 u}{\partial t^2} \tag{3-150}$$

This equation applies over both drill pipe and drill collar sections. Possible boundary conditions for the axial case are illustrated in Fig. 3-21. The same equation will used for both pipe and collars where

$u_1(x,t)$ - drill collar displacements
$u_2(x,t)$ - drill pipe displacements

The determination of mode shapes and corresponding natural frequencies proceeds as follows.

Boundary conditions are

$$u_1(0,t)=0 \tag{3-151a}$$

$$u_1(L_1,t)=u_2(L_1,t) \tag{3-151b}$$

$$A_1E\frac{du_1}{dx}(L_1,t)=A_2E\frac{du_2}{dx}(L_1,t) \tag{3-151c}$$

$$M\frac{\partial^2 u_2}{\partial t^2}+A_2E\frac{\partial u_2}{\partial t}+ku_2=0 \tag{3-151d}$$

The fourth boundary condition is derived by direct application of the second law to the top mass.

Assume a solution of the form

$$u(x,t)=X(x)\sin\omega t \tag{3-152}$$

Substituting into Eq. (3-150) gives

$$\frac{d^2X_1}{dx^2}+\lambda_1{}^2X_1=0 \text{ (drill collars)} \tag{3-153}$$

$$\frac{d^2X_2}{dx^2}+\lambda_2{}^2X_2=0 \text{ (drill pipe)} \tag{3-154}$$

where

$$\lambda^2=\frac{m\omega^2}{AE}$$

but

$$mL = \rho AL$$

$$\lambda^2=\left(\frac{\omega}{c}\right)^2$$

$$c=\sqrt{\frac{E}{\rho}} \text{ (velocity of compression wave)}$$

Therefore, $\lambda_1 = \lambda_2 = \lambda$. Applying Eq. (3-152) to the boundary conditions Eq. (3-151) gives

$$X_1(0) = 0 \tag{3-155a}$$

$$X_1(L_1)=X_2(L_1) \tag{3-155b}$$

$$Area_1E\frac{dX_1}{dx}(L_1)=Area_2E\frac{dX_2}{dx}(L_1) \tag{3-155c}$$

$$Area_2E\frac{dX_2}{dx}(L)=kX_2\left[\left(\frac{\omega}{\Omega}\right)^2-1\right]=0 \tag{3-155d}$$

where $\Omega = \sqrt{\dfrac{k}{M}}$

The solution to both differential equations is of the form

$$X_1(x) = A_1 \cos\lambda x + B_1 \sin\lambda x \tag{3-156}$$

$$X_2(x) = A_2 \cos\lambda x + B_2 \sin\lambda x \tag{3-157}$$

For case 1, $X_1(0) = 0$ and for case 2, $\dfrac{dX_1}{dx}(0) = 0$. The other three boundary conditions are the same for both cases. The first boundary condition gives $A_1 = 0$. The other boundary conditions give

$$B_1 \sin\lambda L_1 = A_2 \cos\lambda L_1 + B_2 \sin\lambda L_1$$

$$B_1 \cos\lambda L_1 = -\frac{Area_2}{Area_1}\left(A_2 \sin\lambda L_1 - B_2 \cos\lambda L_1\right)$$

$$-A_2\lambda \sin\lambda L + B_2\lambda \cos\lambda L = \frac{k}{EArea_2}\left[\left(\frac{\omega}{\Omega}\right)^2 - 1\right]\left(A_2 \cos\lambda L + B_2 \sin\lambda L\right)$$

These three equations are arranged in the matrix equation

$$\begin{bmatrix} a_{11} & a_{12} & a_{13} \\ a_{21} & a_{22} & a_{23} \\ a_{31} & a_{32} & a_{33} \end{bmatrix} \begin{Bmatrix} B_1 \\ A_2 \\ B_2 \end{Bmatrix} = \begin{Bmatrix} 0 \\ 0 \\ 0 \end{Bmatrix}$$

The expansion of the square matrix produces the characteristic equation. The roots to this equation are the eigenvalues, λ's. The ratios of the amplitudes give the mode shapes corresponding to each natural frequency. There are multiple eigenvalues for each set of drillstring parameters.

Figure 3-22 shows how natural frequencies change with hole depth for the following drillstring parameters. Multiple roots to the characteristic equation give the frequencies of each of the modes of vibration.

Drill pipe size (5 in. OD × 4.276 in. ID)
Drill collar size (7 in. OD × 2.937 in. ID)
Length (L_1) = 550 ft
Acoustic velocity = 16,850 ft/sec
Spring constant of drawwork cables (k) = 640,000 lbs/ft (53,333 lbs/in.)
Mass of swivel and traveling block (M) = 466 slugs (15,000 lbs)
Drillstring length (L) is a variable.

Considering a well depth of 5000 ft, the natural frequency of the fourth mode is 6.13 cps and a critical rotary speed of 122.6 rpm (Table 3-6). The fourth mode would resonant with a roller cone bit at this speed. Note also that the frequency

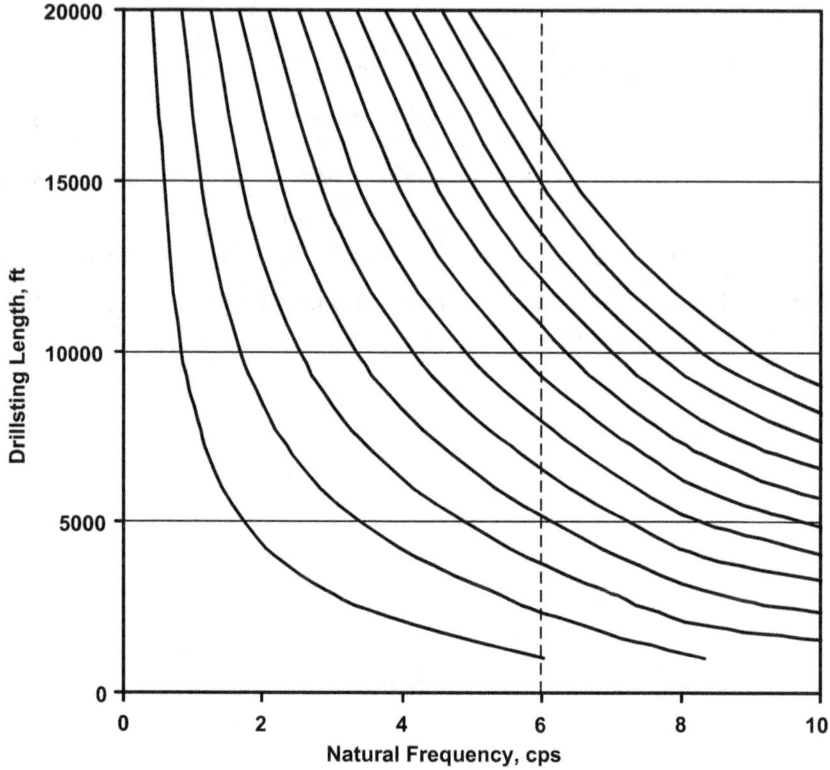

Figure 3-22. Natural frequencies of axial modes (rotary speed equals 20 times frequency).

band width of the response is quite large so any of the axial modes can easily be excited.

On the other hand, a rotary speed of 80 rpm, would fall between critical speeds of the 2nd and 3rd axial modes so a roller cone bit should operate relatively smooth at this speed.

Table 3-6. Natural frequencies of axial modes.

	Length – 5000 ft		Length – 20,000 ft	
Mode	Frequency (cps)	Critical rotary speed* (rpm)	Frequency (cps)	Critical rotary speed* (rpm)
1	1.72	34.4	0.414	8.3
2	3.39	67.8	0.84	16.8
3	4.87	97.4	1.26	25.2
4	6.13	122.6	1.67	33.4
5	7.26	145.2	2.09	41.8
6	8.30	166.0	2.50	50.0
7	9.77	195.4	2.92	58.4

*Based on three cone roller bit

According to Fig. 3-19, the frequency band width of mode resonance is of the order of $\Delta f \sim 0.1$ cps ($\Delta N \sim 2$ rpm). Since the exciting frequency is related directly to bit rotation, rotary speed has to be maintained within a ± 1 rpm variance in order for energy to be fed into the mode. It takes time to put energy into a mode because of drillstring length. On the other hand, the frequency band with of drill collar response is much larger, plus the drill collars are much shorter making it easier to excite drill collar modes as demonstrated in shallow test wells. The frequency band width at 20,000 ft is much smaller, and it is more difficult for bit speeds to latch onto these modes.

Figure 3-22 also shows how natural frequency changes with depth. If, for example, rotary speed is 120 rpm, forcing frequency by a three cone bit is 6 cps. At this frequency the mode number changes about every 1000 ft.

3.6.3.2 *Natural Frequencies of Torsion Modes*

The natural frequencies of torsion modes of vibration are determined in a similar manner. The differential equation of free vibration in this case is

$$GJ\frac{\partial^2\theta}{\partial x^2} = J_m\frac{\partial^2\theta}{\partial x^2} \tag{3-158}$$

This equation applies over both drill pipe and drill collar sections.

$\theta_1(x,t)$ - drill collar displacements
$\theta_2(x,t)$ - drill pipe displacements

These variables are related through boundary conditions. Assuming solutions for both drill collar and drill pipe sections of the form

$$\theta(x,t) = \Theta(x)\sin\omega t$$

and substituting into Eq. (3-158) gives

$$\frac{d^2\Theta_1}{dx^2} + \lambda_1^2\Theta_1 = 0 \quad \text{(drill collars)} \tag{3-159}$$

$$\frac{d^2\Theta_2}{dx^2} + \lambda_2^2\Theta_2 = 0 \quad \text{(drill pipe)} \tag{3-160}$$

where

$$\lambda^2 = \frac{J_m\omega^2}{GJ}$$

but

$$J_m = \frac{1}{2} m R^2$$

$$mL = \rho A L$$

$$J = \frac{\pi}{2} R^4$$

Combining these equations gives

$$\lambda^2 = \left(\frac{\omega}{c_\theta}\right)^2 \tag{3-161}$$

where

$$c_\theta = \sqrt{\frac{G}{\rho}} \quad \text{(velocity of shear wave)}$$

therefore $\lambda_1 = \lambda_2 = \lambda$.

Solutions to both differential equations, Eqs. (3-159) and (3-160), are of the form

$$\Theta_1(x) = A_1 \cos \lambda x + B_1 \sin \lambda x \tag{3-162}$$

$$\Theta_2(x) = A_2 \cos \lambda x + B_2 \sin \lambda x \tag{3-163}$$

Boundary conditions for the case illustrated by Fig. 3-23b (fixed-fixed case) are

$$\Theta_1(0) = 0 \tag{3-164a}$$

$$\Theta_1(L_1) = \Theta_2(L_1) \tag{3-164b}$$

$$J_1 G \frac{d\Theta_1}{dx}(L_1) = J_2 G \frac{d\Theta_2}{dx}(L_1) \tag{3-164c}$$

$$\Theta_2(L) = 0 \tag{3-164d}$$

These boundary conditions give $A_1 = 0$ and

$$B_1 \sin \lambda L_1 = A_2 \cos \lambda L_1 + B_2 \sin \lambda L_1$$

$$J_1 B_1 \cos \lambda L_1 = J_2(-A_2 \sin \lambda L_1 + B_2 \cos \lambda L_1)$$

$$0 = A_2 \cos \lambda L + B_2 \sin \lambda L$$

These equations can also be put in a matrix form yielding a characteristic equation.

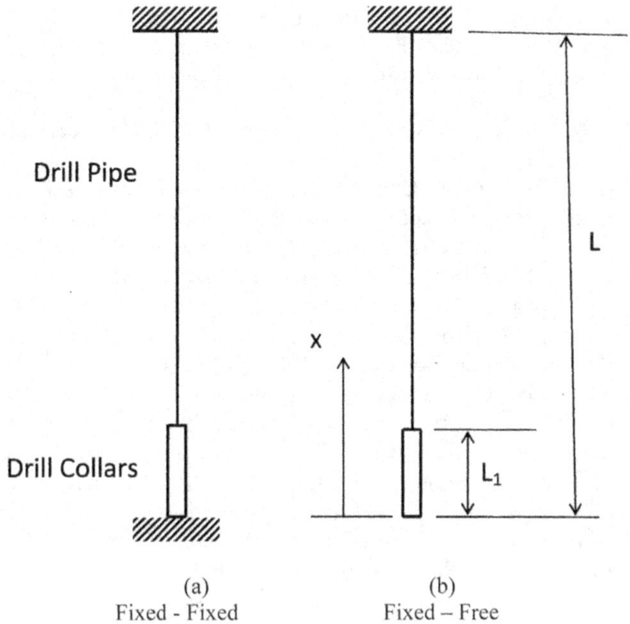

Figure 3-23. Torsion vibration models of drillstring.

3.6.3.3 *Natural Frequencies of Lateral Modes*

Timoshenko [38, p. 454] gives natural frequencies for a simply supported beam under an axial tension load.

$$p_i = \frac{a(n\pi)^2}{L^2}\sqrt{1+\frac{SL^2}{(n\pi)^2 EI}} \tag{3-165}$$

where

$a^2 = \dfrac{EI}{m}$

EI — bending stiffness
L — length of simply support beam
m — mass per unit length
S — axial tension force in beam
n — mode of vibration (n = 1, 2, 3, etc)

This equation can be converted to account for internal pressure and applied tension by use of Eq. (2-76). In this case, Eq. (3-165) becomes

$$\omega = a\left(\frac{n\pi}{L}\right)^2\left[1+\frac{L^2(F+P)}{(n\pi)^2 EI}\right]^{\frac{1}{2}} \tag{3-166}$$

Plus F is tension and $P = -p_i A_1$ accounts for internal pressure.

Zare, Hashemi, Rashed [39] use the finite element method to predict lateral vibrations accounting for well bore contact and drilling mud. Their analyses give frequency response at a different location above the drill bit.

3.6.4 Polycrystalline Diamond Compact (PDC) Drill Bits

The cutting mechanism for PDC drill bits is different from roller cone drill bits. PDC cutters are designed to shear the rock instead of chipping it. Fundamental to these drill bits is a synthetic polycrystalline diamond cutter surface. Tiny synthetic diamonds bonded together to form a small cylindrical disc which is then bonded to a tungsten carbide backing. The tungsten carbide substrate is then attached to a stud. Multiple studs attached to a ridged cutter head make up Polycrystalline Diamond Compact (PDC) drill bits.

Various types of fish tail drag bits were made since the early 1900s but were limited primarily to soft formations along the gulf coast. Tungsten Carbide was applied to the tips of the cutters to reduce wear. PDC drill bits, which were introduced in the 1970s, are typically used in medium hard formations. PDC drill bits have no moving parts and have become a good option for use with high-speed motors and turbines.

Basic understanding of the mechanics of PDC bit chatter has not been understood until recently. Penetration rate with PDC drag bits is sometimes inconsistent in the same formation under the same drilling condition. Tooth breakage is also a problem.

3.6.4.1 Self-Exciting Mechanism

Self-excited vibrations are sustained by a feedback mechanism, which continuously puts energy into the motion. The driving force stops when the vibration stops. Energy that drives the vibration usually comes from a steady flow of energy from sources, such as air, fluid, friction or speed. In the case of PDC drill bits, axial movement modulates a steady bit force into a periodic force, which feeds back and excites the vibrations. By contrast, force of excitation in forced vibrations is independent of the vibration it produces.

When vibrations cannot be explained by frequency tuning there is a good chance the vibration is self-excited. This is the premise of the following discussion. Stick-slip or negative damping explains self-excited vibrations in many situations. There is field evidence that this mechanism may explain vibrations while pulling drillstrings out of the hole or rotating off bottom with a top drive. While this mechanism may play secondary role in drilling, it does not explain longitudinal vibrations often seen at the surface while using drag bits. The key to unlocking the mystery of self-excited vibrations is the feedback mechanism.

The theory presented here is a spin-off from metal tool chatter theory developed by Tlusty and Ismail [40]. Dareing, Tlusty, and Zamudio [41-43] applied this theory to PDC drill bits. The proof of concept as applied to drilling has been advanced by Elsayed [44-46] by further analytical studies and testing at the Sandia National Laboratory. Elsayed shows how drillstring, drill bit design and formation properties all affect bit chatter.

To illustrate the application of this theory, consider the idealized bit having three rows of cutters (Fig. 3-24).

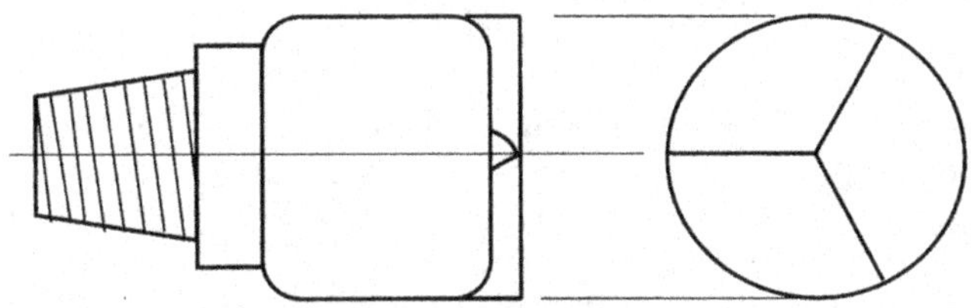

Figure 3-24. Model of PDC drill bit with three cutter rows.

The self-exciting mechanism is based on the regeneration of waviness of the surface, which each cutter edge leaves behind if vibration is superimposed over the otherwise steady motion of the bit. The following edge cuts into the wavy surface and the corresponding variable cutting force excites the vibration, etc. The phase difference between the previous wave and the current wave gives rise to a variable instantaneous cutting depth, and this variation account for the variable force which drives the vibration (see Fig. 3-25).

Whether self-excitation occurs (unstable cutting) or not (stable cutting) depends on the dynamic flexibility of the drillstring as expressed by the transfer function G(f) between the force acting on the bit, length of the cutting edges and on the rotational speed of the bit.

Here, instantaneous bit force depends directly on the instantaneous depth of cut as defined by

$$F(t) = Kbh(t) \tag{3-167}$$

where

F — axial force applied to drill bit
K — specific cutting force (depends on rock strength)
b — length of cutting edge
h — thickness of cut

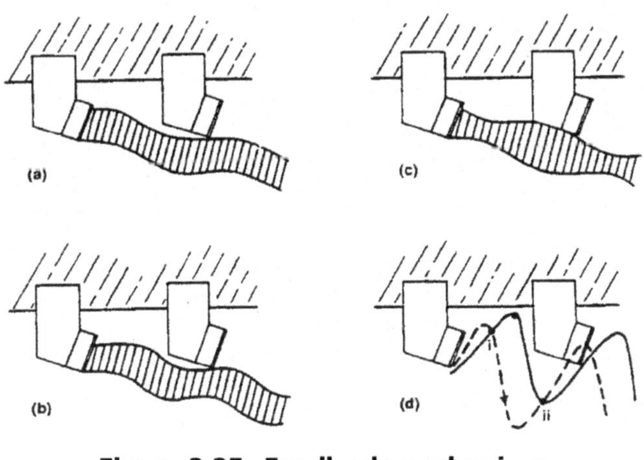

Figure 3-25. Feedback mechanism.

A leading cutter creates a surface undulation and a trailing cutter creates its own undulation. The depth of cut is the key. Vertical force on the cutter increases with thickness of the cut. If the depth of cut varies as in (c) the vertical force varies and creates a time varying vertical driving force.

The variation of the thickness h depends on the phase ε with which the cutter meets the undulation created by the preceding cutter. Various values of the phase angle, are illustrated in Fig. 3-25. The drawing in Fig. 3-25a corresponds to $\varepsilon = 0$ and as the two undulation are in phase, the depth of cut remains constant. The drawing Fig. 3-25b shows a one and a quarter waves between subsequent teeth, $\varepsilon = 90$, and there is thickness variation. Thickness variation is a maximum for $\varepsilon = 180$ (Fig. 3-25c) where there are one and a half waves between subsequent cutters.

3.6.4.2 *Stability Analysis*

The thickness of cut consists of a steady value, c, minus the vibratory displacement of the bit, $x(t)$ plus the displacement $x(t-\tau)$, where τ is the period between two subsequent cutters passing through a point.

Assuming a single degree of freedom case and m blades (edges) in the bit, the equation of motion is

$$M\ddot{x} + C\dot{x} + kx = m\frac{d}{2}K\left[c - x(t) + x(t-\tau)\right] \tag{3-168}$$

Note how the force depends on x and x depends on the force, a typical self excited feedback mechanism. The condition for stability can be determined

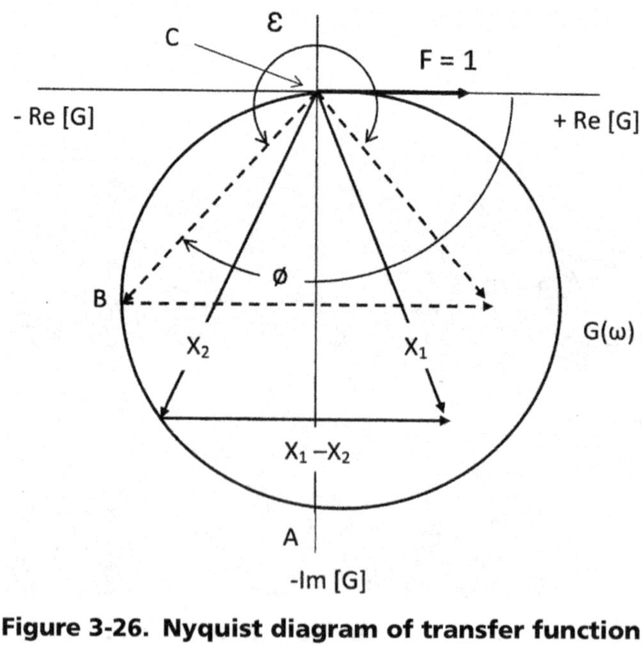

Figure 3-26. Nyquist diagram of transfer function.

without solving this equation directly. This will be illustrated using an SDOF system, assuming the response is simple harmonic.

The transfer function for this case is shown in Fig. 3-26 showing the complex amplitude vector, X, at a frequency, f. The vector representing current bit displacement is shown as X_2 and is phase shifted by ϕ behind the force. For a force whose magnitude is one and phase is zero, the vector representing the displacement will be equal to that representing the transfer function. Therefore X_2 also represents the transfer function G at this instant. This force representation is adopted here where the unit force is shown in the horizontal direction. The vector representing the previous wave is shown as x_1, which is separated by the phase angle ε from x_2.

Considering the vibration frequency to be $f(Hz)$ and the rotational speed to be n rev/s and the number of blades in the bit as m, the number of cycles between the current surface undulation and the previous on is $\frac{f}{mn}$. This can be expressed in terms of complete cycles N and a fraction of cycle $= \frac{\varepsilon}{360}$. Hence the phase angle can be written as

$$\frac{\varepsilon}{360} + N = \frac{f}{mn} \tag{3-169}$$

The vibration which was happening at the time when the current tooth was passing the location of the preceding tooth is shown as vector X_2. If we define the limit of stability as the condition at which vibrations do not grow or diminish but remain constant then

$$|X_0| = |X| \tag{3-170}$$

This equation indicates bit vibration is neither increasing nor decreasing; this is the limit of stability. The right hand side of Eq. (3-168) represents the force and can be written as:

$$m\frac{d}{2}K\left(c - X_2 + X_1\right)$$

and hence the difference, $X_2 - X_1$ must be equal and in phase with force and is shown parallel to the real axis. Considering our previous discussion that in this case is equal to G, Fig. 3-26 shows that

$$X - X_0 = -2\,\mathrm{Re}(G) \tag{3-171}$$

where $Re(G)$ is the real component of the transfer function. Using the force equation and taking the force amplitude as one (1),

$$1 = mK\frac{d_{\lim}}{2}(X_2 - X_1) \tag{3-172}$$

So

$$d_{\text{lim}} = \frac{-1}{mK\,\text{Re}(G)} \tag{3-173}$$

This equation applies to any elastic system including drill strings. Notice the negative sign in Eq. (3-173). Only the negative component of the transfer function affects stability. This insight is often used in manipulating the transfer function to improve stability.

The dynamic stability of the feedback mechanism is shown in a lobbing diagram identifying zones of stability in terms of cutter head dimensions and cutter speeds (see Fig. 3-27).

The limit of stability is defined in terms of d_{lim} as illustrated in the figure. Each lobe or curve corresponds to N = 0, 1, 2, etc. In this example, an SDOF system was used to generate the lobes.

In multi-degree-of freedom system, each degree of freedom will have its own value for d_{lim} and the lobes will appear in levels. The lobbing diagram is then reduced to the envelope of all lobes. Each lobe is generated by:

1. assuming a value for f, the frequency of vibration
2. determining the corresponding value of $Re(G)$ from the transfer function
3. calculate the phase angle, ϕ from

$$\phi = \tan^{-1}\left[\frac{Im(G)}{Re(G)}\right] \tag{3-174}$$

4. calculating d_{lim} from Eq. (3-173)
5. calculate ε from

$$\varepsilon = 180 + 2(180 - \phi)$$

6. assuming $N = 0$ for lobe (0) of the stability diagram, calculate n (rev/s) from Eq. (3-169), hence bit speed = 60 n (rpm)

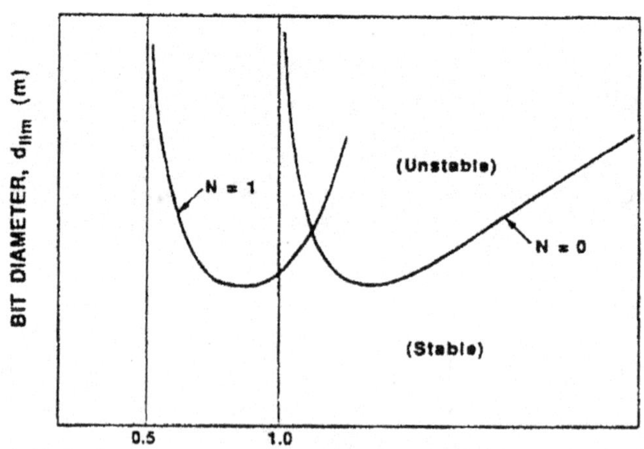

Figure 3-27. Lobbing diagram indicating stable and unstable regions.

7. repeat steps 1 to 6 above for various frequencies and plot d_{lim} versus bit rpm for lobe number (0)
8. increment N by 1 and repeat the above steps 1 to 7 to plot lobe (1) of the stability diagram.
9. The above process is repeated for various values of N to plot the desired number of lobes of the stability diagram

The lobe identifies the limit of stability. Points above each curve correspond to conditions of instability. Points below each curve correspond to conditions of stability. The diagram in Fig. 3-27 is schematic only.

This concept will now be applied to a drillstring modeled with multiple degrees of freedom. The complex direct transfer function, $\overline{G}_{11}$, for this case is (Craig [47], p. 356)

$$\overline{G}_{11} = \sum_{1}^{N} \frac{X_{1r}X_{1r}}{K_r}\left[\frac{1}{(1-r_r^2)+i(2\zeta r_r)}\right] \tag{3-175}$$

The real part of this function is

$$R\left[\overline{G}_{11}\right] = \sum_{1}^{N} \frac{X_{1r}X_{1r}}{K_r}\left[\frac{(1-r_r^2)}{(1-r_r^2)^2+(2\zeta r_r)^2}\right] \tag{3-176}$$

which is labeled $R(G)$ in the earlier discussion. X_{1r} is the displacement of the rth mode at point 1, r_r is the frequency ratio of the rth mode and K_r is the stiffness of the rth mode. The PDC drill bit is located at point 1.

Note that the force is applied to the bottom end of the string so the natural modes of vibration are based on an open ended model (Fig. 3-23b) and not on the pinned boundary condition as previously assumed in the earlier discussion of roller cone bits.

Equation (3-176) was used to determine the limit of stability for a drillstring composed of the following;

Drill pipe = 4.5 in. (16.6 lb/ft)
Length = 7200 ft
Drill collars = 6.5 in. O by 2 in. ID
Length = 780 ft
M = 22,600 lbs (swivel, traveling block)
K - 52,500 lb/in (draw works cables)

The numerical example given in ref. [43] uses this equation where the modes of vibration are determined numerically for a 26 degrees of freedom drill string model. The lobbing diagram for this example is given in Fig. 3-28.

3.6.5 Experimental Verification of Theory

Elsayed [44,45] advanced this theory by addressing important issues such as process damping, axial and torsion vibration coupling and vibration saturation of bit motion. In addition, he tested the validity of the self excitation theory [46].

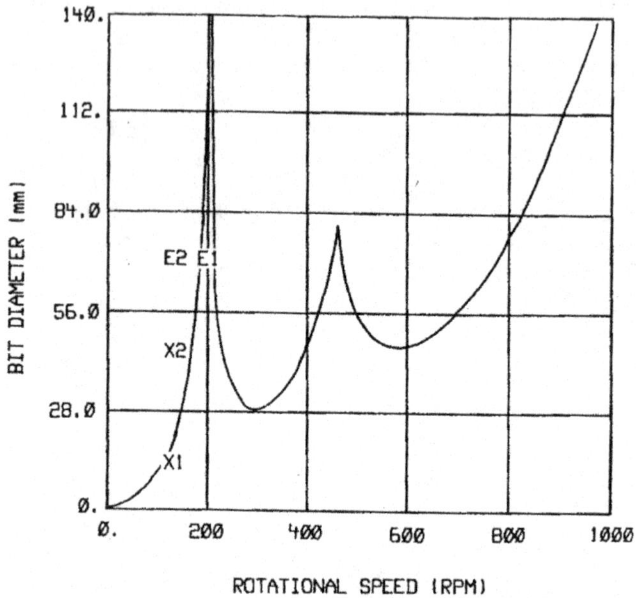

Figure 3-28. Lobbing diagram.

Laboratory tests were conducted at the Sandia National Laboratories using their Hard-Rock Drilling Facility (HRDF).

In order to simulate structural flexibility along the axis of drillstrings, a compliant drill test rig was built at Sandia National Laboratories. This indoor rig, called Hard Rock Drilling Facility (HRDF), had been previously used to study the effects of weight on bit (WOB), rate of penetration (ROP), rotational speed, and vibration characteristics on drill bit performance.

Since the natural frequencies of drillstrings can be very low (say 0.3 Hz) [43], it is important to analyze chatter behavior at low frequencies. The HRDF is designed to make the fundamental frequencies of the test fixture as low as possible to be consistent with drilling natural frequencies. Therefore, chatter effects observed at frequencies of the test rig were representative of the system characteristics encountered in the field.

A rock cube is positioned on a base plate within the HRDF. Drilling was conducted with a 3.25-inch PDC coring bit. The bit incorporates three, half-inch diameter PDC cutters in a coring configuration to simulate PDC drill bits with three rows of cutters. Water was used as a drilling fluid and circulated through the bit. The WOB was applied through a set of springs to introduce compliance in the axial direction. The longitudinal mode of the system is approximately 6 Hz. The HRDF is controlled through a computer and parameters such as WOB, speed, ROP and axial position are recorded. The axial position is obtained through a Center Position Displacement Transducer (CPDT) attached to the drillstring support beam.

Experimental modal analysis was conducted on the system and showed an axial translation frequency of the short drillstring to be about 6 Hz. This multi-mode system allowed us to observe some effects of closely-spaced modal frequencies that are also characteristic of field drillstrings. The displacement transducer on the drillstring centerline (CPDT) provides a response that is a direct measure of vibration amplitude.

Drilling tests were conducted in two rock types. Berea Sandstone, a soft formation typically encountered in oil and gas formations, was used first. Subsequent testing was conducted in Sierra White Granite, a hard, abrasive rock typical of geothermal reservoirs. Tests consisted of drilling a series of holes at constant weight on bit and constant rotary speed and recording drilling parameters for post-processing. Test parameters could easily be used to compare theory predictions of chatter with test results.

Chatter, typically, takes place near the natural frequencies of the system. Its frequency varies little with speed. This distinguishes chatter from other types of vibrations that are often present during drilling, such bit whirl, where chatter frequency is directly proportional to the rotational speed.

The effect of chatter on the topology of the rock surface could be seen at bit-rock interface at the bottom of each hole. Figure 3-29 shows the hole bottom for stable drilling, while Fig. 3-30 shows the surface for unstable drilling with high chatter. These patterns were made while drilling the Berea sandstone. The highly irregular surface shown in Fig. 3-31 is an indication of PDC drill bit instability. Figure 3-30 also gives an indication of impact loading on the bit under chatter conditions.

Similar tests were conducted in Sierra White Granite (SWG). However, chatter was so severe and so destructive to the cutters that the outside cutter, which runs at the highest speed, broke frequently. The bit used for these tests was also used previously for wear testing, and performed well. The difference in cutter life is attributed to the added flexibility incorporated in the current setup in order to approximate actual drilling conditions. The added flexibility allowed for high chatter amplitudes and increased impact loads that led to cutter failure.

Figure 3-29. PDC cutter pattern under stable drilling.

Figure 3-30. Irregular PDC cutter pattern caused by self-excitation.

The essence of these tests is as follows. Chatter appears in drilling of both soft and hard rock. Severity of chatter varies with speed and system dynamics. There is an agreement between measured severity of chatter and calculated system instability. Frequency of chatter is mostly close to the natural frequency of the system. As the operating point crosses the point of lowest stability, frequencies increase at a higher rate. With stability lobes so close, this frequency increase is interrupted as conditions enter the lower frequency range of the next lobe. In a multi-degree-of-freedom system, this lobe may belong to another mode with entirely different frequency. This can lead to a sudden change in frequency as a result of minor changes in speed. This is the "mode hunting" phenomenon observed in the laboratory.

Their work showed that chatter severity correlates with drill string and bit instability as predicted by the self-excited vibration model. It was shown experimentally, that torsion vibrations greatly reduced by limiting axial vibrations [45].

3.6.6 Backward Whirl of PDC Drill Bits

3.6.6.1 Gear Tracking

Another type of PDC drill bit motion is backward whirl. Laboratory testing by Bret, Warren, and Behr [48] show a star shape pattern in rock samples. The resulting whirling motion is similar to a planetary gear train with the bit being the planet gear and well bore being the annular sun gear. Individual cutters on the bit would seem to follow a cycloid path. Backward whirl of these bits creates centrifugal forces, which encourages backward whirl and well bore enlargement. The whirling frequency in this case is

$$\frac{\omega_{bit}}{\omega_{whirl}} = 1 - \frac{r_{hole}}{r_{bit}}$$

$$\omega_{whirl} = -\frac{r_{bit}}{e}\omega_{bit}$$

where "e" is the radial location of the bit center. This equation indicates that whirl velocity is quit high initially.

3.6.6.2 Cutter Impulsive Force

Tooth breakage on PDC drill bits is most likely due to large impulsive forces caused by momentum changes. Consider a disc rotating at a constant angular velocity (ω_1) about its geometric center (Fig. 3-31a). Over a very short time interval, $\Delta t = t_2 - t_1$, the edge of the disc is caught by a fixed pin as shown. At time t_2 the disc sudden rotates about its instantaneous center, point "a", with angular velocity (ω_2) (Fig. 3-31b).

Over this short time interval (Δt) angular momentum with respect to point "a" is conserved. However, linear momentum with respect to the center of gravity is not conserved. Based on the conservation of angular momentum,

$$H_{a2} = H_{a1} \tag{3-177}$$

from which

$$(I_G + Mr^2)\omega_2 = I_G\omega_1 \tag{3-178}$$

and

$$\omega_2 = \frac{I_G}{I_G + Mr^2}\omega_1 = \frac{1}{3}\omega_1 \tag{3-179}$$

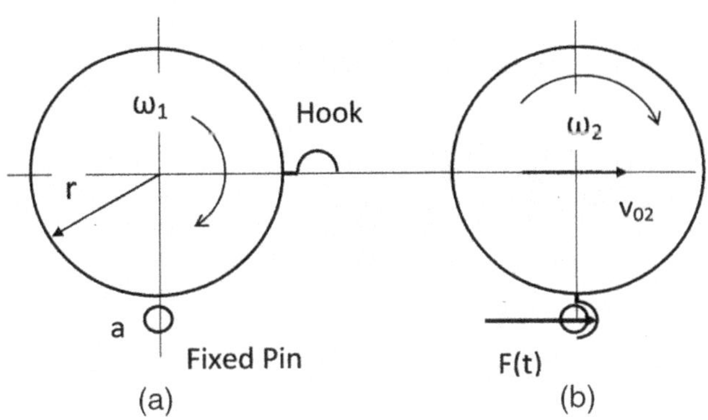

Figure 3-31. Cutter impulsive force.

assuming a solid disc. The velocity of the center of gravity is now $v_{o2} = r\omega_2$. Considering the component of impulse to the right

$$\text{Im}_x = G_{x2} - G_{x1} \tag{3-180}$$

giving

$$\int_1^2 F(t)dt = v_{o2}M - 0 = r\omega_2 M \tag{3-181}$$

This change in linear momentum takes place over a very short period of time. The magnitude of the average impulsive force can be approximated by

$$\int_1^2 F(t)dt = F_{ave}\Delta t \tag{3-182}$$

Dividing the impulse by Δt gives the average impulsive force.

$$F_{ave} = \frac{r\omega_2 M}{\Delta t} \tag{3-183}$$

Consider, for example, a rigid disc weighing 60,000 lbs, having a 7 inch diameter and rotating at a speed of 100 rpm. With a distributed weight of 131 lb/ft, collar length is 458 ft. These numbers were chosen to simulate drill collar. Adjusting the units for substitution into eq. 3-183.

$$r = \frac{3.5}{12} = 0.2917 \text{ ft}$$

$$M = \frac{60{,}000}{32.2} = 1{,}863 \text{ slugs}$$

$$\omega_2 = \frac{1}{3}\frac{2\pi 100}{60} = 3.49 \text{ rad/sec}$$

Putting these numbers into Eq. (3-183) gives

$$F_{ave} = \frac{0.2917(3.49)1{,}863}{\Delta t} = \frac{1{,}896.6}{\Delta t} \text{ lbs}$$

It follows that

Δt, sec	0.1	0.05	0.04	0.03	0.02
F_{ave}, kips	18.97	37.93	47.42	63.22	94.83

The duration of Δt has a large effect on the average impulsive force. The actual force level depends on the strength of both rock and cutter, so these force levels may not be reached in reality.

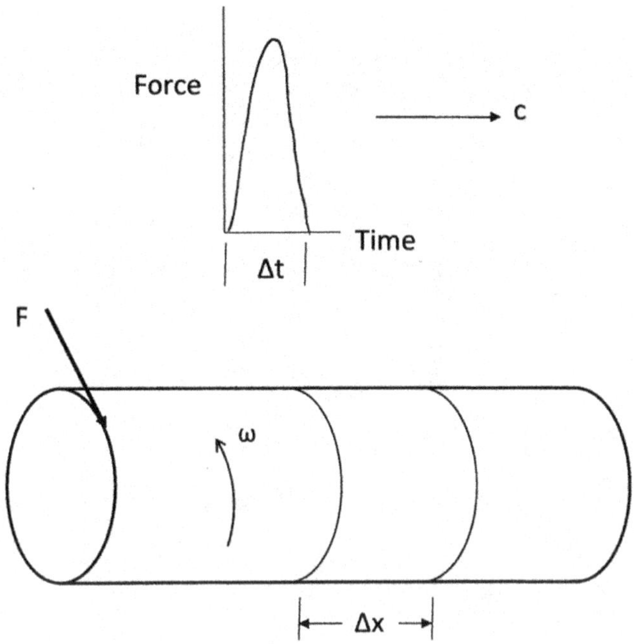

Figure 3-32. Torsion shear wave set up by cutter impact.

3.6.7 Accounting for Torsion Flexibility in Drill Collars

Now consider the torsion flexibility of drill collars of length, L (Fig. 3-32). Assume the collars rotate clockwise as in drilling. Once a cutter makes contact with an obstruction, creating impact force, F(t), a shear wave starts to travel to the right with acoustic velocity c = 10,600 ft/sec $\left(c = \sqrt{\frac{G}{\rho}} = 10{,}600\,fps\right)$. Over a short span of time the leading edge of the shear wave has travel a distance $x = ct$. The shape of the impulse force, F(t), will be sustained within the collars as illustrated. The space interval Δx is related to the time interval, Δt by $\Delta x = c\Delta t$. As the shear wave travels to the right, linear momentum is changed according to

$$\int_0^{\Delta t} F(t)dt = r\omega_2 m\Delta x \tag{3-184}$$

In terms of the average impulse force

$$F_{ave}\Delta t = r\omega_2 mc\Delta t$$

$$F_{ave} = r\omega_2 mc$$

Applying the previous number gives

$$F_{ave} = 0.2917(3.49)4.0676(10{,}600) = 43{,}894 \text{ lbs}$$

$$F_{ave} = r\omega_2 mc \tag{3-185}$$

Note the average force level does not depend on Δt because Δx and Δt are related by $\Delta x = c\ \Delta t$. If Δt is large, Δx is large. If Δt is small, Δx is small. For example if $\Delta t = 0.03$ sec, then $\Delta x = 318$ ft. This interval is less than $L = 458$ ft, explaining why F_{ave} for the rigid case is greater than F_{ave} for the elastic case.

3.6.7.1 Free Torsion Vibration Caused by Cutter Impulse

Since the collar section is long and flexible, its twist varies with both distance x and time t and is represented by $\theta(x,t)$. The differential equation of motion is

$$J_m \frac{\partial^2\theta}{\partial t^2} + \gamma_\theta \frac{\partial\theta}{\partial t} - GJ \frac{\partial^2\theta}{\partial x^2} = T(x,t) \tag{3-186}$$

where

$T(x,t)$ is applied torque per unit length. It can vary with time and location, x.
J_m is mass torsional moment of inertia per unit length
γ_θ is torsional damping per unit length
J is cross section polar moment of inertia

The drill collar section will be modeled as being free at both ends since drill pipe has a minor effect and the torque at both ends are non-impulsive.

Using modal analysis techniques, the solution to Eq. (3-186) will be represented by

$$\theta(x,t) = \sum X_i(x)\eta_i(t) \tag{3-187}$$

where X_i defines the shape of each of the natural torsion modes and

$$\frac{\partial^2\theta}{\partial x^2} = \sum \frac{d^2X_i}{dx^2}\eta_i(t)$$

$$\frac{\partial\theta}{\partial t} = \sum X_i(x)\frac{d\eta_i}{dt}$$

$$\frac{\partial^2\theta}{\partial t^2} = \sum X_i(x)\frac{d^2\eta_i}{dt^2}$$

Substituting these expressions into Eq. (3-186), multiplying through by X_j and integrating each term, assuming linear damping, gives

$$J_{\theta i}\frac{d^2\eta_i}{dt^2} + C_{\theta i}\frac{d\eta_i}{dt} + K_{\theta i}\eta_i = T_{\theta i}(x.t) \tag{3-188}$$

Modal torque is determined by

$$T_{\theta i}(x.t) = \int_0^L X_i T(x,t)dx = X_i(0)T(0,t) \tag{3-189}$$

$T(0,t)$ represents the impulsive torque at $x = 0$ and is equal to the moment of the impulsive force. Equation (3-188) then becomes

$$J_{\theta i}\frac{d^2\eta_i}{dt^2} + C_{\theta i}\frac{d\eta_i}{dt} + K_{\theta i}\eta_i = X_i(0)T(t) \tag{3-190}$$

where $T(t)$ represents an impulsive torque at $x = 0$ due to the shifting of the instantaneous center as explained earlier. If the angular impulse of this torque is represented by I_A then the response of each mode can be solved independently following the classic solution by Duhamel's Integral [38]. For an under damped case

$$\eta_i(t) = \frac{X_i(0)}{J_{\theta i}}\frac{I_A}{\omega_{di}}\exp(-\varsigma_i\omega_{ni}t)\sin\omega_{di}t \tag{3-191}$$

The summing all of the modal responses to the impulsive torque is

$$0(x,t) - \sum X_i(x)\frac{X_i(0)}{J_{\theta i}}\frac{I_A}{\omega_{di}}\exp(-\varsigma_i\omega_{ni}t)\sin\omega_{di}t \tag{3-192}$$

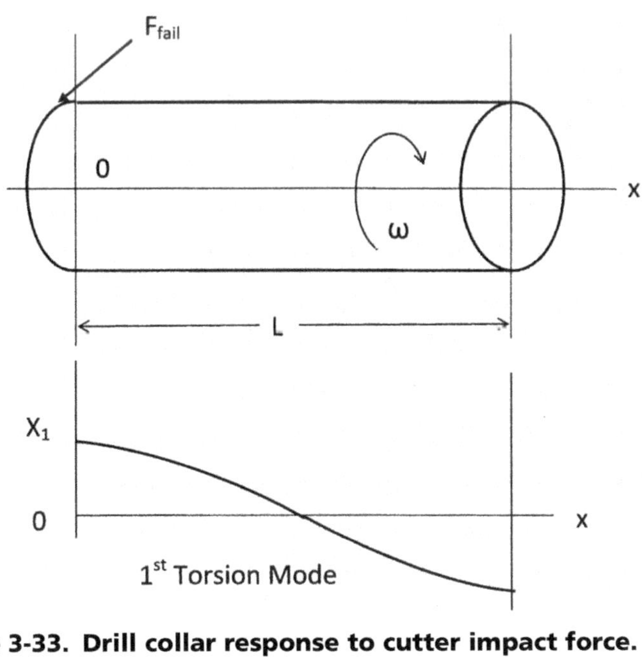

Figure 3-33. Drill collar response to cutter impact force.

The impulsive torque produced by the cutter force is determined as before. These motions could be detected by MWD and used as a marker for tooth failure.

3.7 VIBRATION CONTROL

The primary cause of severe vibrations is frequency tuning between forcing frequency and a natural frequency. This is true of all vibration types, whither axial, torsion, or lateral and for each of their multi-modes.

In most cases, vibrations are driven by applied forces that vary with time. It is useful to identify the possible sources of excitation and quantify their magnitude and frequency. It is also useful to have knowledge of the natural frequencies of each type (axial, torsion, lateral) of the system. Each vibration type will have many vibration modes of their own. Even with this information there are many possibilities for frequency tuning. In practice, active modes will identify themselves during operation. MWD measurements can be most useful in this regard. Vibration control is a matter of identifying active modes and detuning them from driving frequencies [49].

Mode vibration in drilling operations is a reality that has been shown by field measurements. Jogi, Macpherson, and Neubert [50], measured vibration responses for axial, torsion and lateral modes using a downhole tool in a shallow test well at Amoco's test facility in Catoosa, Oklahoma. A 526.8-ft drilling assembly containing a roller cone bit, drill collars (~8.0 in. OD), MWD and stabilizers. The objective of these tests was to validate analytical predictions of natural frequencies for all three modes. Sources of excitation were inherent in collar whirl, imbalance, stabilizer/hole contact and drill bit forces. In nearly every case, natural frequencies predicted by vibration theory agreed with measured frequencies of each of the three modes of vibration. These tests verified that severe vibrations are caused by frequency tuning, even in drillstrings.

Drillstring vibration control is fundamental. The basic options are:

1. Eliminate the source of excitation
2. Change the frequency of the force of excitation (usually linked to rotary speed)
3. Reduce the magnitude of the driving force
4. Change the natural frequencies of the drillstring through design noting that natural frequencies continually change as the well bore advances
   a. It is important to know natural frequencies of all modes of vibration
5. Apply damping
   a. Some damping is inherent
   b. Creating damping is usually costly and should be a last resort

Controlling severe vibrations which are driven by a self excited mechanism is more subtle. Here, the driving mechanism and elastic response are interdependent and feed energy into the vibration on a continuous basis. Self excited vibrations are caused by PDC bits as explained earlier. Stick-slip associated with rotation can cause wide swings in torque as measured and modeled by Lin and Wang [31] and Yigit and Christoforou [32].

The vibration response caused by self excitation depends on the feedback mechanism. The forcing mechanism depends on the vibration response; when one stops, the other stops. It is not always obvious how to untangle the two. A mathematical analysis, based on a proven feedback mechanism, can give useful guidance. A reliable analysis can identify operating regions of stability and instability. PDC chatter is a good example. Rotational stick-slip due to negative damping is another.

Solutions to self excited vibration problems are:

- Alter or eliminate the feedback mechanism.
- Change the Limit of dynamic stability for self excited vibrations or operate within the stable region.

Usually, the design of engineering structures is based on static analyses followed by a dynamic analysis. Once the structure is built and tested. Vibration measurements are made to check for unforeseen vibration problems. If a vibration problem occurs, then one of the control methods listed above is implemented.

3.7.1 Roller Cone Drill Bits

Roller cone bits are known to produce axial and torsion vibrations. The driving frequency is simply three cycles per bit rotation. Care should be taken to avoid tuning natural frequencies of axial and torsion modes to this driving frequency. An example of resonance is shown in Fig. 3-34.

The data was taken near Edmonton while drilling in a sandstone and shale formation with a roller cone drill bit. Well depth was 4370 ft and rotary speed was 110 rpm. This data shows the dynamic axial forces varied from zero (bit came off the bottom) to a maximum value of 150,000 lbs when the indicated (static) bit weight was 40,000 lbs. At the same time, bit torque ranged from 8000 ft/lbs in one direction to 1000 ft/lbs in the opposite directions; typical drilling torque is around 1500 ft/lbs. In each case, the frequency was three cycles per bit rotation.

3.7.1.1 Shock Absorbers

Shock absorbers are commonly used in hard rock areas where drill bit vibrations are often severe [51]. In these areas, more drill collars are needed and this factors into the vibration problem [52]. Shock absorbers basically change the natural frequency of drill collars. A simple, but practical, model is a lump mass setting on a spring (shock absorber). The natural frequency of this SDOF model is simply $f = \frac{1}{2\pi}\sqrt{\frac{k}{M}}$. This effect is shown by the leftward shift in the family of response curves (Fig. 3-35).

A shock absorber located directly above a drill bit changes the frequency response curve from the ones shown in Fig. 3-19 to the ones shown in Fig. 3-35. The response of the drill collar section can still be seen but it has been shifted

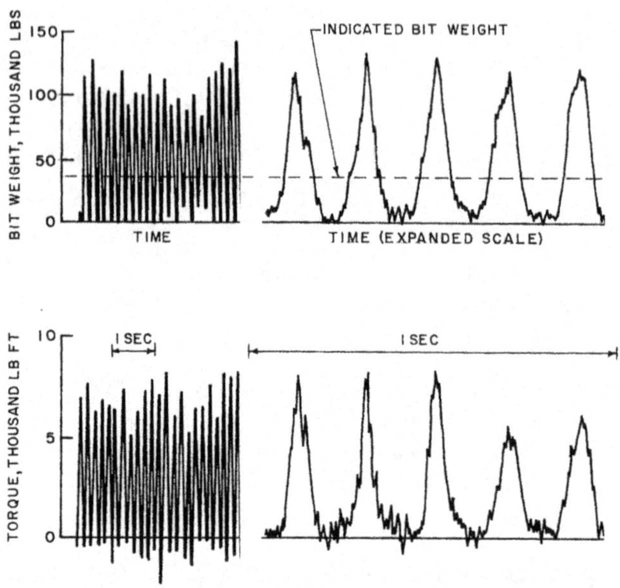

Figure 3-34. Variations in bit forces and torque; ESSO Production Research Data. Taken near Edmonton, Alberta, Canada (1963) [33].

to the left. On the right side of collar resonance, amplitudes of vibration are greatly dimensioned.

3.7.1.2 Drill Collar Design

Drill collars dictate the vibration (axial and torsion) behavior of the entire drill-string. Downhole tools are recipients of these violent forces and motions. If axial and torsion vibrations in the drill collar section are alleviated, dynamic forces in BHA tools will be diminished too.

Earlier analytical results show the drill collars respond directly to bit motion. The drill pipe does too but it's their higher modes that respond to typical rotary speeds. It is easier to resonant the drill collar section than to develop the higher drill pipe modes. Therefore, the drill collar section should be the focus of vibration control [52].

Drill collar length is dictated by required WOB. In hard rock drilling areas more drill collars are carried to generate a high bit force. The irony of this practice is the natural frequency of axial modes of vibration becomes close to frequencies of excitation associated with normal drilling speeds. This situation creates frequency tuning almost automatically. The numbers for this scenario are given below.

The natural frequencies of a drill collars (pinned at the bottom and free at the top) are determined from

$$f_{na} = \frac{n16{,}850}{4L} \qquad \text{(axial mode)} \qquad (3\text{-}193)$$

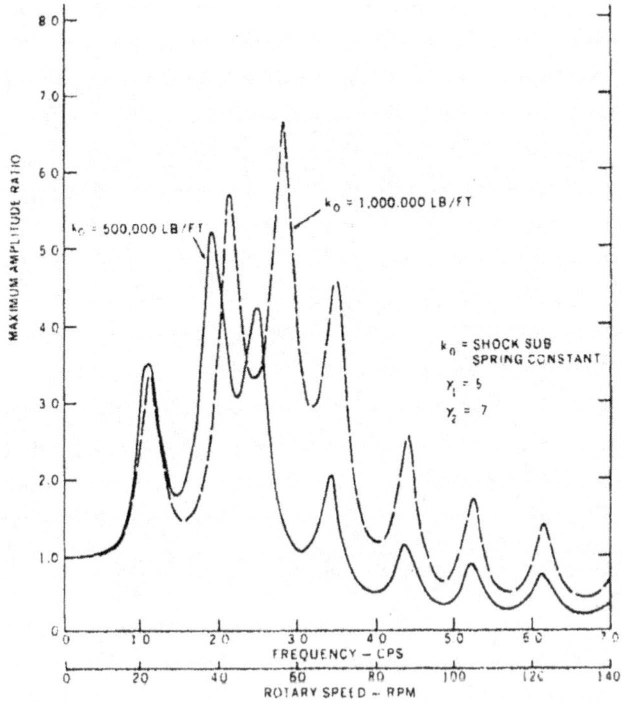

Figure 3-35. Frequency response with shock absorber [35].

For the first mode, $n = 1$ and

$$f_{1a} = \frac{4212}{L}$$

The natural torsion frequencies are found by

$$f_{n\vartheta} = \frac{n10,650}{4L} \qquad \text{(torsion mode)}$$

And for the first torsion mode

$$f_{1\vartheta} = \frac{2662}{L}$$

If, for example, L = 600 ft, $f_{1a} = 7.02$ cps.

Now compare this frequency to that associated with a three roller cone bit rotating at, 140 rpm. The drill bit frequency in this example would be

$$f = \frac{3N}{60} \text{ cps} \tag{3-194}$$

$$f = \frac{140}{20} = 7 \text{ cps}$$

Actual drilling cases may vary but the reality remains. Both natural frequency and driving frequency are close enough to excite the first mode of the drill collars. The same condition could exist for higher drill collar modes when PDMs and turbines are used.

Shock absorbers offer a practical solution for changing natural frequency of drill collars. Drill collar weight (length) can remain the same for a required WOB, while lowering natural frequency.

3.7.2 PDC Rock Bit

There is a fundamental change in the cause of axial vibrations with PDC drill bits. Axial vibrations in this case are caused by dynamic instability and not by resonance.

3.7.2.1 Operation and Design for Regions of Stability

The force between the drill bit cutters creates undulations in the rock which directly affect the interaction between bit and formation. This feedback mechanism causes energy to be continually be fed into axial vibrations. In this case vibration analyses show operating zones of dynamic stability and instability. Bit chatter results from operating within zones of instability. Instability can be avoided through: bit design, rotary speed, and drillstring design (see Fig. 3-28).

3.7.2.2 Shock Absorber Design for PDC Drill Bit Applications

Shock absorbers do not help in this case. Figure 3-36 gives a lobbing diagram for the case of a shock absorber placed directly above the PDC drill bit. The drillstring is the same one used to generate Fig. 3-28. The shock absorber in this example has the following design features:

k = 43,000 lbs/in.
M = 704 lbs
ζ = 0.2

M represents the mass of the drill bit and a short section of drill collars, i.e., the shock absorber in not located directly above the PDC drill bit. The amount of damping is expressed in terms of the damping factor. It is clear that the zone of instability has been increased (compared with Fig. 3-28) meaning that shock absorbers may create unstable vibrations when used with PDC bits.

Shock absorbers can be designed to expand the zone of stability as shown in Fig. 3-37. This is done by incorporating substantial damping within shock absorbers. While this feature is not helpful for roller cone bit applications, it

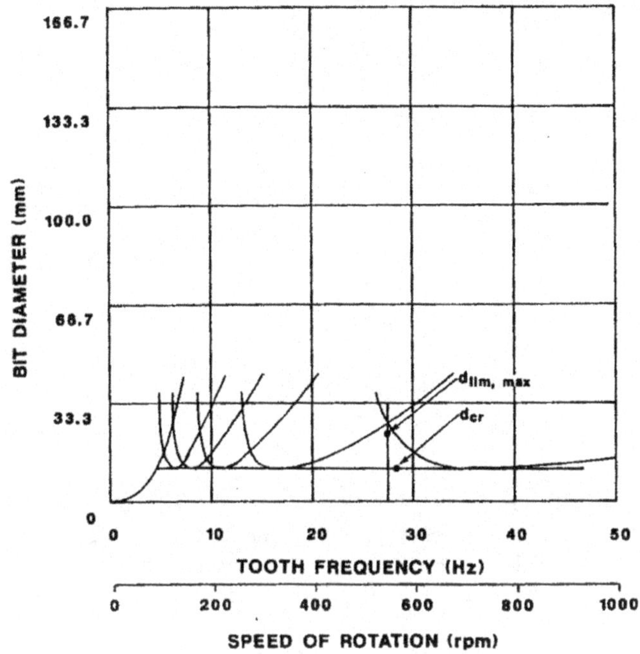

Figure 3-36. Limit of stability with shock sub [42].

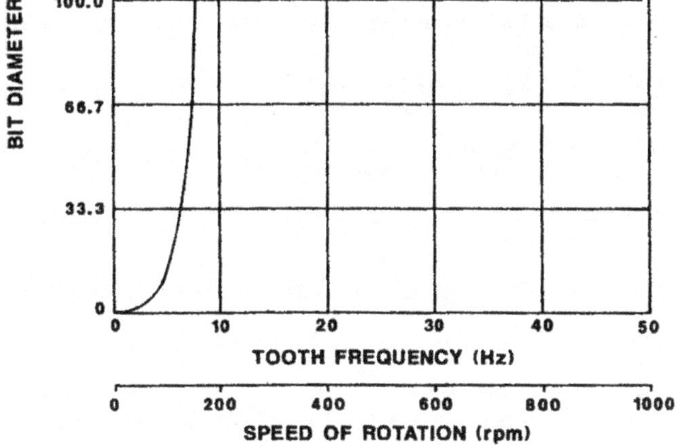

Figure 3-37. Stability diagram with shock absorber [42].

does expand the region of stability when used with PDC drill bits as shown in Fig. 3-37. The shock absorber design features are:

k = 215,000 lbs/in.
M = 141 lbs
ζ = 0.4

This shock absorber has a very stiff spring and a high damping factor. M represents the assumed mass of the PDC drill bit.

3.7.2.3 Drill Bit Design

Warren, Brett, and Sinor [53] propose a PBC drill bit design containing a relative smooth pad integral to the bit and opposite to rows of cutters. The purpose of the pad is to keep the center of rotation of the bit always inline with the center of the well bore.

Laboratory drilling tests clearly show a star pattern in rock samples drilled with PDC bits caused by backward whirl. Hole enlargement and severe lateral bit vibrations and cutter damage were evident. The very nature of PDC drill bits, having individual cutters protruding from a bit body, encourages backward whirl. Large impact forces resulting from sudden changes in linear momentum of the drill bit, as explained earlier, causes cutter breakage. When this happens, the instantaneous center of rotation suddenly changes from the geometric center of the collars (bit) to the point of contact between drill bits and well bore. The whirl resistant bit design does not allow the instantaneous center to switch to the walls of the well bore, thus there is no change in linear momentum and therefore no linear impulsive forces on the cutters.

3.8 FRICTION IN DIRECTIONAL WELLS

Directional wells may reach laterally several thousand feet. The well plan may call for defined paths to be drilled to reach a specific target in a reservoir. Complex well paths can now be drilled and navigated by rotary steerable tools and MWD monitoring. Friction forces created by these complex well paths can create high drilling torque and pull out forces. These loads can sometimes exceed the capability of the surface equipment and strength of the drillstring. Through experience, it has been determined that friction forces can be greatly reduced by rotating the pipe and maintaining mud circulation during tripping. Top drives have been designed for this purpose.

3.8.1 Coefficient of Friction

Johancsik, Friesen, and Dawson [54] discuss a method of determining axial pipe force directly below the Kelly sub and rotary torque imparted to the Kelly. The method is based on a load sensor and torque sensor designed by the authors. The load cell could be used only in the static mode. The torque transducer could be operated in the dynamic mode. Measured surface forces and torque,

Table 3-7. Coefficients of frictions [54].

	Well #1	Well #2	Well #3*
Pick-up	0.28	0.31	0.40
Slack-off	0.27	0.31	0.41
Rotating	0.27	0.29	0.41

*Well #3 had several severe dog legs in the lower portion of the build zone.

along with a computer model of drillstring friction, were used to determine the coefficient of friction in three different wells (9790 ft, 15,573 ft and 12,200 ft). The coefficient of frictions for each well is given in Table 3-7.

The friction model for computational purposes is based on Coulomb type frictions where the normal force is the resultant contact force over a 100 ft drilled interval. Torque predictions were tangential friction multiplied by pipe radius. Bending stiffness was not considered.

An interesting aspect of the measured data is that the total friction torque as measured at the surface is about 10 times the magnitude of typical drilling torque (~1500 ft-lb). In the third well, the pick up force is about 75% of the calculated string weight. While these numbers will vary from well to well, they do show the significance of friction in directional wells.

3.8.2 Soft Pipe Model

The soft string model does not include the bending stiffness of the pipe. The free body diagram for this model is shown in Fig. 3-38.

To account for the effects of hydrostatic pressure, let $T = F_\theta + p_\theta A_c$, where F_θ is the internal force, p_θ the local hydrostatic pressure and A_c the cross sectional

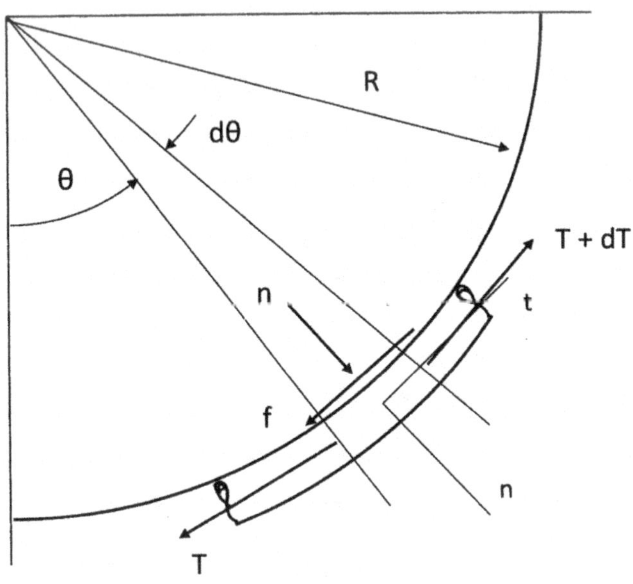

Figure 3-38. Freebody diagram of a soft drillstring.

area; T is actually effective tension. Summation of the forces in the tangent direction gives

$$dT = fRd\vartheta + (w - b)Rd\vartheta \sin\vartheta \tag{3-195}$$

Summing forces in the normal direction gives

$$nR = T - (w - b)R\cos\vartheta \tag{3-196}$$

Note that the internal shear force at any cross section can be shown to be equal to zero by summing moments about the center of curvature and combining this equation with the tangent force component.

Noting that $f = \mu n$ and combining Eqs. (3-195) and (3-196) gives

$$\frac{dT}{d\vartheta} - \mu T = RwBF[\sin\vartheta - \mu\cos\vartheta] \tag{3-197}$$

The complimentary solution is

$$T_c = Ce^{\mu\vartheta} \tag{3-198}$$

where C is an arbitrary constant.

The particular solution can be determined by using complex variables.

$$\overline{T_p} = \overline{A}e^{i\vartheta} \tag{3-199}$$

and replacing

$$\sin\theta = -ie^{i\theta}$$

$$\cos\theta = -e^{i\theta}$$

gives

$$\overline{T_p} = RwBF\,\frac{-(i+\mu)}{i-\mu}\,e^{i\theta} \tag{3-200}$$

The numerator is equal to $re^{i\phi}$ while the denominator is equal to $re^{i(\pi-\phi)} = -re^{-i\phi}$, where $\tan\phi = \frac{1}{\mu}$. Bringing these terms together gives

$$\overline{T_p} = RwBFe^{i2\phi}e^{i\theta} \tag{3-201}$$

Giving a particular solution of

$$T_p = RwBF\cos(\theta + 2\phi) \tag{3-202}$$

An alternative approach to the particular solution gives

$$T_p = \frac{RwBF}{(1+\mu^2)^{\frac{1}{2}}}\left[-\sin(\theta+\phi) + \mu\cos(\theta+\phi)\right] \tag{3-203}$$

Combining Eqs. (3-198) and (3-202) gives the total solution

$$T = Ce^{\mu\vartheta} + RwBF\cos(\theta + 2\phi) \tag{3-204}$$

Imposing the boundary condition that $T = T_0$ at $\theta = \theta_0$ gives

$$C = e^{-\mu\theta_0}\left[T_0 - RwBF\cos\left(\theta_0 + 2\phi\right)\right] \tag{3-205}$$

Equation (3-204) represents the solution and defines tension in the pipe at each θ location.

When the weight term is dropped, Eq. (3-204) gives the belt friction solution [55].

$$T = T_0 e^{\mu(\theta-\theta_0)} \tag{3-206}$$

Consider the following example.

R = 800 ft
w = 20 lb/ft
BF = 0.847 (10 ppg)
T_0 = 40,000 lbs
θ_0 = 30 deg
μ = 0.3

Effective tension for the pipe between θ_0 = 30 deg and θ = 90 deg. is given in Fig. 3-39 by the solid line. If pipe weight is ignored the solution resorts to a belt problem, whose solution is shown by the dashed line.

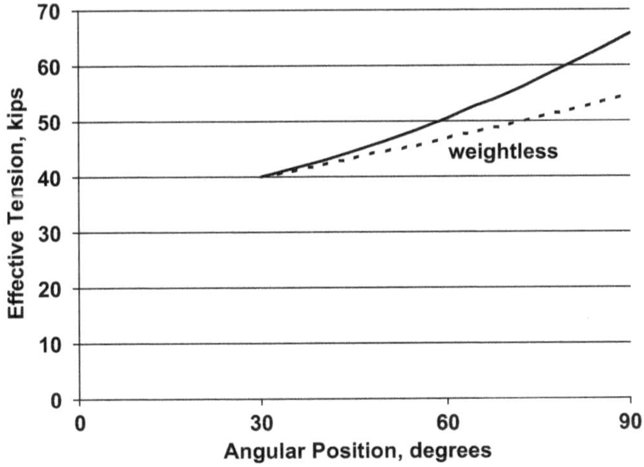

Figure 3-39. Effective tension around a curved well path.

3.8.3 Elastically Flexible Model

In reality, bending stiffness affects pipe bending deflections and friction within curved portions of directional wells. The magnitude of contact forces effects friction in rotation and pull (push). Friction affects are usually predicted by Coulomb's law. One approach taken by Dareing and Ahlers [56] and Rocheleau and Dareing [57] is to assume contact is made by each tool joint within a curved portion of the well bore. This may not be exactly true, but the solutions give insight as to how bending stiffness enters in to the calculation.

The set for the differential equation of bending is similar to the bending between stabilizers discussed earlier, except that they are expressed in polar coordinates. The method of solving these equations and determining contact force between tool joints and well bore, is the same method as used to determine bit and stabilizer contact forces in our discussion of stabilizer placement. The procedure is explained in detail in Appendix A at the end of this chapter. The basic equations for 1) pulling out of the well bore and 2) going in to the well bore are explained below.

3.8.3.1 Pulling Out of Well Bore [56]

With reference to Fig. 3-40, a freebody diagram is shown for a differential element between support points i and $i + 1$. The support points are assumed to be tool joints making point contact with the well bore. The radial location of these support points is arbitrary in the derivation.

The differential equation for the bending for an element of length $ds = R\,d\theta$ located between support points i and $i + 1$.

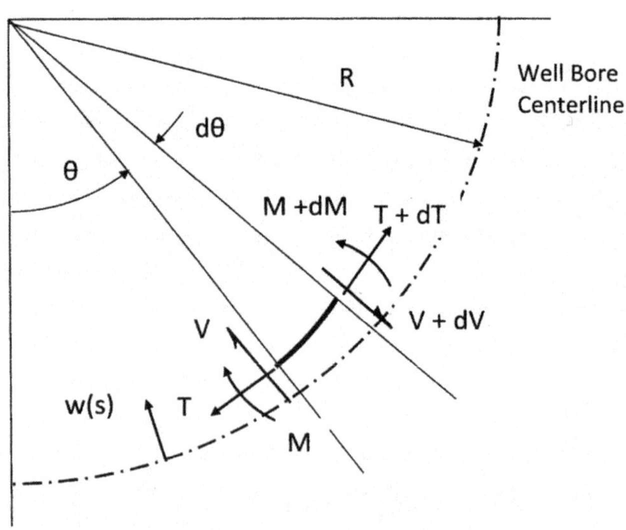

Figure 3-40. Freebody diagram of stiff pipe in a curved well bore.

Summing forces in the n direction gives

$$\frac{dV}{ds} = q + \frac{T}{R} \tag{3-207}$$

Summing forces in the t direction gives

$$\frac{dT}{ds} = -\frac{V}{R} \tag{3-208}$$

Summing moments about the center of curvature of the well bore (point 0) gives

$$\frac{dM}{ds} - T\frac{dw}{ds} + R\frac{dT}{ds} = 0 \tag{3-209}$$

By substitution

$$\frac{dM}{ds} - V - T\frac{dw}{ds} = 0 \tag{3-210}$$

The curvature of the displaced beam is expressed in terms of beam deflection by (see Timoshenko and Woinowsky-Krieger [58, pp. 503–504]

$$\frac{1}{\rho} = \frac{1}{R} + \frac{w}{R^2} + \frac{d^2w}{ds^2} \tag{3-211}$$

Assuming Euler bending

$$M = EI\frac{1}{\rho} \tag{3-212}$$

and combining the above equations gives

$$\frac{d^4w}{ds^4} - \lambda^2\frac{d^2w}{ds^2} = \zeta \tag{3-213}$$

where

$$\lambda^2 = \frac{T}{EI} - \frac{1}{R^2}$$

$$\zeta = \frac{Rq + T}{REI}$$

The solution to Eq. (3-213) is

$$w(s) = -\frac{1}{2}\chi_i s^2 + A_i s + B_i + C_i \cosh \lambda_i s + D_i \sinh \lambda_i s \tag{3-214}$$

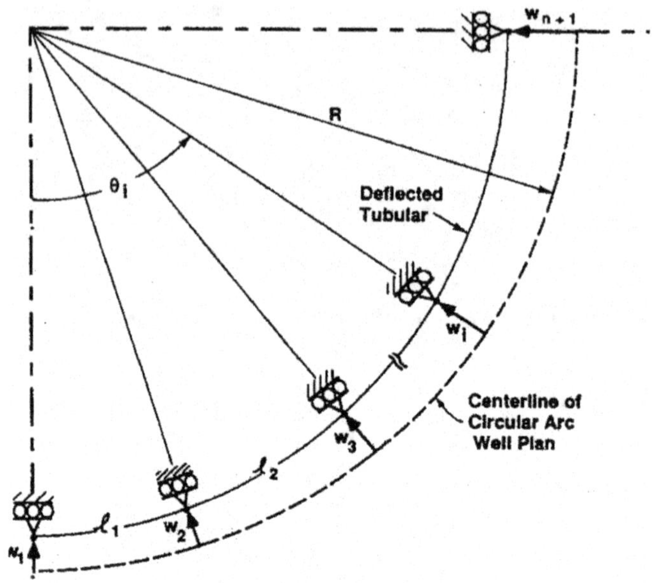

Figure 3-41. Pipe deflection defined in curvilinear coordinates [56].

where

$$\chi_i = \frac{Rq + T_i}{RT_i - \frac{EI_i}{R}}$$

This equation is now applied between a typical *span* between two *support* points as illustrated in Fig. 3-41.

When $s = 0$, $w(0) = w_i$, giving

$$B_i = w_i - C_i \tag{3-215}$$

When $s = \ell$, $w(\ell) = w_{i+1}$, giving

$$A_i = \left(\psi_i + \frac{1}{2}\chi_i \ell_i\right) - \frac{(\cosh \lambda_i \ell_i - 1)}{\ell_i} C_i - \frac{\sinh \lambda_i \ell_i}{\ell_i} D_i \tag{3-216}$$

where

$$\psi_i = \frac{w_{i+1} - w_i}{\ell_i}$$

Also, at $s = 0$

$$\phi_i = A_i + D_i \lambda_i \tag{3-217}$$

and

$$M_i = EI\left[\frac{1}{R} + \frac{1}{R^2}(B_i + C_i) - \chi_i + C_i \lambda_i^2\right] \tag{3-218}$$

Substituting for A and B

$$\phi_i = \left(\psi_i + \frac{1}{2}\chi_i\ell_i\right) + \frac{(1-\cos h\lambda_i\ell_i)}{\ell_i}C_i + \left(\lambda_i - \frac{\sin h\lambda_i\ell_i}{\ell_i}\right)D_i \tag{3-219}$$

and

$$M_i = EI\left(\frac{1}{R} + \frac{w_i}{R^2} - \chi_i\right) + C_i\lambda_i^2 EI \tag{3-220}$$

These two equations can be written in matrix form as

$$\begin{Bmatrix}\phi_i\\ M_i\end{Bmatrix} = [\alpha]_i^{-1}\begin{Bmatrix}C_i\\ D_i\end{Bmatrix} + [a]_i \tag{3-221}$$

where

$$[\alpha]_i^{-1} = \begin{bmatrix}\dfrac{(1-\cos h\lambda_i\ell_i)}{\ell_i} & \lambda_i - \dfrac{\sin h\lambda_i\ell_i}{\ell_i}\\ \lambda_i^2 EI & 0\end{bmatrix} \tag{3-222}$$

and

$$[a]_i = \begin{Bmatrix}\psi_i + \dfrac{1}{2}\chi_i\ell_i\\ EI\left(\dfrac{1}{R} + \dfrac{w_i}{R^2} - \chi_i\right)\end{Bmatrix} \tag{3-223}$$

By inversion

$$\begin{Bmatrix}C_i\\ D_i\end{Bmatrix} = [\alpha]_i\left(\begin{Bmatrix}\phi_i\\ M_i\end{Bmatrix} - \{a\}_i\right) \tag{3-224}$$

Similarly, at $s = \ell_i$,

$$\begin{Bmatrix}\phi_{i+1}\\ M_{i+1}\end{Bmatrix} = [\beta]_i\begin{Bmatrix}C_i\\ D_i\end{Bmatrix} + \{b\}_i \tag{3-225}$$

where

$$[\beta]_i = \begin{bmatrix}\dfrac{1-\cosh\lambda_i\ell_i}{\ell_i} + \lambda_i\sin h\lambda_i\ell_i & \lambda_i\cos h\lambda_i\ell_i - \dfrac{\sinh\lambda_i\ell_i}{\ell_i}\\ EI\lambda_i^2\cosh\lambda_i\ell_i & EI\lambda_i^2\sinh\lambda_i\ell_i\end{bmatrix} \tag{3-226}$$

and

$$\{b\}_i = \begin{Bmatrix} \psi_i - \frac{1}{2}\chi_i \ell_i \\ \frac{EI}{R}\left(1 + \frac{w_{i+1}}{R} - R\chi_i\right) \end{Bmatrix} \tag{3-227}$$

Substituting Eq. (3-224) into Eq. (3-225) gives

$$\begin{Bmatrix} \phi_{i+1} \\ M_{i+1} \end{Bmatrix} = [\beta]_i[\alpha]_i\left(\begin{Bmatrix} \phi_i \\ M_i \end{Bmatrix} - \{a\}_i\right) + \{b\}_i \tag{3-228}$$

According to Eq. (3-222), $\begin{Bmatrix} \phi_i \\ M_i \end{Bmatrix}$ at the ith support can be expressed in terms of $\begin{Bmatrix} \phi_1 \\ M_1 \end{Bmatrix}$ at the first support. This relationship can be put in the form of

$$\begin{Bmatrix} \phi_i \\ M_i \end{Bmatrix} = [\gamma]_i \begin{Bmatrix} \phi_1 \\ M_1 \end{Bmatrix} + [\delta]_i \tag{3-229}$$

where $i = 1, 2, 3, 4, \ldots\ldots n + 1$. Note that

$$[\gamma]_1 = \begin{bmatrix} 1 & 0 \\ 0 & 1 \end{bmatrix} \text{ and } \{\delta\}_1 = \begin{Bmatrix} 0 \\ 0 \end{Bmatrix} \tag{3-230}$$

Substituting Eq. (3-229) into Eq. (3-228) gives

$$\begin{Bmatrix} \phi_{i+1} \\ M_{i+1} \end{Bmatrix} = [\gamma]_{i+1} \begin{Bmatrix} \phi_1 \\ M_1 \end{Bmatrix} + [\delta]_{i+1} \tag{3-231}$$

where

$$[\gamma]_{i+1} = [\beta]_i[\alpha]_i[\gamma]_i \tag{3-232}$$

and

$$\{\delta\}_{i+1} = [\beta]_i[\alpha]_i(\{\delta\}_i - \{a\}_i) + \{b\}_i \tag{3-233}$$

Equation (3-231) can now be written as

$$\begin{Bmatrix} \phi_{n+1} \\ M_{n+1} \end{Bmatrix} = [\gamma]_{n+1} \begin{Bmatrix} \phi_1 \\ M_1 \end{Bmatrix} + [\delta]_{n+1} \tag{3-234}$$

which relates the slope and moments at each end of the multi-spanned curved beam. This matrix equation gives two algebraic equations from with to determine two unknowns. The two unknowns can be any two of the following: ϕ_1, M_1, ϕ_{n+1}, M_{n+1}. For example, if ϕ_1 and ϕ_{n+1} are given then

$$M_1 = \frac{\phi_{n+1} - (\lambda_{11})_{n+1}\phi_1 - (\delta_1)_{n+1}}{(\gamma_{12})_{n+1}} \tag{3-235}$$

M_1 and ϕ_1 are then used in Eq. (3-229) to determine all M_i and ϕ_i. Constants A,B,C,D can be determined from Eqs. (3-215), (3-216), and (3-224). Once these constants have been determined for each span, the deflection, slope, moment, and shear can be determined for each span.

The normal reaction at each contact point is determined from internal shear on each side of the contact point. Friction force between each contact point and the well bore can then be determined.

Bending of the tubular depends on tension. Tension in turn depends on bending. The calculation starts with an assumed tension in each span. Following an iterative procedure, pipe tension over each span converges in about three or four calculation cycles.

As an example, consider the following:

- radius of hole curvature is 200 ft (DLS = 28.65 deg/100 ft)
- total arc of dog leg is 90 deg.
- drill pipe is 4.5 in. (16.6 lb/ft)
- numbers of spans over the arc length

Following the above procedure, pull out forces corresponding to six pull back forces were calculated and the results given in Table 3-8. Two sets of boundary conditions were considered. The results were then compared with pull forces assuming no stiffness.

The two sets of boundary conditions had little effect on required pull out force. Also, tension for both sets of boundary conditions is slightly higher than the tension curve based on the "soft" model. Table 3-8 shows that pull out force accounting for elastic bending approaches the soft model prediction as back pull tension is increased.

3.8.3.2 *Putting Pipe Into Well Bore [57]*

The assumption made here is that the stabilizers make contact on the outside of the centerline instead of the inside as earlier discussed. The freebody diagram

Table 3-8. Pull out Force over a 90-deg Turn; R = 200 ft (4 ½ in. 16.5 lb/ft drill pipe).

Back pull (lbs)	Top tension free-free (lbs)	Top tension fixed-fixed (lbs)	Top tension no stiffness (lbs)
2000	6162	5544	4234
4000	9421	8773	7438
6000	12,794	12,361	10,642
8000	16,164	15,849	13,846
10,000	19,614	19,337	17,050
12,000	23,117	22,826	20,254

in Fig. 3-41 still applies except the axial force is a compressive force instead of a tension force. The differential equation of bending becomes

$$\frac{d^4w}{ds^4} + \lambda^2 \frac{d^2w}{ds^2} = \zeta \tag{3-236}$$

where

$$\lambda^2 = \frac{P}{EI} + \frac{1}{R^2}$$

$$\zeta = \frac{Rq - P}{REI}$$

The solution to Eq. (3-236) is

$$w(s) = -\frac{1}{2}\chi_i s^2 + A_i s + B_i + C_i \cos\lambda_i s + D_i \sin\lambda_i s \tag{3-237}$$

At the beginning of the ith section (where the ith support is located), $s = 0$ and $w(0) = w_i$, giving

$$B_i = w_i - C_i \tag{3-238}$$

and at the end of the ith section (where the $i + 1$ support is located), $s = \ell_i$, and $w(\ell_i) = w_{i+1}$, giving

$$w_{i+1} = -\frac{1}{2}\chi_i \ell_i^2 + A_i \ell_i + B_i + C_i \cos\lambda_i \ell_i + D_i \sin\lambda_i \ell_i \tag{3-239}$$

Solving Eq. (3-239) for A_i gives

$$A_i = \psi_i + \frac{1}{2}\chi_i \ell_i - \frac{(\cos\lambda_i \ell_i - 1)}{\ell_i} C_i - \frac{\sin\lambda_i \ell_i}{\ell_i} D_i \tag{3-240}$$

where

$$\psi_i = \frac{w_{i+1} - w_i}{\ell_i}$$

Also, at $s = 0$, the slope at the beginning of the ith section is

$$\phi_i = A_i + D_i \lambda_i \tag{3-241}$$

and the moment at the beginning of the ith section is

$$M_i = EI\left[\frac{1}{R} + \frac{1}{R^2}(B_i + C_i) - \chi_i - C_i \lambda_i^2\right] \tag{3-242}$$

Substituting A_i as defined by Eq. (3-240) into Eq. (3-241) gives

$$\phi_i = \psi_i + \frac{1}{2}\chi_i \ell_i - \frac{(\cos\lambda_i \ell_i - 1)}{\ell_i} C_i - \left(\lambda_i - \frac{\sin\lambda_i \ell_i}{\ell_i}\right) D_i \qquad (3\text{-}243)$$

and placing B_i as defined by Eq. (3-238) into Eq. (3-242) gives

$$M_i = EI\left[\frac{1}{R} + \frac{w_i}{R^2} - \chi_i\right] - C_i \lambda_i^2 EI \qquad (3\text{-}244)$$

Note that the derivation is similar to the previous derivation except that hyperbolic functions have been replaced by trigonometric functions. The rest of the solution proceeds as before (also see ref. [57]).

REFERENCES

1. Huang, T. and Dareing, D. W., "Buckling and Lateral Vibration of Drill Pipe," *Trans. ASME, Journal of Engineering for Industry*, Vol. 90, Series B., No. 4, 1968, 613–619.
2. A. Lubinski, "A study of the buckling of rotary drilling string", *API Drilling and Production Practice*, **17** (1950), pp. 178–214.
3. *Tables of Bessel Functions of Fractional Orders*, prepared by the Computation Laboratory of the National Applied Mathematics Laboratories, National Bureau of Standards, Columbia University Press, New York 1948, Vol. I; 1949, Vol. II.
4. Lubinski, A., Althouse, W. S., and Logan, J. L.,"Helical Buckling of Tubing Sealed in Packers", *Journal of Petroleum Technology* (June 1962) 655–70.
5. Mitchell, R. F., "Buckling Behavior of Well Tubing: The Packer Effect", S. Pet. Engrs. J., October, 1982, pp. 616–624.
6. Mitchell, R. F., "New Concepts for Helical Buckling", *SPE Drilling Engineering*, September 1988, pp. 303–310.
7. Cheatham, John B. Jr, and Pattillo, Phillip D., "Helical Postbuckling Configuration of a Weightless Column Under the Action of an Axial Load", *S. Pet, Engrs. J.*, August 1984, pp. 467–472.
8. Sorenson, K. G. and Cheatham J. B. Jr., "Post-Buckling Behavior of a Circular Rod Constrained Within a Circular Cylinder", *ASME Journal of Applied Mechanics*, December 1986, Vol. 53, pp. 929–934.
9. Hoch, R. S, "A Review of the Crooked-Hole Problem and an Analysis of Packed Bottom-Hole Drill-Collar Assemblies", Paper 926-7-K, Div. of Production, API, Dallas, Tex. (1962).
10. Wai-Fah Chen and Toshio Atsuta, Theory of Beam-Columns, Vol. 1: In-Plane Behavior and Design, McGraw-Hill, 1976.
11. Millheim K. K., Jordan S. and Ritter C. J., "Bottom-hole assembly analysis using the finite-element method", *Journal of Petroleum Technology*. **30** (1978), pp. 265–274.
12. Pestal, I. E. C. and Leckie, F. A.; *Matrix Methods in Elastomechanics*, McGraw-Hill, 1963.

13. Bogy, D. B. and Paslay, P. R.: "Buckling of Drill Pipe in an Inclined Hole", *Trans., ASME, Series B* (May 1964) 214–220.
14. Paslay, P. R. and Bogy, D. B., "The Stability of a Circular Rod Laterally Constrained to be in Contact with an Inclined Circular Cylinder," *Journal of Applied Mechanics* (Dec 1964) **31**, 605–610.
15. Rapier Dawson and P. R. Paslay, "Drill Pipe Buckling in Inclined Holes", *Journal of Petroleum Technology*, October 1984, pp. 1734–1738.
16. Mitchell, R. G., "Effects of Well Deviation on Helical Buckling," SPE 29462, SPE Production Operations Symposium, Oklahoma City, April 1995.
17. Shuh, F.J., "The Critical Buckling Force and Stresses for Pipe in Inclined Curved Boreholes," SPE/IADC 21942, SPE/IADC Drilling Conference, Amsterdam, March 11–14, 1991.
18. Kyllingstad, Aage, "Buckling of Tubular Strings in Curved Wells," *Journal of Petroleum Science and Engineering*, 12, 1995, pp. 209–218.
19. Suryanarayana, P.V.R., McCann, R. C., "An Experimental Study of Buckling and Post-Buckling of Laterally Constrained Rods," *(ASME) Journal of Energy Resources Technology*, Vol. 177, June 1995, pp. 115–124.
20. API RP 7G; API Recommended Practice for Drill Stem Design and Operating Limits.
21. Kuru, Ergun, Alexander Martinez, Stefan Miska and Weiyong Qiu, "The Buckling Behavior of Pipes and Its Influences on the Axial Force Transfer in Directional Wells", *Journal of Energy Resources Technology*, September 2000, Vol. 122, pp. 129–135.
22. Martinez, Alexander, Stefan Miska, Ergun Kuru, and James Sorem, "Experimental Evaluation of the Lateral Contact Force in Horizontal Wells", *Journal of Energy Resources Technology*, September 2000, Vol. 122, pp. 123–128.
23. Amjad A. Bersalsow and Mike L. Payne, "A Study of Excitation Mechanisms and Resonances Inducing Bottomhole-Assembly Vibrations", *SPE Drilling Engineering*, March 1988, pp. 93–101.
24. J. Kim Vandiver, James W. Nicholson, Rong-Juin Shyu, "Case Studies of the Bending Vibration and whirling Motion of Drill Collars", *SPE Drilling Engineering*, December 1990 pp. 282–290.
25. M. W. Dykstra, D. C.-K. Chen, T. M. Warren, J. J. Azar, "Drillstring component Mass Imbalance: A Major source of Downhole Vibrations", *SPE Drilling & Completion*, December 1996, pp. 234–241.
26. G. Greenhill, *Proc. Cambridge Phil. Soc.*, vol. 4, 1881.
27. R. V. Southwell and B. S. Dough; "On the Stability of a Rotating Shaft Subjected Simultaneously to End Thrust and Twist", *British Association for the Advancement of Science*, Vol. 8 (1921) pp. 345–355.
28. Huang, T. and Dareing, D. W., "Predicting the Stability of Long Vertical Pipe Transmitting Torque in a Viscous Medium," Trans., *ASME, Journal of Engineering for Industry*, Vol. 88, Series B, No. 2, May 1966, 191–200.
29. P. D. Spanos, M. L. Payne, C. K. Secora; "Bottom-Hole Assembly Modeling and Dynamic response Determination", *Journal of Energy Resources Technology*, Sept. 1997, Vol. 119, pp. 153–158.
30. V. A. Dunayevsky, Fereldoun Abbassian, Arnis Judzis, "Dynamic Stability of Drillstrings Under Fluctuating Weight on Bit", *SPE Drilling & Completion*, June 1993.

31. Yao-Qun Lin and Yu-Hwa Wang; "Stick-Slip Vibration of Drill Strings", *ASME Journal of Engineering for Industry*, Vol. 113, February 1991, pp. 38–43.
32. Yigit, Ahmet S. and Andreas P. Christoforou, "Stick-Slip and Bit Bounce Interaction in Oil-Well Drillstrings", *Journal of Energy Resources Technology*, December 2006, Vol. 128, pp. 268–274.
33. Dareing, D. W.: "Downhole Measurements of Drill-String Forces and Motions", Esso Production Research Report, No. EPR65-PR12, October 28, 1965.
34. P. D. Spanos, A. K. Sengupta, R. A. Cunningham, P. R. Paslay; "Modeling of Roller Cone Bit lift-Off Dynamics in Rotary Drilling", *Journal of Energy Resources Technology*, Sept. 1995, Vol. 117, pp. 197–20.
35. Dareing, D. W. and Livesay, B. J., "Longitudinal and Angular Drillstring Vibration with Damping," *Trans. ASME Journal of Engineering for Industry*, Vol. 90, Series B, No. 4, 1968, 671–679.
36. Kreisle, Leonardt F. and John M. Vance; "Mathematical Analysis of the Effect of a Shock Sub on the Longitudinal Vibrations of an Oilwell Drill String" Society of Petroleum Engineers, Dec. 1970.
37. Sengupta, A. K.; Numerical Simulation of the Roller Cone Drill Bit Lift-Off Phenomenon, M .S. Thesis, Rice University (1993), Houston, TX.
38. S. Timoshenko, D. W. Young and W. Weaver, Jr.; *Vibration Problems in Engineering*, 4th ed., John Wiley & Sons, 1974.
39. Zare, J., Hashemi, S.J., and Rashed G., "Finite element analysis of drill-string lateral vibration", Scientific Research and Essays Vol. 6(13), July 2011, pp. 2682–2694.
40. Tlusty, J. and Ismail F., "Special Aspects of Chatter in Milling", Trans ASME, Journal of Vibration Stress and Reliability in Design, Vol. 105, Jan 1983, pp. 24–32.
41. Zamudio, C. A., J. Tlusty, D. W. Dareing, "Self Excited Vibrations in Drillstrings", SPE 16661, Presented at the 62nd Annual Technical Conference, Dallas, TX, September 27–30, 1987.
42. Zamudio, C. A., J. Tlusty, D. W. Dareing, "Effect of Shock Absorber on Drag Bit Chatterr", SPE 17194, Presented at 1988 IADC/SPE Drilling Conference, Dallas, TX, February 28–March 2, 1988.
43. Dareing, D. W., Tlusty, J., Zamudio, C., "Self-excited vibrations induced by drag bits," *ASME, Journal of Energy Resources Technology*, Vol. 112, March 1990, pp. 54–61.
44. Elsayed, M. A., D. W. Dareing, and Dupuy C. A., "Effect of downhole assembly and polycrystalline diamond compact (PDC) bit geometry on stability of drillstrings", (1997) *ASME, Journal of Energy Resources Technology*, Vol. 119, pp. 159–163.
45. Elsayed, M. A. and David W. Raymond, "Measurement and analysis of chatter in a compliant model of a drillstring equipped with a PDC Bit", ETCE2000/ER-10280, Proceedings of ETCE/OMAE2000 Joint Conference, February 14–17, 2000, New Orleans, LA.
46. Elsayed, M. A., Yalamanchili A., and Aissi C., "Visualization of Surface Topology in Unstable Drilling with PDC Bits", Presented at the international Instrumentation Symposium, May 5–8, 2008, Pensacola Beach, FL.

47. Craig, Roy R., Jr; *Structural Dynamics*, 1981, John Wiley & Sons, NY.
48. Ford Brett J., Thomas M. Warren, Suzanne M. Behr, "Bit whirl—a new theory of PDC bit failure", *SPE Drilling Engineering*, December 1990.
49. Dareing, D. W., "Guidelines for Controlling Drill String Vibrations," *Trans. ASME, Journal of Energy Resources Technology*, Vol. 106, June 1984, 272–277.
50. Pushkar N. Jogi, John D. Macpherson, and Neubert M., "Field Verification of Model-Derived Natural Frequencies of a Drill String", *(ASME) Journal of Energy Resources Technology*, Vol. 124 Sept 2002.
51. Garrett, W. R., "Effect of a Downhole Shock Absorber on Drill Bit and Drill Stem Performance, Paper ASME 62-Pet-21 presented at the ASME 1962 Petroleum and Mechanical Engineering Conference, Dallas, Sept 23–26, 1962.
52. Dareing, Don W., "Drill Collar Length is a Major Factor in Vibration Control", *Journal of Petroleum Technology*, April 1984, pp. 637–644.
53. Thomas M. Warren, J. Ford Brett, L. Allen Sinor: "Development of a Whirl-Resistant Bit", *SPE Drilling Engineering*, Dec 1990, pp. 267–274.
54. C. A. Johancsik, D. B. Friesen, Rapier Dawson, "Torque and Drag in Directional Wells—Prediction and Measurement", *Journal of Petroleum Technology*, June 1984, pp. 987–992.
55. Shigley, J. E. and Mitchell, L. D.; *Mechanical Engineering Design*, McGraw-Hill, 1983.
56. Dareing, D. W., and Ahlers, C. A., "Tubular Bending and Pull Out Forces in High Curvature Well Bores," *ASME, Journal of Energy Resources Technology*, Vol. 113, pp. 133–139, June 1991.
57. Rocheleau, David N. and Dareing, Don W., "Effect of Drag Forces on Bit Weight in High Curvature Well Bores," *Trans ASME, Journal of Energy Resources Technology*, Vol. 114, No. 3, pp. 175–180, Sept. 1992.
58. Timoshenko, S. and Woinowsky-Krieger, S.; *Theory of Plates and Shells*, McGraw-Hill, 1959 (see pp. 503–504).

Appendix 3A

COMPUTER MODEL OF DRILL COLLAR BENDING BETWEEN MULTIPLE STABILIZERS

It is worth going through the calculation steps for stabilized drill collars. The same method and procedure is applied to friction in curved well bores (Chapter 3) and bundled tubulars attached to marine risers (Chapter 6) with some modifications.

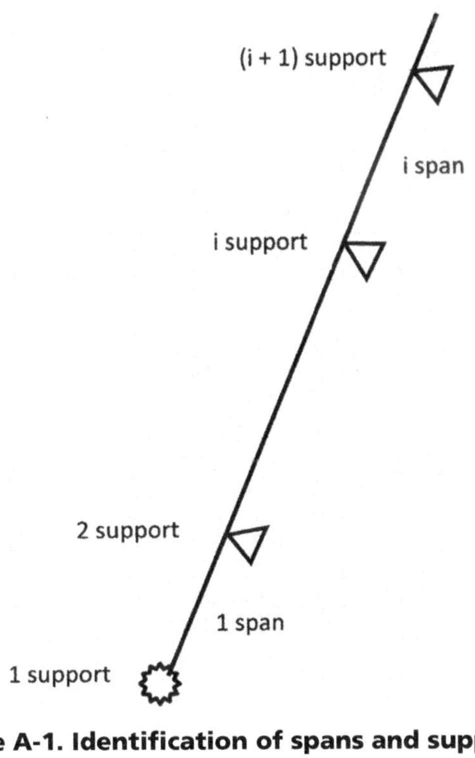

Figure A-1. Identification of spans and supports.

COMPUTER PROGRAM LOGIC

1. Calculate all $[\alpha]_i$, $[\beta]_i$, ψ_i for each ***span***.

$$[\alpha]_i = \frac{1}{-Q(\lambda_i L_i - \sin\lambda_i L_i)}\begin{bmatrix} 0 & \lambda_i L_i - \sin\lambda_i L_i \\ -QL_i & \cos\lambda_i L_i - 1 \end{bmatrix}$$

$$[\beta]_i = \frac{1}{L_i}\begin{bmatrix} 1-\cos\lambda_i L_i - \lambda_i L_i \sin\lambda_i L_i & \lambda_i L_i \cos\lambda_i L_i - \sin\lambda_i L_i \\ -QL_i\cos\lambda_i L_i & -QL_i \sin\lambda_i L_i \end{bmatrix}$$

$$\psi_i = \frac{y_{i+1} - y_i}{L_i}$$

Note: The product $[\beta]_i\,[\alpha]_i$ is dimensionless.

2. Calculate all $[\gamma]_i$ and $[\delta]_i$ for each ***support***, noting that

$$[\gamma]_1 = \begin{bmatrix} 1 & 0 \\ 0 & 1 \end{bmatrix} \text{ and } (\delta)_1 = \begin{pmatrix} 0 \\ 0 \end{pmatrix}$$

$$[\gamma]_i = [\beta]_i\,[\alpha]_i\,[\gamma]_i$$

$$(\delta)_{i+1} = [\beta]_i[\alpha]_i\left((\delta)_i - \begin{pmatrix} \psi_i - \dfrac{q}{Q}\dfrac{L_i}{2} \\ \dfrac{q}{\lambda^2} \end{pmatrix}\right) + \begin{pmatrix} \dfrac{q}{Q}\dfrac{L_i}{2} + \psi_i \\ \dfrac{q}{\lambda^2} \end{pmatrix}$$

3. Set $M_{n+1} = 0$. Through trial and error vary L_n until $\theta_{n+1} = 0$

$$\begin{pmatrix} \theta_{n+1} \\ M_{n+1} \end{pmatrix} = [\gamma]_{n+1}\begin{pmatrix} \theta_1 \\ M_1 \end{pmatrix} + (\delta)_{n+1}$$

$$\theta_{n+1} = (\gamma_{11})_{n+1}\,\theta_1 + (\gamma_{12})_{n+1}M_1 + (\delta_1)_{n+1}$$

$$M_{n+1} = (\gamma_{21})_{n+1}\,\theta_1 + (\gamma_{22})_{n+1}M_1 + (\delta_2)_{n+1}$$

These two equations are the only ones that are solved simultaneously. At least sixteen algebraic equations would have to be solved simultaneously using other approaches. Note that M_1 and M_{n+1} are both assumed to be zero. The two unknowns are L_n and θ_1. Must vary L_n until θ_{n+1} is zero. In the process of finding L_n, θ_1 is also determined.

4. Calculate θ_i, M_i, A_i, B_i, C_i, D_i

$$\begin{pmatrix} \theta_i \\ M_i \end{pmatrix} = [\gamma]_i\begin{pmatrix} \theta_1 \\ M_1 \end{pmatrix} + (\delta)_i$$

$$\begin{pmatrix} C_i \\ D_i \end{pmatrix} = [\alpha]_i \left(\begin{pmatrix} \theta_i \\ M_i \end{pmatrix} - \begin{pmatrix} \psi_i - \frac{q}{Q}\frac{L_i}{2} \\ \frac{q}{\lambda^2} \end{pmatrix} \right)$$

Solve for A

$$A_i = \psi_i - \frac{q}{Q}\frac{L_i}{2} + C_i \frac{1 - \cos\lambda_i L_i}{L_i} - D_i \frac{\sin\lambda_i L_i}{L_i}$$

Solve for B

$$y_i = B_i + C_i$$

5. Calculate y, θ, M, V distributions over each span.

$$y(x) = \frac{q}{Q}\frac{x^2}{2} + A_i x + B_i + C_i \cos\lambda x + D_i \sin\lambda x$$

6. Determine resultant bit force direction and magnitude.

$$V_1 = \frac{dM}{dx}(0) + Q\theta_1$$

Example Calculations

In applying the above equations, pay special attention to the units. They must be uniform. Also note that the rate of side loading, w, was taken plus in the plus y direction; downward is minus.

Consider, for example, a BHA with two stabilizers. Assuming

Hole size 8 ½ in.
Collar size 6 ¾ × 2 13/16 in.

$I = 71.83$ in.^{4}
$A = 24.47$ in.^{2}
$w = 83.27$ lb/ft

Mud weight = 12 lb/gal (BF = 0.817)
Stabilizer = 8 15/32 in.
WOB = 40,000 lbs
Hole angle = 30 deg.

There are three sections (spans) and four fixed points (supports). Input parameters for each span are:

$L_1 = 20$ ft
$L_2 = 40$ ft
$L_3 = L_n$ (initially unknown)
$Q = WOB - BFwx\cos 30$
$Q_1 = WOB$
$Q_2 = WOB - BFwL_1\cos 30$
$Q_3 = WOB - BFw(L_1 + L_2)\cos 30$

Input parameters for the supports are:

$y_1 = 0$
$y_2 = 0$
$y_3 = 0$
$y_4 = -0.875$ in. ($y_4 = y_{n+1}$)

For span i = 1

$$\psi_1 = 0 \quad [\alpha]_1 = \begin{bmatrix} 0 & -0.000025 \\ 1308 & 0.0000687 \end{bmatrix} \quad [\beta]_1 = \begin{bmatrix} -0.00171 & -0.00144 \\ -19844 & -34730 \end{bmatrix}$$

For span i = 2

$$\psi_1 = 0 \quad \psi_2 = 0 \quad [\alpha]_2 = \begin{bmatrix} 0 & -0.0000258 \\ 401.58 & 0.0000319 \end{bmatrix} \quad [\beta]_2 = \begin{bmatrix} -0.000701 & -0.00390 \\ 18659 & -34043 \end{bmatrix}$$

For span i = 3

$$\psi_3 = \frac{y_4}{L_3} = \frac{-0.875}{L_3} \quad [\alpha]_3 = \begin{bmatrix} 0 & -0.0000274 \\ 542 & 0.00004196 \end{bmatrix} \quad [\beta]_3 = \begin{bmatrix} -0.001289 & -0.00312 \\ 6760 & -35832 \end{bmatrix}$$

Note that L_3 is input to these parameters.

For support 1

$$[\gamma]_1 = \begin{bmatrix} 1 & 0 \\ 0 & 1 \end{bmatrix} \quad (\delta)_1 = \begin{pmatrix} 0 \\ 0 \end{pmatrix}$$

For support 2

$$[\gamma]_2 = \begin{bmatrix} -1.889 & 0 \\ -454{,}417{,}810 & -1.889 \end{bmatrix} \quad (\delta)_2 = \begin{pmatrix} -0.0007840 \\ -40{,}054 \end{pmatrix}$$

For support 3

$$[\gamma]_3 = \begin{bmatrix} 7.797 & 0 \\ 97{,}031{,}121 & 3.733 \end{bmatrix} \quad (\delta)_3 = \begin{pmatrix} -0.000764 \\ -77{,}372 \end{pmatrix}$$

Table A-1. Support Constants.

Support i	θ, rad.	M, lb-in
1	−4.19 (10⁻⁵)	0
2	−0.0007	−38,149
3	−0.0011	−81,441
4	0	0

Table A-2. Span Constants.

Span i	A	B	C	D
1	0.0045	3.690	−3.690	−1.043
2	0.0152	2.935	−2.935	−3.684
3	0.0195	2.207	−2.207	−4.935

For support 4 (n+1)

$$[\gamma]_4 = \begin{bmatrix} -22.426 & 0 \\ -315{,}398{,}777 & -11.93 \end{bmatrix} \qquad (\delta)_4 = \begin{pmatrix} -0.001127 \\ -13{,}225 \end{pmatrix}$$

Substituting into Eq. (3-58) and iterating on L_3 we obtain L_3 = 35 ft and $\theta_1 = -4.193 \times 10^{-5}$ rad. Recall that $M_1 - 0$. Tables A-1 and A-2 give the calculated values for the support and span constants.

The shear force at point 1 (drill bit) is determined from

$$V_1 = \left[EI \frac{d^3 y_1}{dx^3} + Q_1 \frac{dy_1}{dx} \right]_{x=0}$$

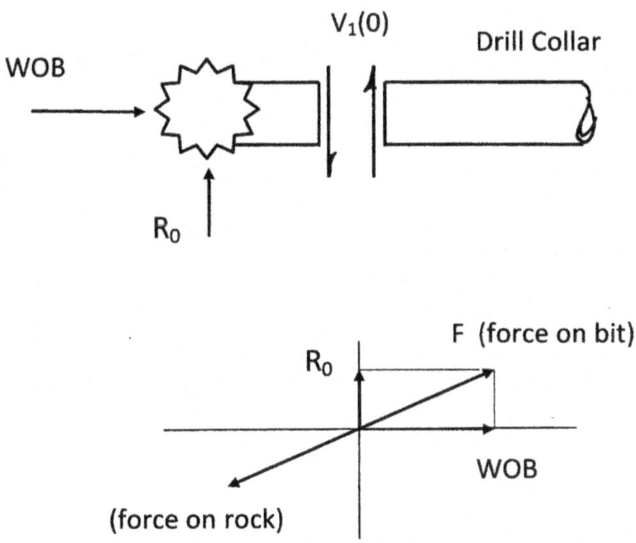

Figure A-2. Force on drill bit and reaction force on formation.

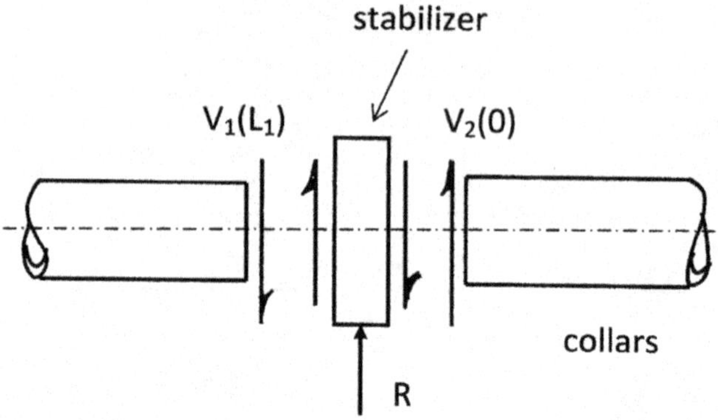

Figure A-3. Freebody diagram of stabilizer support and top end.

The resultant of this shear force and weight on bit is the reaction between bit and formation (Fig. A-2). In this case the bit is pushing down and forward indicating a dropping assembly. The shear force at the bit, R_1, indicates the bit tends to move back toward the vertical. The resultant bit force reaction then is the vector sum of WOB and R_1. This force vector makes an angle of 0.0043 rad. (0.25 deg) with the well bore axis.

The side reaction at each support is determined from (see Fig. A-3)

From the freebody diagram in Fig. A-3

$$R_2 = V_2(0) - V_1(L_1)$$

$$R_2 = \left[EI\frac{d^3y_2}{dx^3} + Q_2\frac{dy_2}{dx} \right]_{x=0} - \left[EI\frac{d^3y_1}{dx^3} + Q_1\frac{dy_1}{dx} \right]_{x=L_1}$$

Since the Q force and slope at support #2 and #3 are the same, this equation simplifies to

$$R_2 = \left[EI\frac{d^3y_2}{dx^3} \right]_{x=0} - \left[EI\frac{d^3y_1}{dx^3} \right]_{x=L_1}$$

The full expression for shear has to be considered in determining shear at the drill bit and upper contact point with the well bore.

Values of the reactions at each support are given in Table A-3.

Table A-3. Transverse support reactions.

	Support	Reaction, lbs
(bit)	1	181
	2	1088
	3	1481
(tangent)	4	471*

*Shear is actually minus at the upper tangent location.

Chapter 4

MECHANICS OF MARINE RISERS

Marine risers are tubulars that extend from the ocean floor to drill ships or production platforms. They are attached to a blowout preventer (BOP stack), which in turn is attached to surface casing. Risers are supported at the top by a motion compensation system which maintains tension and compensates for vessel heave. The top and bottom of drilling risers are attached to flex/ball joint connections with rotational stiffness. Tapered stress joints are often used in production risers to alleviate bending stresses and provide a smooth transition into the well. Loads to marine risers may come from rig offset, ocean currents, or ocean waves. The amount of flexing and bending is controlled mainly by tension. A guide to design and operation of drilling risers is given in API RP 16Q [1].

There are two types of marine risers: drilling and production. Drilling risers are used in exploratory drilling. They connect the drill ship to mud-line equipment and allow access to the well bore and serve as a conduit for drilling mud. Production risers protect production tubing which extend deep into the well. They represent an extension of the well bore and accommodate downhole production equipment. Drilling risers are on location throughout drilling and completion of a well, which usually takes around four month or so. On the other hand, production risers are meant to stay on location throughout the life of the production system.

Lateral deflections and bending stresses in marine risers can be viewed as the sum of static and dynamic effects. Each is usually considered separately and their effects combined. In most cases, static analyses give stress levels that dictate riser design. Cyclic stress can sometimes be determined from a quasi-static analysis, provided frequency of the loading does not match any of the natural frequencies. However, in many cases a complete dynamic analysis is required.

4.1 STATIC ANALYSIS OF MARINE RISERS

4.1.1 Buckling of Marine Risers

Riser stability depends on tension/compression and hydrostatic pressure created by fluids inside (drilling mud) and outside (sea water). Drilling mud, because of its high density, encourages riser buckling. These factors are included in a buckling analysis by Huang and Dareing [2]. Their paper, however, addresses drill pipe buckling where densities of fluid inside and outside are the same. The solution in ref. [2] is adapted here to predict riser buckling. The differential equations of bending for both problems are the same. This means the mathematical solution to both is the same. Only the content of the αs and βs are different.

The results are modified as explained below to predict riser buckling for which the fluid inside is typically drilling mud and the fluid outside is sea water. Results from ref. [2] are given in Tables 3-1 and 3-2 (Chapter 3). In applying these results, remember that the buckling model is based on a structurally fixed boundary condition at the lower end, such as tapered flex joints. Other boundary conditions, such as flex ball joints, are different, and should be considered separately.

Using the general differential equation of bending Eq. (3-2) and adjusting the sign on β to account for the negative sign in Eq. 3-5 gives

$$\alpha = \frac{(w - A_o\gamma_o + A_i\gamma_i)L^3}{EI}\left|\frac{144\text{ in}^2}{1\text{ ft}^2}\right| \tag{4-1}$$

$$\beta = -\frac{(F_B + LA_o\gamma_o - LA_i\gamma_i)L^2}{EI}\left|\frac{144\text{ in}^2}{1\text{ ft}^2}\right| \tag{4-2}$$

Critical values of β are related to α by

$$\phi_{cr} = \frac{\beta_{cr}}{\alpha^{\frac{2}{3}}} \tag{4-3}$$

or

$$\beta_{cr} = \phi_{cr}\alpha^{\frac{2}{3}} \tag{4-4}$$

When $\alpha > 4{,}000$, ϕ_{cr} becomes constant for each mode. Note that F_B is the actual internal force in the riser at the bottom end. It is related to top tension by

$$T_{top} = wL + F_B \tag{4-5}$$

Equation (4-5) assumes F_B is a plus number or tension. It follows that

$$\beta_{cr} = -\frac{(F_{B,cr} + LA_o\gamma_o - LA_i\gamma_i)_{cr}L^2}{EI}\left|\frac{144\text{ in}^2}{1\text{ ft}^2}\right| \tag{4-6}$$

The minus sign adjusts β for tension. The derivation in ref. [2] assumes β to be negative. So to account for tension the following adjustments are made.

$$-(F_B + LA_o\gamma_o - LA_i\gamma_i)_{cr} = \phi_{cr}\left(\frac{EI}{144}\right)^{\frac{1}{3}}(w - A_o\gamma_o + A_i\gamma_i)^{\frac{2}{3}} \tag{4-7}$$

The critical internal force at the lower end of the riser is

$$F_{B,cr} = -\phi_{cr}\left(\frac{EI}{144}\right)^{\frac{1}{3}}(w - A_o\gamma_o + A_i\gamma_i)^{\frac{2}{3}} - (LA_o\gamma_o - LA_i\gamma_i) \tag{4-8}$$

$F_{B,cr}$ can be either plus (tension) or minus (compression). For values of $\alpha >$ 4,000, ϕ_{cr} are fairly unchanged and approach the following values for the first two modes of buckling

$$(\phi_{cr})_1 = 3.29 \qquad \text{(1st mode)}$$

$$(\phi_{cr})_2 = 4.55 \qquad \text{(2nd mode)}$$

For values of $\alpha < 4000$, ϕ_{cr} values should be taken from Table 3-2. If the fluid is the same inside and out, say w_m, then according to Eq. (3-4)

$$F_{B,cr} = -(F_{cr} + w_m L) \tag{4-9}$$

where F_{cr} is the compressive force above local hydrostatic pressure and by substitution Eq. (4-8) reduces to

$$F_{cr} = \phi_{cr}(EI)^{\frac{1}{3}}(w - w_m)^{\frac{2}{3}} \tag{4-10}$$

This is the same as Eq. (3-19). However, for marine risers, it is best to define buckling in terms of the actual internal force at the well head for which

$$T_{top} = wL + F_{B,cr} \tag{4-11}$$

In this case, $F_{B,cr}$ does depend on riser length and drilling mud density. As an example, consider the following case:

Riser pipe OD — 22 in.
Riser pipe ID — 20.75 in.
w — 142.81 lb/ft
I — 2398 in.^{4}
E — 29 (10^6) psi
γ_0 — 64 lb/ft^3 (sea water)
γ_m — variable, ppg
L — variable, ft

Values of α for the range of fluid densities and lengths of 1000 ft and 2000 ft are less than 4000 so ϕ_{cr} should be taken from Table 3-2. Values of α for riser lengths of 3000, 4000, and 5000 ft are greater than 4000 so $(\phi_{cr})_1 = 3.29$ and $(\phi_{cr})_2 = 4.55$.

Critical forces for the first buckling mode are displayed in Fig. 4-1 for five different riser lengths. The lower end is fixed and the top end pinned. The buckling force, $F_{B,cr}$, is the actual internal tension/compression at the lower end as determined from $T_{top} - wL$. This is different from the F_{cr} discussed in Chapter 3.

All five lines intersect at one point. According to Eq. (4-8), $F_{B,cr}$ is independent of riser length, L, when $A_0\gamma_0 = A_i\gamma_i$. For the example case, this point corresponds to

$$\gamma_i = \left(\frac{d_o}{d_i}\right)^2 64 = \left(\frac{22}{20.75}\right)^2 64 = 71.94 lb/ft^3 \quad \text{(9.62 ppg)}$$

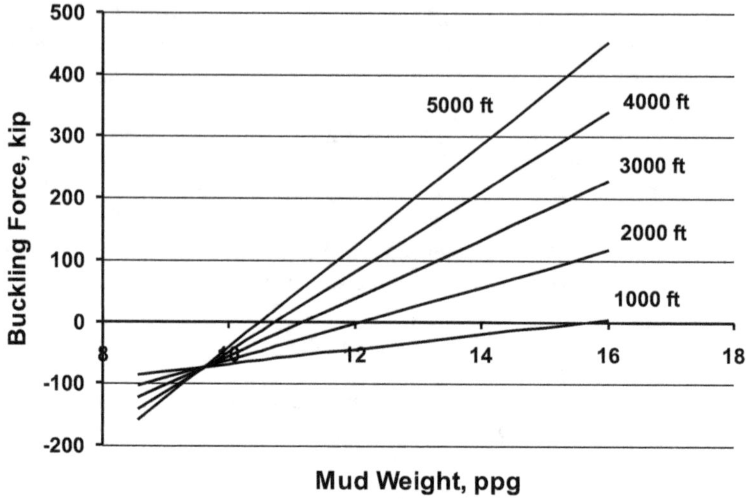

Figure 4-1. Critical buckling force ($F_{B,cr}$) for marine riser; 1st Mode (ϕ = 3.29) (22 in. OD – 20.75 in. ID).

Under this condition

$$F_{Bcr} = -\phi_{cr}\left(\frac{EI}{144}\right)^{\frac{1}{3}} w^{\frac{2}{3}} = -70{,}516 lb$$

which agrees with Fig. 4-1.

A separate calculation showed that the critical buckling force for each case in air is also 70.52 kips (compression). This number is determined from Eq. (4-10) with $w_m = 0$.

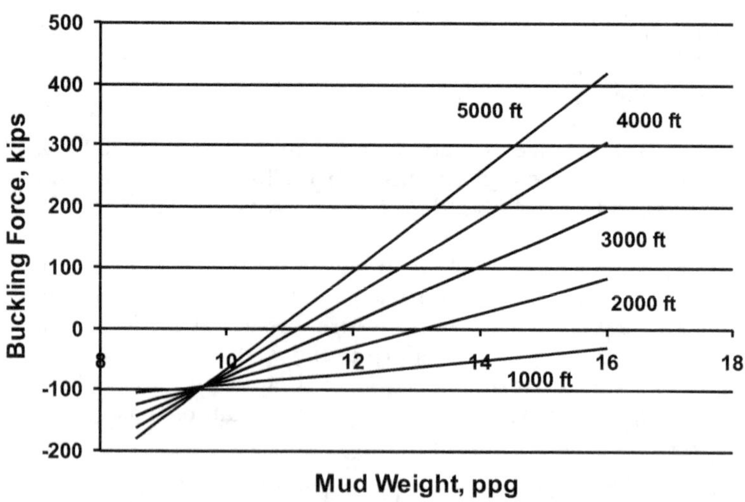

Figure 4-2. Critical buckling force ($F_{B,cr}$) for marine riser; 2nd mode (ϕ = 4.55) (22 in. OD – 20.75 in. ID).

Numbers that are minus represent the amount of compressive force at the bottom end that initiate buckling. Plus numbers define conditions of tension at the lower end that correspond to the onset of buckling. This data, which was generated using Eq. (4-8), shows the effects of higher mud weight on buckling especially in long riser pipes. Top tension in each case is determined from $T_{top} = F_B + wL$.

Results given in Fig. 4-1 show how fluid density inside risers influences buckling. The effect is huge. Even at mud weights of 12 ppg, buckling will occur when riser pipe is in tension at the bottom. Under extreme cases of 5000 ft and 16 ppg mud, bottom tension must exceed 450 kips to prevent buckling. In each case, top tension is the sum of riser air weight plus critical buckling force as shown in this figure.

Critical forces for the second buckling mode are given in Fig. 4-2. The buckling force (tension) reduces for all mud weights; all lines shift downward slightly.

4.1.2 Static Displacement of Risers in One Plane

This section covers lateral deflections of marine risers subjected to rig offset and current. Water depth is assumed to be 1500 ft. Calculations are made for different boundary constraints.

4.1.2.1 Differential Equation of Bending

The differential equation of bending was developed in Chapter 2. The simplified equation for these examples is

$$EI\frac{d^4y}{dx^4} - \frac{d}{dx}\left[(ax+b)\frac{dy}{dx}\right] = q(x) \tag{4-12}$$

where

$$a = w - A_0\gamma_0 + A_i\gamma_i$$
$$b = F_B + LA_0\gamma_0 - LA_i\gamma_i$$

The densities of sea water, γ_w and drilling mud, γ_m, are recognized in these expressions. In dimensionless form, Eq. (4-12) reverts to

$$\frac{d^4y}{d\zeta^4} - \frac{d}{d\zeta}\left[(\beta+\alpha\zeta)\frac{dy}{d\zeta}\right] = f(x) \tag{4-13}$$

where

$$\zeta = \frac{x}{L}$$

$$f(x) = \frac{q(x)L^4}{EI}$$

Equation (4-13) is expanded into

$$\frac{d^4y}{d\zeta^4} - \alpha\zeta\frac{d^2y}{d\zeta^2} - \beta\frac{d^2y}{d\zeta^2} - \alpha\frac{dy}{d\zeta} = f(x) \tag{4-14}$$

For riser response due solely to rig offset

$$\frac{d^4y}{d\zeta^4} - \alpha\zeta\frac{d^2y}{d\zeta^2} - \beta\frac{d^2y}{d\zeta^2} - \alpha\frac{dy}{d\zeta} = 0 \tag{4-15}$$

The solution to Eq. (4-15) is assumed to be

$$y(\zeta) = \sum_{n=0}^{\infty} a_n \zeta^n \tag{4-16}$$

Substituting Eq. (4-16) and its derivatives into Eq. (4-15) gives a power series in ζ. The coefficients of each ζ must be equal to zero in order to satisfy Eq. (4-15), giving the recurrence equation.

$$a_n = \frac{\beta}{n(n-1)} a_{n-2} + \frac{\alpha(n-3)}{n(n-1)(n-2)} a_{n-3} \tag{4-17}$$

where $n \geq 4$. By repeated applications of Eq. (4-17), starting with $n = 4$, a_n can be expressed as a linear combination of a_0, a_1, a_2, and a_3. Thus in general

$$a_n = F_n a_0 + G_n a_1 + H_n a_2 + I_n a_3 \tag{4-18}$$

Following the procedure given in Chapter 2,

$$F_n = \frac{\beta}{n(n-1)} F_{n-2} + \frac{(n-3)\alpha}{n(n-1)(n-2)} F_{n-3}, \text{ etc.} \tag{4-19}$$

for $n \geq 4$. These equations are determined by substituting Eq. (4-18) for a_{n-2} and a_{n-3} into Eq. (4-17) and then equating Eq. (4-17) to the resulting equation.

The coefficients in the power series (Eq. (4-16)) can now be expressed in terms of a_0, a_1, a_2, and a_3, which are determined from the four boundary conditions. These equations are applied to four different bottom end boundary constraints. In each case, the rig is offset by distance, δ.

4.1.2.2 Boundary Conditions

Four different boundary conditions will now be considered. In each case, the top boundary condition is assumed to be pinned. The four boundary conditions at the lower end are:

1. Ball Joint
2. Flex Ball Joint
3. Fixed Constraint
4. Tapered Flex Joint

4.1.3 Deflections Caused by Rig Offset

As an example, consider a drilling riser having the following properties;

L = 1500 ft
Riser size = 22 in. OD × 20.75 in. ID
w = 142.8 lb/ft (based on nominal cross section dimensions)

Mud weight = 12 ppg
Sea water = 64 lb/ft^3
$E = 29(10)^6$ psi
$I = 2399$ in.^{4} (based on nominal cross section dimensions)
$F_B = 100{,}000$ lbs
Flex/ball joint ($K_\phi = 0$)

4.1.3.1 *Ball Joint at Lower End*

Ball joints in this case are assumed to have no rotational bending resistance. Boundary conditions are

$$y(0) = 0$$

$$\frac{d^2y}{d\zeta^2}(0) = 0$$

$$y(1) = \delta$$

$$\frac{d^2y}{d\zeta^2}(1) = 0$$

Applying these boundary conditions to Eq. (4-16) gives $a_0 = a_2 = 0$. Following the steps given in Chapter 2, a_1 and a_2 are determined from

$$\left[\sum_{n=1}^{\infty} G_n\right] a_1 + \left[\sum_{n-1}^{\infty} I_n\right] a_3 = \delta \tag{4-20}$$

$$\left[\sum_{n=1}^{\infty} n(n-1)G_n\right] a_1 + \left[\sum_{n-1}^{\infty} n(n-1)I_n\right] a_3 = 0 \tag{4-21}$$

The solution to these equations for this particular input gives

$$a_0 = 0$$

$$a_1 = 126.35$$

$$a_2 = 0$$

$$a_3 = -1485.66$$

All a_n's have length units. The remaining coefficients, a_n, are determined by repeated application of Eq. 4-17, with $\alpha = 1289.9$ and $\beta = 173.4$. Lateral deflections, slope, bending moments and shear can now be determined. Figure 4-3 shows deflections for this case.

$$\frac{dy}{d\varsigma} = \sum_{1} a_n n \varsigma^{n-1} \tag{4-22}$$

$$\frac{d^2y}{d\varsigma^2} = \sum_{2} a_n n(n-1)\varsigma^{n-2} \tag{4-23}$$

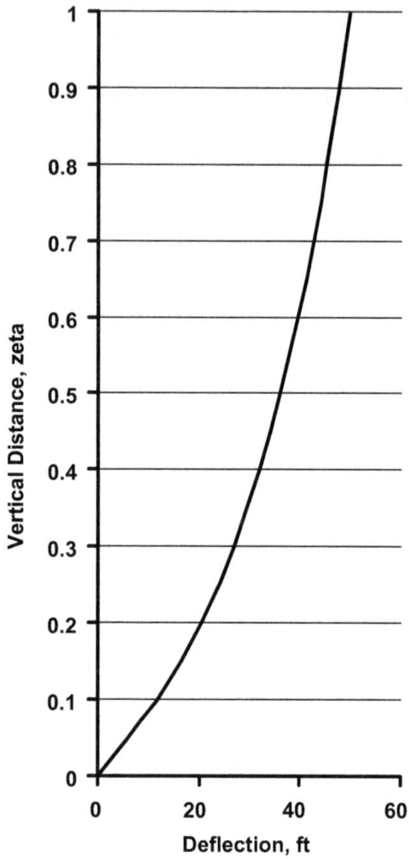

Figure 4-3. Riser deflection due to rig offset; bottom support is pinned connected. *L* = 1500 ft (ζ = 1).

Curvature at any location is

$$\frac{1}{\rho} = \frac{d^2y}{dx^2} = \frac{1}{L^2}\frac{d^2y}{d\varsigma^2} \tag{4-24}$$

and bending stress is

$$\sigma_b = E\frac{c}{\rho} \tag{4-25}$$

The slope and curvature of beam deflection at $\zeta = 0$ are

$$\frac{dy}{d\varsigma}(0) = a_1 \tag{4-26a}$$

$$\frac{dy}{dx}(0) = \frac{a_1}{L} \tag{4-26b}$$

and

$$\frac{d^2y}{d\varsigma^2}(0) = 2a_2 \tag{4-26c}$$

$$\frac{d^2y}{dx^2}(0) = \frac{2a_2}{L^2} \tag{4-26d}$$

According to Eq. (4-8) (Fig. 4-1) buckling for the first mode occurs when F_B is near 0 lbs, so the riser is safe from buckling.

Equation (4-23) was used to calculate bending moment and bending stress distribution (see Fig. 4-4).

Shear force between bottom of riser and BOP stack is

$$V(0) = EI\frac{d^3y}{dx^3}(0) - F_B\frac{dy}{dx}(0)$$

$$V(0) = \frac{EI}{L^3}\frac{d^3y}{d\zeta^3}(0) - \frac{F_B}{L}\frac{dy}{d\zeta}(0)$$

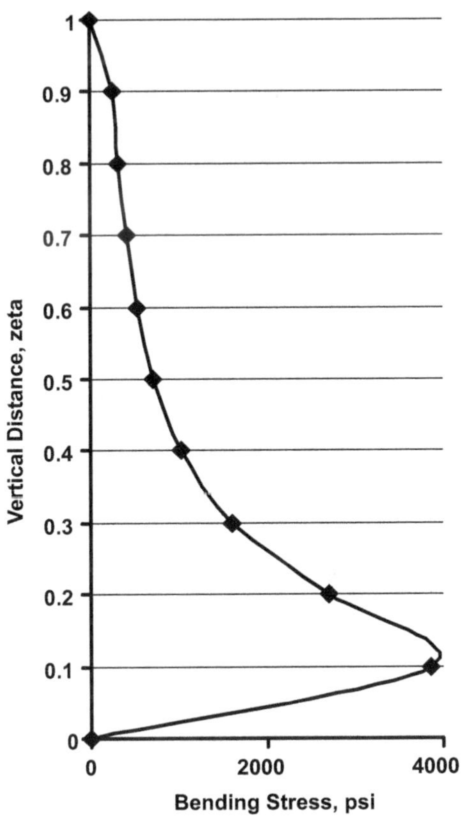

Figure 4-4. Bending stress distribution (L = 1500 ft, ζ = 1).

$$V(0) = \frac{EI}{L^3}(6a_3) - \frac{F_B}{L}(a_1)$$

$$V(0) = \frac{29(10)^6 2399}{1500^3} 6(-1485.66)\left|\frac{1\ \text{ft}^2}{144\ \text{in}^2}\right| - \frac{10^5}{1500}(126.35)$$

$$V(0) = -9699\ \text{lbs}$$

The angular rotation across the ball joint is

$$\Delta\theta = \frac{dy}{dx}(0) = \frac{a_1}{L},\ \text{radians}$$

$$\Delta\theta = \frac{126.35}{1500} = 0.0842\ \text{radians}$$

or

$$\Delta\theta = 4.83\ \text{deg}$$

4.1.3.2 Flex/Ball Joint

Flex ball joints are commonly used at the lower end of drilling risers to control bending and angle of rotation. Guidelines from API RP 16Q suggest that the built-in rotational stiffness should range from 10,000 to 30,000 ft-lb/deg of rotation. The rotational stiffness is usually assumed to be linear to accommodate the mathematics. Accordingly, the lower boundary condition is written as

$$EI\frac{d^2y}{dx^2}(0) = K_\theta \frac{dy}{dx}(0) \tag{4-27}$$

The set of boundary conditions for this case is

$$y(0) = 0 \tag{4-28a}$$

$$EI\frac{d^2y}{d\varsigma^2}(0) = K_\phi L\frac{dy}{d\varsigma}(0) \tag{4-28b}$$

$$y(1) = \delta \tag{4-28c}$$

$$\frac{d^2y}{d\varsigma^2}(1) = 0 \tag{4.28d}$$

The differential equation of bending (Eq. (4-15)) still applies as well as the recurrence formula, Eq. (4-17).

Consider again the power series solution (Eq. (4-16)). The first boundary condition yields $a_0 = 0$. The second boundary condition yields

$$a_2 = \frac{K_\phi L}{2EI}a_1 \tag{4-29}$$

where the term, $\frac{K_\phi L}{2EI}$, is dimensionless. The third boundary condition yields

$$\sum_{n=1}^{\infty} G_n a_1 + \sum_{n=1}^{\infty} H_n a_2 + \sum_{n=1}^{\infty} I_n a_3 = \delta \tag{4-30}$$

Using boundary condition Eq. (4-28b)

$$\left(\sum_{n=1}^{\infty} G_n + \frac{K_\phi L}{2EI}\sum_{n=1}^{\infty} H_n\right)a_1 + \left[\sum_{n=1}^{\infty} I_n\right]a_3 = \delta \tag{4-31}$$

The fourth boundary condition gives

$$\sum_{n=1}^{\infty} n(n-1)G_n a_1 + \sum_{n=1}^{\infty} n(n-1)H_n a_2 + \sum_{n=1}^{\infty} n(n-1)I_n a_3 = 0 \tag{4-32}$$

and with the second boundary condition

$$\left(\sum_{n=1}^{\infty} n(n-1)G_n + \frac{K_\phi L}{2EI}\sum_{n=1}^{\infty} n(n-1)H_n\right)a_1 + \left[\sum_{n=1}^{\infty} n(n-1)I_n\right]a_3 = 0 \tag{4-33}$$

a_1 and a_3 are determined from Eq. (4-31) and Eq. (4-33). a_2 is determined from Eq. (4-29).

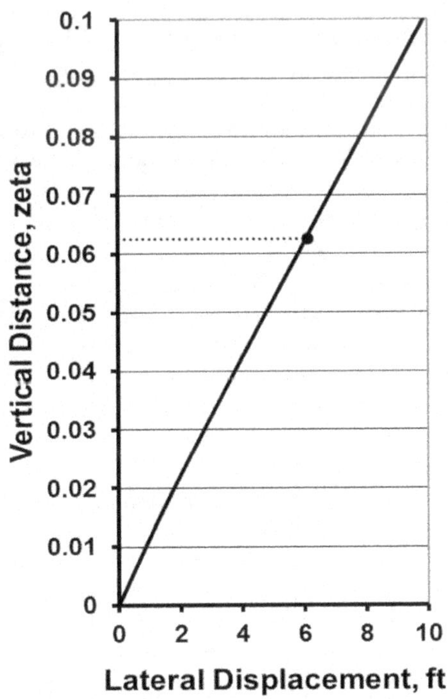

Figure 4-5. Flex ball joint at lower end, K_ϕ = 50,000 ft-lb/degree.

$$a_0 = 0$$
$$a_1 = 83.376$$
$$a_2 = 370.79$$
$$a_3 = -3040.31$$

The remaining values of a_n are with the recurrence formula (Eq. (4-17)) using α =1289.9 and β = 173.4.

Riser deflections were calculated using the same input data given earlier with the exception of a flex ball joint located at the bottom. Figure 4-5 shows deflections near the flex ball joint having a rotational spring constant of 50,000 ft-lb/deg. Input to these calculations is

L = 1500 ft
Riser size = 22 in. OD × 20.75 in. ID
w = 142.8 lb/ft (based on nominal dimensions)
Mud weight = 12 ppg (89.76 lb/ft^3)
Sea water = 64 lb/ft^3
E = 29 (10)^6 psi
I = 2399 in.^{4} (based on nominal dimensions)
F_B = 100,000 lbs
Flex/ball joint stiffness = 50,000 ft-lb/degree

Note the change in curvature caused by the flex ball joint. There is a point of inflection located at $\zeta = 0.0627$. When the flex joint stiffness factor, $\frac{K_\phi L}{2EI}$, is zero, the algebraic pair (Eqs. (4-31) and (4-33)) revert back to the pinned bottom end algebraic pair (Eqs. (4-20) and (4-21)). The rotation across the flex ball joint is

$$\frac{dy}{dx}(0) = \frac{dy}{Ld\zeta}(0) = \frac{a_1}{L}, \text{ radians}$$

Using the value of a_1 = 83.376 (ft) as determined from the solution, the angle of rotation across the flex/ball joints is determined by

$$\frac{dy}{dx}(0) = \frac{83.376}{1500}\frac{180}{\pi}\text{ deg}$$

$$\Delta\theta = \frac{dy}{dx}(0) = 3.1847 \text{ deg}$$

Note the slight decrease in $\Delta\theta$ compare with the previous example ($K_\phi = 0$). The bending moment applied to the well head is

$$M(0) = K_\phi \Delta\theta$$
$$M(0) = 50{,}000(3.187) = 159{,}237 \text{ ft-lbs}$$

Bending moment is also calculated by

$$M(0) = EI \frac{d^2y}{dx^2}(0)$$

$$M(0) = EI \frac{1}{L^2} \frac{d^2y}{d\zeta^2}(0)$$

$$M(0) = \frac{EI}{L^2} 2a_2$$

$$M(0) = 159{,}237 \text{ ft} - \text{lbs}$$

This number agrees with the bending moment determined by flex joint stiffness. The corresponding bending stress in the riser at the bottom end is 8750 psi.

The shear force at the lower end is determined as follows.

$$V(0) = EI \frac{d^3y}{dx^3}(0) - F_B \frac{dy}{dx}(0)$$

$$V(0) = \frac{EI}{L^3} \frac{d^3y}{d\zeta^3}(0) - \frac{F_B}{L} \frac{dy}{d\zeta}(0)$$

$$V(0) = \frac{EI}{L^3} 6a_3 - F_B a_1$$

$$V(0) = -8{,}169 \text{ lbs}$$

The minus sign means the shear force on the riser is to the left and the shear force on the well head is to the left.

4.1.3.3 *Fixed Lower End*

It is well known that structural stiffness attracts bending moment, so it is unlikely that a riser pipe would be attached directly to a well head without special connections. This unlikely case is illustrated next to show stress levels. The boundary conditions in this case are

$$y(0) = 0 \tag{4-34a}$$

$$\frac{dy}{dx}(0) = 0 \tag{4-34b}$$

$$y(1) = \delta \tag{4-34c}$$

$$\frac{d^2y}{dx^2}(1) = 0 \tag{4-34d}$$

The recursion formula for the coefficients for this case is

$$a_n = \frac{\beta}{n(n-1)}a_{n-2} + \frac{\alpha(n-3)}{n(n-1)(n-2)}a_{n-3} \tag{4-35}$$

Values of a_n can be expressed by linear combinations of a_0, a_1, a_2, a_3 by

$$a_n = F_n a_0 + G_n a_1 + H_n a_2 + I_n a_3 \tag{4-36}$$

It follows that the values of $F_n G_n H_n I_n$ for $n = 0,1,2,3$ are given by Eq. (2-110). For $n \geq 4$,

$$F_n = \frac{\beta}{n(n-1)}F_{n-2} + \frac{(n-3)\alpha}{n(n-1)(n-2)}F_{n-3}, \text{ etc.} \tag{4-37}$$

which can be grouped as follows

$$\begin{Bmatrix} F_n \\ G_n \\ H_n \\ I_n \end{Bmatrix} = \frac{\beta}{n(n-1)}\begin{Bmatrix} F_{n-2} \\ G_{n-2} \\ H_{n-2} \\ I_{n-2} \end{Bmatrix} + \frac{(n-3)\alpha}{n(n-1)(n-2)}\begin{Bmatrix} F_{n-3} \\ G_{n-3} \\ H_{n-3} \\ I_{n-3} \end{Bmatrix} \tag{4-38}$$

The first two boundary conditions given in Eq. (4-34) give $a_0 = a_1 = 0$. Following the procedure of Chapter 2, the other two are determined from.

$$\left[\sum_{n=2}^{\infty} H_n\right]a_2 + \left[\sum_{n-2}^{\infty} I_n\right]a_3 = \delta \tag{4-39}$$

$$\left[\sum_{n=2}^{\infty} n(n-1)H_n\right]a_2 + \left[\sum_{n-2}^{\infty} n(n-1)I_n\right]a_3 = 0 \tag{4-40}$$

a_0, a_1, a_2, and a_3 are now known.

$$\begin{aligned} a_0 &= 0 \\ a_1 &= 0 \\ a_2 &= 1090.13 \\ a_3 &= -6056.36 \end{aligned}$$

All other a's can be determined by repeated application of the recurrence formula, Eq. (4-35) with α and β the same as before. The set of input data used in the previous example is used in this case, giving riser displacements as shown in Fig. 4-6.

The maximum curvature is at $x = 0$.

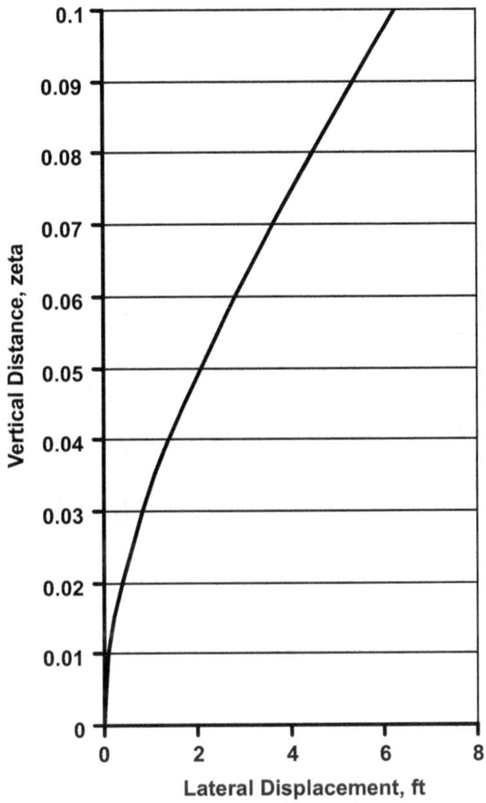

Figure 4-6. Riser displacement; bottom support fixed (x = 150 ft, ζ = 0.1).

$$\frac{1}{\rho} = \frac{d^2y}{dx^2}(0) = \frac{2a_2}{L^2} \tag{4-41}$$

$$\sigma_b = E\frac{c}{\rho} = Ec\frac{d^2y}{dx^2}(0) = Ec2\frac{a_2}{L^2}\left|\frac{1\text{ ft}}{12\text{ in}}\right|$$

The value of a_2 = 1090.13 (ft) is established from the determination of the a_n s.

$$\sigma_b = Ec2\frac{1090.13}{L^2}\left|\frac{1\text{ ft}}{12\text{ in}}\right| = 29(10)^6 1 1(2)\frac{1090.13}{(1500)^2}\frac{1}{12} = 25{,}759\text{ psi}$$

Parameters, L and a_2, have units of feet. The corresponding axial stress is σ_a = 2383 psi.

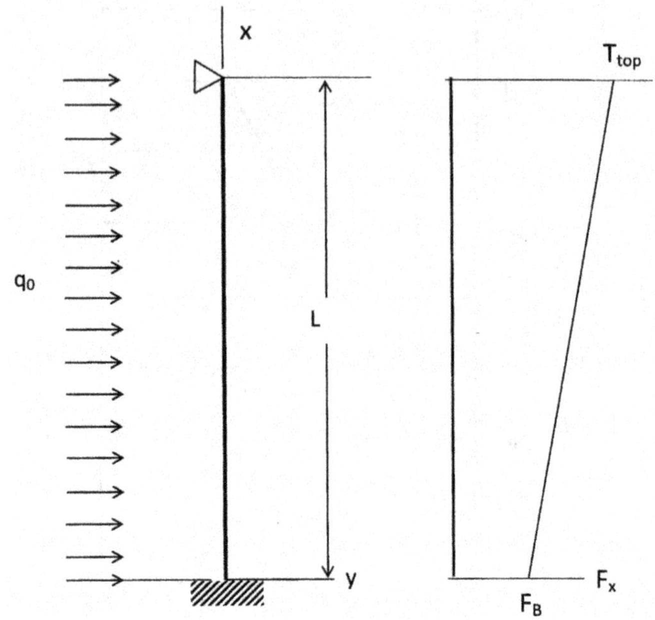

Figure 4-7. Schematic of current loaded riser (bottom end fixed).

The shear force at $x = 0$ is

$$V(0) = EI\frac{d^3y}{dx^3}(0) - F_B\frac{dy}{dx}(0) \tag{4-42}$$

With the value of $a_3 = -6056.36$ from the solution

$$V(0) = \frac{EI}{L^3}\frac{d^3y}{d\zeta^3}(0) = \frac{EI}{L^3}(6a_3)$$

$$V(0) = \frac{29(10)^6 2399}{1500^3}6(-6056.36)\left|\frac{1\text{ ft}^2}{144\text{ in}^2}\right|$$

$$V(0) = -5{,}202\text{ lb}$$

The minus sign means the shear force is applied to the end of the riser in the minus y direction. The force applied to the well head is to the right. Well head equipment has to accommodate both moment and shear loads.

4.1.4 Deflections Caused by Uniform Current Loading

Ocean currents apply significant side load to risers. This type of loading will now be considered. A schematic of the mathematical model is shown in Fig. 4-7.

4.1.4.1 Differential Equation of Bending

The riser is assumed to be structurally fixed (zero slope) at the bottom end and simply supported at the top, with no rig offset. The following boundary conditions apply.

$$y(0) = 0 \tag{4-43a}$$

$$\frac{dy}{dx}(0) = 0 \tag{4-43b}$$

$$y(L) = 0 \tag{4-43c}$$

$$\frac{d^2y}{dx^2}(L) = 0 \tag{4-43d}$$

The differential equation of bending is

$$EI\frac{d^4y}{dx^4} - \frac{d}{dx}\left[(b+ax)\frac{dy}{dx}\right] = q(x) \tag{4-44}$$

where

$$q(x) = C_d \frac{\gamma_0 d_0}{2g} v_0^2(x)$$

For example, if v = 1.5 fps and assuming C_d = 1.33, the rate of side loading is

$$q = 1.33\frac{64(22)}{2(32.2)}(1.5)^2 \left|\frac{1\,ft}{12\,in}\right| = 5.47 \text{ lb/ft}$$

For the special case of uniform rate of side loading, $q(x)$ will be taken as q_0, having units of lb/ft. Eq. (4-44) can now be written as

$$\frac{d^4y}{d\varsigma^4} - \alpha\zeta\frac{d^2y}{d\varsigma^2} - \beta\frac{d^2y}{d\varsigma^2} - \alpha\frac{dy}{d\varsigma} = f(x) \tag{4-45}$$

where

$$f(x) = \frac{q(x)L^4}{EI}$$

where $f(x)$ has units of length. Assuming uniform current velocity with depth

$$\frac{d^4y}{d\zeta^4} - \alpha\zeta\frac{d^2y}{d\zeta^2} - \beta\frac{d^2y}{d\zeta^2} - \alpha\frac{dy}{d\zeta} = f_0 \tag{4-46}$$

where

$$f_0 = \frac{q_0L^4}{EI}$$

Applying the above set of numbers gives

$$f_0 = \frac{5.47(1500)^4}{29(10)^6 2399}\left|\frac{144 \text{ in}^2}{1 \text{ ft}^2}\right| = 57{,}320 \text{ ft} \tag{4-47}$$

4.1.4.2 Method of Solution

A particular solution to Eq. (4-46) is

$$y_p = -\frac{f_0}{\alpha}\zeta \tag{4-48}$$

The complimentary solution is the solution to

$$\frac{d^4y}{d\zeta^4} - \alpha\zeta\frac{d^2y}{d\zeta^2} - \beta\frac{d^2y}{d\zeta^2} - \alpha\frac{dy}{d\zeta} = 0 \tag{4-49}$$

As before, the solution to Eq. (4-49) is taken as

$$y_c = \sum_{n=0}^{\infty} a_n\zeta^n \tag{4-50}$$

The total solution to Eq. (4-46) is therefore

$$y(\zeta) = \sum_{n=0}^{\infty} a_n\zeta^n - \frac{f_0}{\alpha}\zeta \tag{4-51}$$

Using the first two boundary conditions, $a_0 = 0$ and $a_1 = \frac{f_0}{\alpha}$. The other two constants, a_2 and a_3, are determined as follows:

The third boundary condition gives

$$\sum_{2} a_n = 0 \tag{4-52}$$

The fourth boundary condition gives

$$\sum_{2} a_n n(n-1) = 0 \tag{4-53}$$

Consider now Eq. (4-49) and its solution Eq. (4-50). After substitution, each coefficient must be equal to zero to satisfy the Eq. (4-49) giving the following recurrence equation.

$$a_n = \frac{\beta}{n(n-1)}a_{n-2} + \frac{\alpha(n-3)}{n(n-1)(n-2)}a_{n-3} \tag{4-54}$$

where $n \geq 4$. By repeated applications of this recurrence equation, starting with $n = 4$, a_n can be expressed as linear combinations of a_0, a_1, a_2, and a_3. Thus in general

$$a_n = F_n a_0 + G_n a_1 + H_n a_2 + I_n a_3 \tag{4-55}$$

Applying Eq. (2-110) for $n = 0,1,2,3$ and for $n \geq 4$

$$\begin{Bmatrix} F_n \\ G_n \\ H_n \\ I_n \end{Bmatrix} = \frac{\beta}{n(n-1)} \begin{Bmatrix} F_{n-2} \\ G_{n-2} \\ H_{n-2} \\ I_{n-2} \end{Bmatrix} + \frac{(n-3)\alpha}{n(n-1)(n-2)} \begin{Bmatrix} F_{n-3} \\ G_{n-3} \\ H_{n-3} \\ I_{n-3} \end{Bmatrix} \tag{4-56}$$

Equation 4-56 provides a means of determining all values of F_n, G_n, H_n, I_n which are needed in Eq. (4-55). The coefficients in Eq. (4-50) are expressed in terms of a_0, a_1, a_2, and a_3, as shown by Eq. (4-55).

Substituting Eq. (4-55) into Eq. (4-52) and noting that $a_0 = 0$ gives

$$\sum_{n=2}^{\infty} H_n a_2 + \sum_{n=2}^{\infty} I_n a_3 = -\sum_{n=2}^{\infty} G_n a_1 \tag{4-57}$$

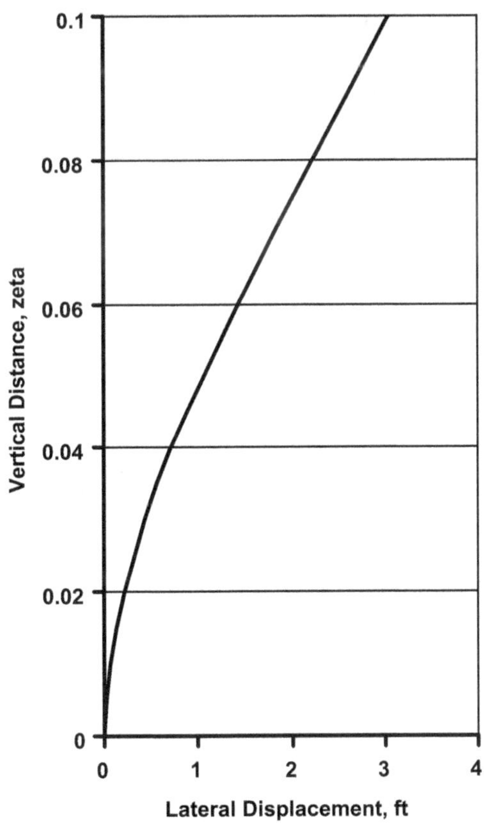

Figure 4-8. Riser deflection due to uniform current. (x = 150 ft, ζ = 0.1).

and from Eq. (4-53)

$$\sum_{n=2}^{\infty} n(n-1)H_n a_2 + \sum_{n=2}^{\infty} n(n-1)I_n a_3 = -\sum_{n=2}^{\infty} n(n-1)G_n a_1 \tag{4-58}$$

where

$$a_1 = \frac{f_0}{\alpha}$$

The algebraic Eqs. (4-57) and (4-58) are solved to obtain the remaining two constants, a_2 and a_3. The set is now complete.

$$\begin{aligned} a_0 &= 0 \\ a_1 &= 44.43 \\ a_2 &= 585.42 \\ a_3 &= -3774.82 \end{aligned}$$

Figure 4-8 gives a plot of the deflection at the lower end of the riser were bending stresses are greatest. The input data is the same as before with q_0 = 5.47 lb/ft (f_0 = 57,300).

Bending stress is determined from local curvature

$$\sigma_b = E\frac{c}{\rho} = Ec\frac{d^2y}{dx^2} = E\frac{c}{L^2}\frac{d^2y}{d\varsigma^2}\left|\frac{1 \text{ ft}}{12 \text{ in}}\right| \tag{4-59}$$

where

$$\frac{d^2y}{d\varsigma^2} = \sum_2 n(n-1)a_n\varsigma^{n-2} \tag{4-60}$$

At the bottom end of the riser.

The bending moment is

$$M(0) = EI\frac{d^2y}{dx^2}(0) = \frac{EI}{L^2}\frac{d^2y}{d\zeta^2}(0) \tag{4-61}$$

where

$$\frac{d^2y}{d\zeta^2}(0) = 2a_2 \tag{4-62}$$

For this particular example

$$M(0) = \frac{29(10)^6 2399}{1500^2} 2(585.4)\left|\frac{1 \text{ ft}^2}{144 \text{ in}^2}\right| \text{lb-ft}$$

$$M(0) = 251{,}400 \text{ lb-ft}$$

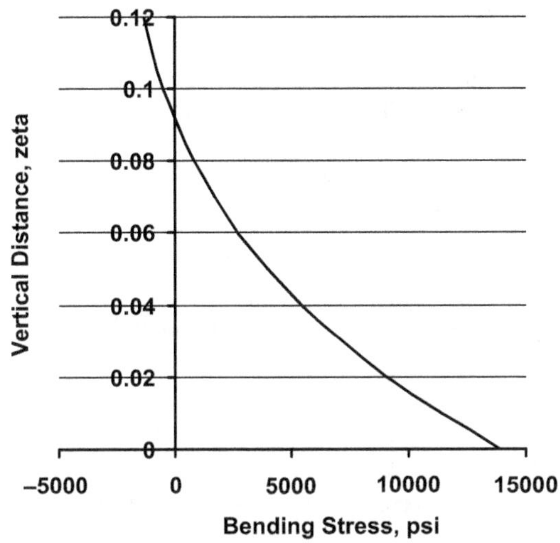

Figure 4-9. Bending stress due to uniform current. ζ = 0.12 corresponds to *x* = 180 ft.

and all a_n's have units of length (ft). Shear force at the base of the riser pipe is

$$V(0) = EI\frac{d^3y}{dx^3}(0) - F_B\frac{dy}{dx}(0) \tag{4-63}$$

$$V(0) = \frac{EI}{L^3}\frac{d^3y}{d\zeta^3}(0) = \frac{EI}{L^3}(6a_3) \tag{4-64}$$

$$V(0) = \frac{29(10)^6 2399}{1500^3}6(-3775)\left|\frac{1\ \text{ft}^2}{144\ \text{in}^2}\right|\text{lbs}$$

$$V(0) = -3242\ \text{lbs}$$

The minus sign means the shear on the riser pipe is to the left, which means the shear force on the well head equipment is to the right. The total side force on the entire 1500 ft long riser is Q = 5.47(1500) = 8,205 lbs. The difference between Q and $V(0)$ is 8205–3242 = 4,963 lbs. This horizontal force is taken by the rig support to the riser.

The bending stress distribution over the lower 180 ft is shown in Fig. 4-9. Note the point of inflection around 140 ft.

Riser deflection due to other current distribution and be determined in a similar way. Also combinations of current and rig offset can be superimposed. The challenge in each case is to establish particular solutions that satisfied Eq. (4-45).

4.1.4.3 Effective Tension vs Average Effective Tension

It is of interest to compare bending stresses corresponding to variable tension and uniform average tension. This comparison provides some insight into the expected accuracy by assuming uniform average tension. Deflection and bending stresses will now be determined based on the assumption of uniform tension equal to the average effective tension.

Recall the expression for effective tension for the variable tension case is

$$T_{\text{eff}} = \left[(F_B + wx) + (L - x)(A_0\gamma_0 - A_i\gamma_i)\right] \tag{4-65}$$

The assumed uniform tension will be

$$T_{\text{ave}} = (T_{eff})_{\text{ave}} = \frac{(T_{eff})_{\text{top}} + (T_{eff})_{\text{bottom}}}{2} \tag{4-66a}$$

$$T_{\text{ave}} = \left[\left(F_B + w\frac{L}{2}\right) + \frac{L}{2}(A_0\gamma_0 - A_i\gamma_i)\right] \tag{4-66b}$$

As a side note, if $\gamma_0 = \gamma_i$

$$T_{eff} = F_B + wL - \frac{1}{2}LwBF \tag{4-67}$$

Using Eq. (4-65) as tension, Eq. (4-24) simplifies to

$$EI\frac{d^4y}{dx^4} - T_{\text{ave}}\frac{d^2y}{dx^2} = q(x) \tag{4-68}$$

and

$$\frac{d^4y}{d\zeta^4} - \beta_{\text{ave}}\frac{d^2y}{d\zeta^2} = f_0 \tag{4-69}$$

where, in this case,

$$\alpha = 0$$

$$\beta_{\text{ave}} = \beta = \frac{T_{\text{ave}}L^2}{EI}$$

$$f_0 = \frac{q_0L^4}{EI} \text{ (units of length)}$$

The solution to Eq. (4-69) can be solved by standard methods. Direct integration of Eq. (4-69) gives

$$\frac{d^2y}{d\zeta^2} - \beta y = \frac{1}{2} f_0 \zeta^2 + C_1 \zeta + C_2 \qquad (4\text{-}70)$$

A particular solution is

$$y_p = -\frac{1}{\beta}\left[\frac{1}{2} f_0 \zeta^2 + C_1 \zeta + C_2 + \frac{f_0}{\beta}\right] \qquad (4\text{-}71)$$

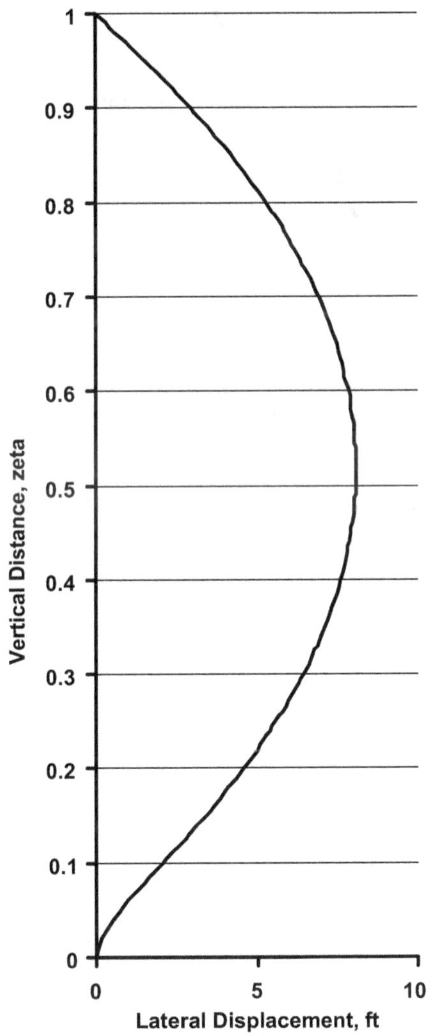

Figure 4-10. Riser displacement based on average tension. $\zeta = 1$ corresponds to L = 1500 ft.

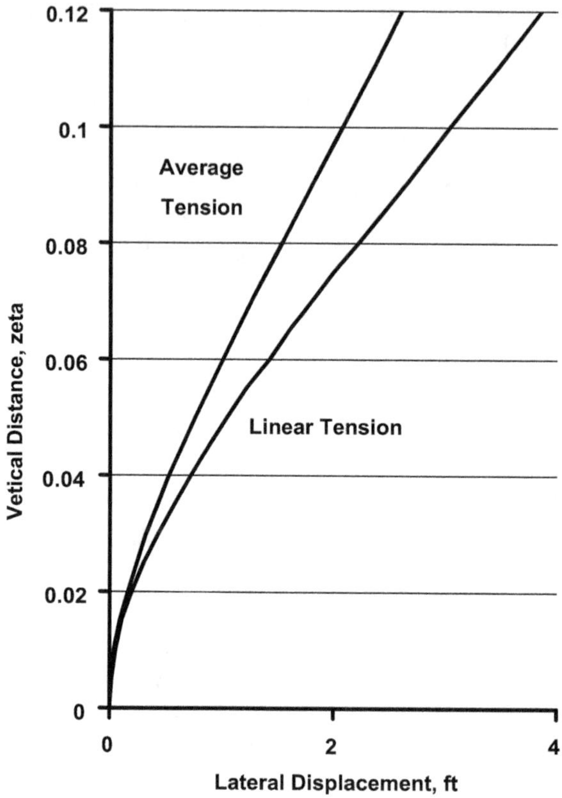

Figure 4-11. Comparison of riser displacement predictions.

The complimentary solution is

$$y_c = C_3 e^{\sqrt{\beta}\zeta} + C_4 e^{-\sqrt{\beta}\zeta} \tag{4-72}$$

The total solution becomes

$$y = y_p + y_c \tag{4-73}$$

The boundary conditions given by Eq. (4-43) also apply here and lead to the following set of algebraic equations.

$$\begin{bmatrix} 0 & -\dfrac{1}{\beta} & 1 & 1 \\ -\dfrac{1}{\beta} & 0 & \sqrt{\beta} & -\sqrt{\beta} \\ -\dfrac{1}{\beta} & -\dfrac{1}{\beta} & e^{\sqrt{\beta}} & e^{-\sqrt{\beta}} \\ 0 & 0 & \beta e^{\sqrt{\beta}} & \beta e^{-\sqrt{\beta}} \end{bmatrix} \begin{Bmatrix} C_1 \\ C_2 \\ C_3 \\ C_4 \end{Bmatrix} = \begin{Bmatrix} \dfrac{f_0}{\beta^2} \\ 0 \\ \dfrac{f_0}{2\beta} + \dfrac{f_0}{\beta^2} \\ \dfrac{f_0}{\beta} \end{Bmatrix} \tag{4-74}$$

Table 4-1. Comparison of bending stresses due to current side loading.

Zeta	Linear tension bending stress (psi)	Average tension bending stress (psi)
0	13833	11407
0.01	11309	8372
0.02	9079	6093
0.03	7127	4384
0.04	5434	3102
0.05	3978	2141
0.06	2738	1421
0.07	1693	882
0.08	821	479
0.09	103	179
0.1	–481	–45
0.11	–950	–211
0.12	–1318	–333

Once the *C*'s are determined for a given set of input numbers, the displacement function and its derivatives are defined.

Displacements over the entire length of the 1500-ft riser are shown in Fig. 4-10. The critical bending region is at the lower end where the riser is structurally fixed (zero slope). This is the area of highest bending stress. Deflections for both average tension and linear tension over the lower 180 ft of the riser are shown in Fig. 4-11. Riser deflections based on linear tension predict higher curvature and bending stresses at the bottom connection as shown in Table 4-1.

The numbers show a point of inflection in the deflection of both cases near $\zeta \sim 0.095$.

4.1.5 Tapered Flex Joints

It is fundamental that structural stiffness attracts bending moment. This makes the top and bottom connections critical areas in the design and operation of risers. Ball joints eliminate bending but the rotation angle across ball joints could create problems for drillstrings and production tubulars and equipment. Most of the riser is relatively flexible. Bending (static and dynamic) over much of the riser is controlled by top tension.

Tapered flex-joints provide a more balanced transition between the flexible riser and mudline equipment. Taper flex joints are typically truncated cones with constant inside diameters but the outside diameter varies linearly from top to bottom (see Fig. 4-12). The method of analysis developed by Dareing [3] and Gordon and Dareing [4] allows tapered stress joints to be analyzed by

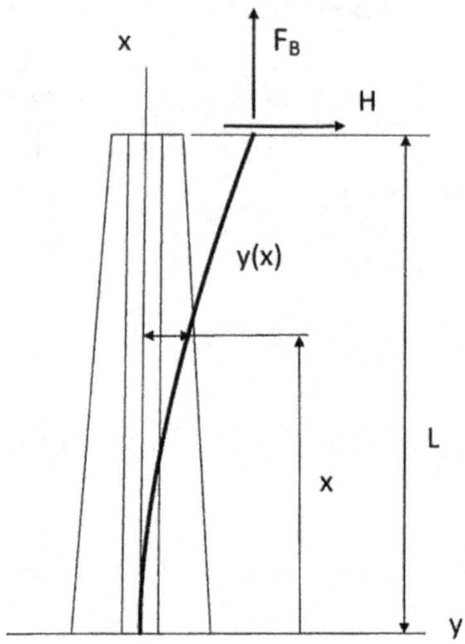

Figure 4-12. Tapered flex joint.

solving one differential equation. Solutions to deflections and bending are obtained in close form.

4.1.5.1 Equation of Bending

Consider a tapered cantilever beam subjected to a transverse force, H and a pull full, F_B at the free end as shown in Fig. 4-12. The differential equation of bending for the beam is

$$EI(x)\frac{d^2y}{dx^2} - F_B y = (L - x)H - F_B\delta \tag{4-75}$$

Gravity and hydrostatic forces are dropped because their effects are small by comparison with forces applied at the top (F_B and H). The solution to this equation is straight forward if there is no pull force, but this force is central to riser operations.

The solution to Eq. (4-75) is also complicated by the variable coefficient, $I(x)$, which accounts for the variation of the cross sectional moment of inertia. By approximating the straight-tapered stress joint with a tubular having a parabolic moment of inertia distribution, the bending equation can be cast into an Euler type differential equation that is amenable to a closed form solution.

4.1.5.2 Parabolic Approximation to Moment of Inertia

If the distribution of the moment of inertia along the column is assumed to be of the form

$$I(x) = I_P\left(\frac{h - x}{a}\right)^2 \tag{4-76}$$

where

$$h = \frac{L}{1 - \sqrt{\frac{I_p}{I_{sJ}}}}$$

and

$$a = h\sqrt{\frac{I_p}{I_{sJ}}}$$

where

I_p — moment of inertia of riser pipe
I_{SJ} — moment of inertia of bottom end of tapered stress joint

the differential equation of bending becomes

$$\frac{EI_p}{a^2}(h-x)^2\frac{\mathrm{d}^2y}{\mathrm{d}x^2} - F_By = (L-x)H - F_B\delta \tag{4-77}$$

Because $I(x)$ is assumed to vary parabolically, the associated stress joint will be referred to as a "parabolic stress joint". Table 4-2 compares the parabolic moment of inertia with the moment of inertia of a straight taper over a 30-ft length.

4.1.5.3 Solution to Differential Equation

The solution to Eq. (4-77) has particular and complimentary components

$$y(x) = y_P(x) + y_C(x) \tag{4-78}$$

The particular solution is

$$y_P(x) = -(L-x)\frac{H}{F_B} + \delta \tag{4-79}$$

To determine the complementary solution it is necessary to perform certain substitutions. By definition

$$\gamma^2 = \frac{a^2F_B}{EI_P} \tag{4-80}$$

The differential equation giving the complementary solution is

$$(h-x)^2\frac{\mathrm{d}^2y}{\mathrm{d}x^2} + \gamma^2y = 0 \tag{4-81}$$

This is a linear differential equation containing a variable coefficient and is of the Euler type.

The earlier approximation for $I(x)$, using a parabolic function (Eq. (4-76)), was made to take advantage of the classic method of solving the Euler type differential equation.

Table 4-2. Comparison of wall thickness and moment of inertia over a 30-ft flex joint. (24" OD at base; 20.5" at interface with riser pipe having 22" OD).

	Wall thickness in.		Moment of inertia, in^4	
Joint height ft	Parabolic	Straight	Parabolic	Straight
0	1.500	1.500	6301	6301
1	1.473	1.475	6164	6174
2	1.446	1.450	6027	6048
3	1.419	1.425	5892	5923
4	1.392	1.400	5759	5798
5	1.365	1.375	5627	5674
6	1.339	1.350	5497	5551
7	1.313	1.325	5369	5429
8	1.286	1.300	5241	5308
9	1.260	1.275	5116	5187
10	1.234	1.250	4992	5067
11	1.208	1.225	4869	4948
12	1.183	1.200	4748	4830
13	1.157	1.175	4629	4712
14	1.132	1.150	4511	4596
15	1.106	1.125	4394	4480
16	1.081	1.100	4279	4365
17	1.056	1.075	4166	4250
18	1.032	1.050	4054	4136
19	1.007	1.025	3943	4023
20	0.983	1.000	3835	3911
21	0.959	0.975	3727	3800
22	0.935	0.950	3621	3689
23	0.911	0.925	3517	3579
24	0.887	0.900	3414	3470
25	0.864	0.875	3313	3361
26	0.841	0.850	3213	3254
27	0.818	0.825	3115	3147
28	0.795	0.800	3018	3040
29	0.772	0.775	2923	2935
30	0.750	0.750	2830	2830

The solution starts by eliminating the variable coefficient as follows.

$$(h-x)=e^{z} \tag{4-82}$$

Differentiating, gives

$$-\mathrm{d}x = e^{z}\mathrm{d}z$$

and

$$\frac{\mathrm{d}z}{\mathrm{d}x}=-\frac{1}{e^{z}} \tag{4-83}$$

Note that

$$\frac{\mathrm{d}y}{\mathrm{d}x}=\frac{\mathrm{d}y}{\mathrm{d}z}\frac{\mathrm{d}z}{\mathrm{d}x}=-\frac{1}{e^{z}}\frac{\mathrm{d}y}{\mathrm{d}z} \tag{4-84}$$

$$\frac{\mathrm{d}^2y}{\mathrm{d}x^2}=-\frac{1}{e^{2z}}\left(\frac{\mathrm{d}^2y}{\mathrm{d}z^2}-\frac{\mathrm{d}y}{\mathrm{d}z}\right) \tag{4-85}$$

and

$$(h-x)^2=e^{2z} \tag{4-86}$$

Combining Eqs. (4-84) and (4-85) with Eq. (4-81) gives

$$\frac{\mathrm{d}^2y}{\mathrm{d}z^2}-\frac{\mathrm{d}y}{\mathrm{d}z}-\gamma^2 y=0 \tag{4-87}$$

The variable coefficient is gone and the solution to Eq. (4-87) can be obtained by assuming $y = Ce^{Dz}$. It follows that

$$(D^2-D-\gamma^2)Ce^{Dz}=0 \tag{4-88}$$

which leads to

$$D_{1,2}=\frac{1}{2}\left(1\pm\sqrt{1+4\gamma^2}\right) \tag{4-89}$$

The complementary solution is therefore

$$yc(x)=C_1e^{D_1z}+C_2e^{D_2z}=C_1(h-x)^{D_1}+C_2(h-x)^{D_2} \tag{4-90}$$

The total solution to Eq. (4-75) is now complete.

$$y(x)=C_1(h-x)^{D_1}+C_2(h-x)^{D_2}-(L-x)\frac{H}{F_B}+\delta \tag{4-91}$$

Equation (4-91) defines the deflection in the tapered flex joint. This closed form solution simplifies the simultaneous solution of stress joints coupled with marine risers as well as other offshore pipe lines and flow lines as will be shown later.

4.1.5.4 Example Calculation

As an example, consider the tapered flex joint defined in Table 4-2. The pull force and horizontal force at the top are assumed to be F_B = 75,000 lb and H = 50,000 lb. Parameters that enter into the deflection and bending stress calculations are

$$h = \frac{L}{1-\sqrt{\dfrac{I_p}{I_{SJ}}}} = 90.94$$

$$a = h\sqrt{\frac{I_p}{I_{sJ}}} = 60.94$$

$$\gamma^2 = \frac{a^2F_B}{EI_p} = \frac{60.94^2(75000)}{29(10)^6(2830)}\left|\frac{144in^2}{1ft^2}\right|$$

$$\gamma = 0.699$$

Applying boundary conditions

$$y(0) = 0 \tag{4-91a}$$

$$\frac{dy}{dx}(0) = 0 \tag{4-91b}$$

$$y(L) = \delta \tag{4-91c}$$

gives

$$h^{D_1}C_1 + h^{D_2}C_2 + \delta = L\frac{H}{F_B} \tag{4-92}$$

$$D_1h^{D_1-1}C_1D_2h^{D_2-1}C_2 = \frac{H}{F_B} \tag{4-93}$$

$$(h-L)^{D_1}C_1 + (h-L)^{D_2}C_2 = 0 \tag{4-94}$$

The solution to these three algebraic equations gives; C_1 = 0.0856, C_2 = –100.079 and δ = 0.417 ft. The stress distributions for his special case are given in Table 4-3. These numbers apply to an assumed stand-alone tapered flex joint (*L* = 30 ft) with assumed vertical and horizontal forces at the top. These equations will be used to interface tapered flex joints with uniform riser pipes.

Table 4-3. Deflection and Stress in Tapered Flex Joint. (Corresponds to data in Table 4-2; H = 50,000 lbs, F_B = 75,000 lbs).

Deflection $y(x)$, ft	Bending moment ft-lbs	Bending stress psi	Axial stress psi	Total tension stress* psi
0.0000	1468742	32865	723	33588
0.0006	1418785	32331	737	33068
0.0023	1368914	31778	750	32528
0.0051	1319127	31203	764	31967
0.0091	1269424	30607	779	31386
0.0141	1219803	29988	794	30782
0.0203	1170262	29344	809	30153
0.0275	1120801	28674	826	29500
0.0357	1071418	27977	842	28820
0.0449	1022110	27251	860	28111
0.0551	972877	26495	878	27373
0.0663	923717	25705	897	26602
0.0785	874627	24881	917	25798
0.0915	825605	24019	937	24957
0.1054	776650	23118	959	24077
0.1202	727759	22175	981	23156
0.1358	678930	21187	1005	22191
0.1522	630160	20150	1029	21179
0.1694	581446	19061	1055	20116
0.1873	532786	17917	1082	18999
0.2058	484175	16712	1110	17822
0.2249	435612	15442	1140	16582
0.2447	387092	14103	1172	15274
0.2649	338612	12687	1205	13892
0.2857	290168	11189	1240	12429
0.3069	241756	9602	1276	10878
0.3284	193370	7917	1316	9232
0.3502	145007	6125	1357	7482
0.3723	96662	4216	1401	5617
0.3945	48328	2179	1448	3627
0.4168	0	0	1498	1498

* Straight taper geometry used in stress calculations.

4.1.5.5 Hydrostatic Effects

Hydrostatic effects are not included in the above derivation. The justification for this is explained below. Starting with Eq. (2-85), the differential equation for the tapered flex joint is

$$\frac{d^2}{dx^2}\left[EI(x)\frac{d^2y}{dx^2}\right]-\frac{d}{dx}\left\{\left[(F_B+wx)+(L-x)(A_0\gamma_0-A_i\gamma_i)\right]\frac{dy}{dx}\right\}=0 \quad (4\text{-}95)$$

Within the effective tension term w and A_0 are both functions of x because the outside radius varies with x. The hydrostatic fluid effects as well as the weight are small in comparison with the direct pull force, F_B, and will be dropped for the sake of simplicity. Therefore,

$$\frac{d^2}{dx^2}\left[EI(x)\frac{d^2y}{dx^2}\right]-F_B\frac{d^2y}{dx^2}=0 \quad (4\text{-}96)$$

The first integration of Eq. (4-96) gives

$$EI(x)\frac{d^3y}{dx^3}-F_B\frac{dy}{dx}=C_1 \quad (4\text{-}97)$$

At $x = 0$ the slope is zero so $C_1 = -H$; see sign convention in Fig. 2-12. The second integration gives

$$EI(x)\frac{d^2y}{dx^2}-F_By=-Hx+C_2 \quad (4\text{-}98)$$

At $x = 0$ the bending moment is $LH - \delta F_B$ so $C_2 = LH - \delta F_B$ and

$$EI(x)\frac{d^2y}{dx^2}-F_By=(L-x)H-\delta F_B \quad (4\text{-}99)$$

which is the same equation as Eq. (4-75).

4.1.6 Interfacing Tapered Flex Joints with Uniform Riser Pipe

The closed form solution to the taper flex joint has enormous advantages over numerical methods when used in connection with riser pipes. This is illustrated with the following example. In this example, a riser pipe is connected to a well head by means of a tapered flex joint. The two solutions for: 1) tapered flex joints and 2) uniform riser pipes are merged for the special case of rig offset. Figure 4-13 illustrates the two components and coordinate system, where

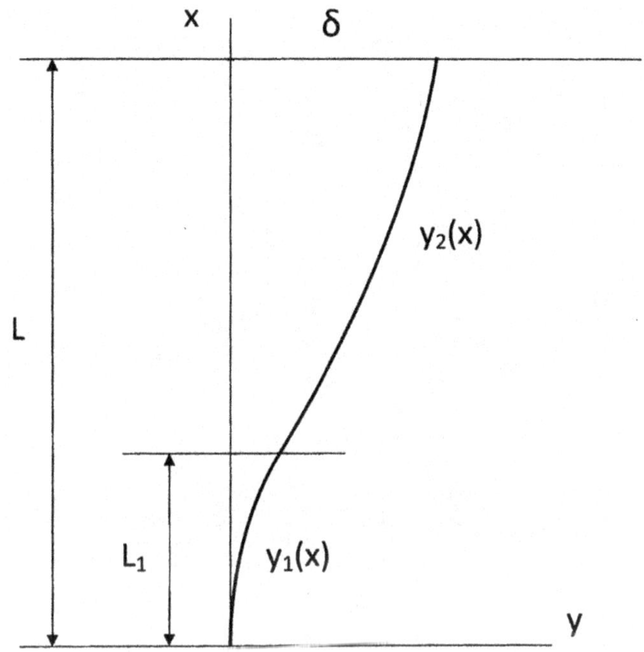

Figure 4-13. Riser pipe (y_2) connected to tapered flex joint (y_1).

$y_1(x)$ — tapered flex joint

$y_2(x)$ — uniform riser pipe

4.1.6.1 *Boundary Conditions*

The boundary conditions which link these solutions are

$$y_1(0) = 0 \tag{4-100a}$$

$$\frac{dy_1}{dx}(0) = 0 \tag{4-100b}$$

$$y_1(L_1) = y_2(\varsigma_1) \tag{4-100c}$$

$$\frac{dy_1}{dx}(L_1) = \frac{1}{L}\frac{dy_2}{d\zeta}(\varsigma_1) \tag{4-100d}$$

$$\frac{d^2y_1}{dx^2}(L_1) = \frac{1}{L^2}\frac{d^2y_2}{d\varsigma^2}(\varsigma_1) \tag{4-100e}$$

$$\frac{d^3y_1}{dx^3}(L_1) = \frac{1}{L^3}\frac{d^3y_2}{d\varsigma^3}(\varsigma_1) \tag{4-100f}$$

$$y_2(1) = \delta \tag{4-100g}$$

$$\frac{d^2y_2}{d\varsigma^2}(1) = 0 \tag{4-100h}$$

where

$$\zeta_1 = \frac{L_1}{L}$$

Note that for pipe under tension

$$\frac{dM}{dx} - F_x \frac{dy}{dx} - V = 0 \tag{4-101}$$

$$EI\frac{d^3y}{dx^3} - (wx + F_B)\frac{dy}{dx} - V = 0 \tag{4-102}$$

$$\frac{EI}{L^3}\frac{d^3y_2}{d\varsigma^3} - \frac{(wx + F_B)}{L}\frac{dy_2}{d\varsigma} - V = 0 \quad \text{(uniform riser pipe)} \tag{4-103}$$

$$EI_p\frac{d^3y_1}{dx^3} - F\frac{dy_1}{dx} - V = 0 \quad \text{(tapered flex joint)} \tag{4-104}$$

Since the tension, slope and both EI's are all the same at $x = L_1$, then

$$\frac{d^3y_1}{dx^3}(L_1) = \frac{1}{L^3}\frac{d^3y_2}{d\varsigma^3}(\varsigma_1) \tag{4-105}$$

Subscripts 1 refer to the tapered stress joint and subscripts 2 refer to the uniform riser pipe. The top boundary condition, δ, is the rig offset.

Earlier discussions considered separate solutions to tapered stress joints and to uniform riser pipes. The two will now be combined. Figure 4-13 illustrates the unified riser string.

4.1.6.2 *Tapered Flex Joint Section*

The differential equation of bending for the tapered flex joint is

$$\frac{d^2}{dx^2}\left[EI(x)\frac{d^2y_1}{dx^x}\right] - F_B\frac{d^2y_1}{dx^2} = 0 \tag{4-106}$$

The hydrostatic fluid terms in the effective tension have been dropped for simplicity. The solution to this differential equation is

$$y_1(x) = C_1(h-x)^{D_1} + C_2(h-x)^{D_2} + C_3x + C_4 \tag{4-107}$$

The four constants of integration are determined from boundary conditions.

4.1.6.3 *Uniform Riser Pipe Section*

Recall that the equation of bending for marine risers with uniform cross section is

$$\frac{d^4y_2}{d\zeta^4} - \frac{d}{d\zeta}\left[(\beta + \alpha\zeta)\frac{dy_2}{d\zeta}\right] = 0 \tag{4-108}$$

where

$$\zeta = \frac{x}{L}$$

Equation (4-108) is expanded to give

$$\frac{d^4 y_2}{dx^4} - \alpha\zeta \frac{d^2 y_2}{dx^2} - \beta \frac{d^2 y_2}{dx^2} - \alpha \frac{dy_2}{dx} = 0 \tag{4-109}$$

Assuming a solution of the form

$$y_2(\zeta) = \sum_{n=0}^{\infty} a_n \zeta^n \tag{4-110}$$

Following the steps outlined in Chapter 2 gives the following recurrence equation.

$$a_n = \frac{\beta}{n(n-1)} a_{n-2} + \frac{\alpha(n-3)}{n(n-1)(n-2)} a_{n-3} \tag{4-111}$$

where $n \geq 4$. By repeated applications of this recurrence equation, starting with $n = 4$, we can express a_n as linear combinations of a_0, a_1, a_2, and a_3. Thus, in general,

$$a_n = F_n a_0 + G_n a_1 + H_n a_2 + I_n a_3 \tag{4-112}$$

The values of $F_n G_n H_n I_n$ for $n = 0, 1, 2, 3$ are given by Eq. (2-110). These numbers are used in the subsequent calculations of the higher-order values of F_n, G_n, H_x, I_n, i.e. $n \geq 4$. The substitution of Eq. (4-112) into Eq. (4-111) for a_n, a_{n-3}, a_{n-4} gives

$$F_n = \frac{\beta}{n(n-1)} F_{n-2} + \frac{(n-3)\alpha}{n(n-1)(n-2)} F_{n-3}, \quad \text{etc.} \tag{4-113}$$

which can be grouped as follows

$$\begin{Bmatrix} F_n \\ G_n \\ H_n \\ I_n \end{Bmatrix} = \frac{\beta}{n(n-1)} \begin{Bmatrix} F_{n-2} \\ G_{n-2} \\ H_{n-2} \\ I_{n-2} \end{Bmatrix} + \frac{(n-3)\alpha}{n(n-1)(n-2)} \begin{Bmatrix} F_{n-3} \\ G_{n-3} \\ H_{n-3} \\ I_{n-3} \end{Bmatrix} \tag{4-114}$$

Equation (4-114) provides a means of determining all values of F_n, G_n, H_n, I_n which are needed in Eq. (4-112). Values for $F_n G_n H_n I_n$ less than $n = 4$ are given by Eq. (2-110).

The coefficients in the power series (Eq. (4-110)) can now be expressed in terms of a_0, a_1, a_2, and a_3. These four constants as well as four from the tapered flex joint solution are determined from the boundary conditions.

The solutions for y_1 and y_2 are combined through the boundary conditions to obtain overall deflection and stress distributions.

4.1.6.4 Merging the Two Solutions

The solution and derivatives for the tapered stress joint deflection are

$$y_1(x) = C_1(h-x)^{D_1} + C_2(h-x)^{D_2} + C_3x + C_4 \tag{4-115}$$

$$\frac{dy_1}{dx} = -C_1D_1(h-x)^{D_1-1} - C_2D_2(h-x)^{D_2-1} + C_3 \tag{4-116}$$

$$\frac{dy_1^2}{dx^2} = C_1D_1(D_1-1)(h-x)^{D_1-2} + C_2D_2(D_2-1)(h-x)^{D_2-2} \tag{4-117}$$

$$\frac{dy_1^3}{dx^3} = C_1D_1(D_1-1)(D_1-2)(h-x)^{D_1-3} + C_2D_2(D_2-1)(D_2-2)(h-x)^{D_2-3} \tag{4-118}$$

The solution and derivatives for the riser pipe deflection are

$$\begin{aligned} y_2(\zeta) &= a_0 + a_1\zeta + a_2\zeta^2 + a_3\zeta^3 + \left(\sum_{n=4}^{\infty} F_n\zeta^n\right)a_0 + \left(\sum_{n=4}^{\infty} G_n\zeta^n\right)a_1 \\ &+ \left(\sum_{n=4}^{\infty} H_n\zeta^n\right)a_2 + \left(\sum_{n=4}^{\infty} I_n\zeta^n\right)a_3 \end{aligned} \tag{4-119}$$

$$\begin{aligned} \frac{dy_2}{d\varsigma} &= a_1 + 2a_2\zeta + 3a_3\zeta^2 + \left(\sum_{n=4}^{\infty} nF_n\zeta^{n-1}\right)a_0 + \left(\sum_{n=4}^{\infty} nG_n\zeta^{n-1}\right)a_1 \\ &+ \left(\sum_{n=4}^{\infty} nH_n\zeta^{n-1}\right)a_2 + \left(\sum_{n=4}^{\infty} nI_n\zeta^{n-1}\right)a_3 \end{aligned} \tag{4-120}$$

$$\begin{aligned} \frac{d^2y_2}{d\varsigma^2} &= 2a_2 + 6a_3\zeta + \left(\sum_{n=4}^{\infty} n(n-1)F_n\zeta^{n-2}\right)a_0 + \left(\sum_{n=4}^{\infty} n(n-1)G_n\zeta^{n-2}\right)a_1 \\ &+ \left(\sum_{n=4}^{\infty} n(n-1)H_n\zeta^{n-2}\right)a_2 + \left(\sum_{n=4}^{\infty} n(n-1)I_n\zeta^{n-2}\right)a_3 \end{aligned} \tag{4-121}$$

$$\begin{aligned} \frac{d^3y_2}{d\varsigma^3} &= 6a_3 + \left(\sum_{n=4}^{\infty} n(n-1)(n-2)F_n\zeta^{n-3}\right)a_0 + \left(\sum_{n=4}^{\infty} n(n-1)(n-2)G_n\zeta^{n-3}\right)a_1 \\ &+ \left(\sum_{n=4}^{\infty} n(n-1)(n-2)H_n\zeta^{n-3}\right)a_2 + \left(\sum_{n=4}^{\infty} n(n-1)(n-2)I_n\zeta^{n-3}\right)a_3 \end{aligned} \tag{4-122}$$

All eight constants can now be determined from boundary conditions. Once C_1, C_2, C_3, C_4 are known, deflections and stresses in the tapered flex joint can be determined. Also, once a_0, a_1, a_2, and a_3 are determined all other a's can be determined by repeated application of the recurrence formula, Eq. (4-111) and the deflections and stresses can be determined throughout the riser pipe. Eight algebraic equations, which are based on boundary conditions, are listed below.

$$h^{D_1}C_1 + h^{D_2}C_2 + C_4 = 0$$

$$-D_1 h^{D_1-1}C_1 - D_2 h^{D_2-1}C_2 + C_3 = 0$$

$$(h-L_1)^{D_1}C_1 + (h-L_1)^{D_2}C_2 + L_1C_3 + C_4 - \left(1+\sum_4 F_n\varsigma_1^n\right)a_0 - \left(\varsigma_1 + \sum_4 G_n\varsigma_1^n\right)a_1$$
$$-\left(\varsigma_1^2 + \sum_4 H_n\varsigma_1^n\right)a_2 - \left(\varsigma_1^3 + \sum_4 I_n\varsigma_1^n\right)a_3 = 0$$

$$-D_1(h-L_1)^{D_1-1}C_1 - D_2(h-L_1)^{D_2-1}C_2 + C_3 - \left(\sum_4 nF_n\varsigma_1^{n-1}\right)\frac{a_0}{L} - \left(1+\sum_4 nG_n\varsigma_1^{n-1}\right)\frac{a_1}{L}$$
$$-\left(2\varsigma_1 + \sum nH_n\varsigma_1^{n-1}\right)\frac{a_2}{L} - \left(3\varsigma_1^n + \sum nI_n\varsigma_1^{n-1}\right)\frac{a_3}{L} = 0$$

$$D_1(D_1-1)(h-L_1)^{D_1-2}C_1 + D_2(D_2-1)(h-L_1)^{D_2-2}C_2 - \left(\sum_4 n(n-1)F_n\varsigma_1^{n-2}\right)\frac{a_0}{L^2}$$
$$-\left(\sum_4 n(n-1)G_n\varsigma_1^{n-2}\right)\frac{a_1}{L^2} - \left(2+\sum_4 n(n-1)H_n\varsigma_1^{n-2}\right)\frac{a_2}{L^2} - \left(6\varsigma_1 + \sum_4 n(n-1)I_n\varsigma_1^{n-2}\right)\frac{a_3}{L^2} = 0$$

$$D_1(D_1-1)(D_1-2)(h-L_1)^{D_1-3}C_1 + D_2(D_2-1)(D_2-2)(h-L_1)^{D_2-3}C_2$$
$$-\left(\sum_4 n(n-1)(n-2)F_n\varsigma_1^{n-3}\right)\frac{a_0}{L^3} - \left(\sum n(n-1)(n-2)G_n\varsigma_1^{n-3}\right)\frac{a_1}{L^3}$$
$$-\left(\sum n(n-1)(n-2)H_n\varsigma_1^{n-3}\right)\frac{a_2}{L^3} - \left(6+\sum n(n-1)(n-2)I_n\varsigma_1^{n-3}\right)\frac{a_3}{L^3} = 0$$

$$\left(1+\sum_4 F_n\right)a_0 + \left(1+\sum_4 G_n\right)a_1 + \left(1+\sum_4 H_n\right)a_2 + \left(1+\sum_4 I_n\right)a_3 = \delta$$

$$
\begin{bmatrix}
h^{D_1} & h^{D_2} & 0 & 1 & 0 & 0 & 0 & 0 \\
-D_1 h^{D_1-1} & -D_2 h^{D_2-1} & 1 & 0 & 0 & 0 & 0 & 0 \\
(h-L_1)^{D_1} & (h-L_1)^{D_2} & L_1 & 1 & 1 & \left(\varsigma_1+\sum G_n\varsigma_1^n\right) & \left(\varsigma_1^2+\sum H_n\varsigma_1^n\right) & \left(\varsigma_1^3+\sum I_n\varsigma_1^n\right) \\
-D_1(h-L_1)^{D_1-1} & -D_2(h-L_1)^{D_2-1} & 0 & 1 & 0 & \left(1+\sum nG_n\varsigma_1^{n-1}\right) & \left(2\varsigma_1+\sum nH_n\varsigma_1^{n-1}\right) & \left(3\varsigma_1^2+\sum nI_n\varsigma_1^{n-1}\right) \\
D_1(D_1-1)(h-L_1)^{D_1-2} & D_2(D_2-1)(h-L_1)^{D_2-2} & 0 & 0 & 0 & \sum n(n-1)G_n\varsigma_1^{n-2} & \left(2+\sum n(n-1)H_n\varsigma_1^{n-2}\right) & \left(6\varsigma_1+\sum n(n-1)I_n\varsigma_1^{n-2}\right) \\
D_1(D_1-1)(D_1-2)(h-L_1)^{D_1-3} & D_2(D_2-1)(D_2-2)(h-L_1)^{D_2-3} & 0 & 0 & 0 & \sum n(n-1)(n-2)G_n\varsigma_1^{n-3} & \sum n(n-1)(n-2)H_n\varsigma_1^{n-3} & \left(6+\sum n(n-1)(n-2)I_n\varsigma_1^{n-3}\right) \\
0 & 0 & 0 & 0 & 1 & \left(1+\sum G_n\right) & \left(1+\sum H_n\right) & \left(1+\sum I_n\right) \\
0 & 0 & 0 & 0 & 0 & \sum n(n-1)G_n & \left(2+\sum n(n-1)H_n\right) & \left(6+\sum n(n-1)I_n\right)
\end{bmatrix}
$$

$$\left(\sum_4 n(n-1)F_n\right)a_0+\left(\sum_4 n(n-1)G_n\right)a_1+\left(2+\sum_4 n(n-1)H_n\right)a_2$$
$$+\left(6+\sum_4 n(n-1)I_n\right)a_3=0$$

These equations are cast in a matrix equation, shown below, for ease of computation.

$$\left[\quad\quad\quad\quad\quad\quad\right]\begin{Bmatrix} C_1\\ C_2\\ C_3\\ C_4\\ a_0\\ a_1\\ a_2\\ a_3 \end{Bmatrix}=\begin{Bmatrix} 0\\ 0\\ 0\\ 0\\ 0\\ 0\\ \delta\\ 0 \end{Bmatrix} \tag{4-123}$$

The square matrix in Eq. (4-123) is expanded on page 238.

4.1.6.5 *Example Calculation*

A sample calculation was made to illustrate the outcome of this procedure. The numerical input is

L_1 =	30
L =	1500 ft
w =	135 lb/ft
γ_o =	64 lb/ft^3
mud wt	12 ppg
riser OD	22 in
riser ID	20.5 in
A_o =	2.6398 ft^2
A_i =	2.2921 ft^2
γ_i =	89.76 lb/ft^3
E =	2.90E + 07 psi
I =	2829.7 in^4
F_B =	150000 lb (plus is tension)
δ =	50 ft (rig offset)

See Table 4-2 for dimensions of tapered flex joint.

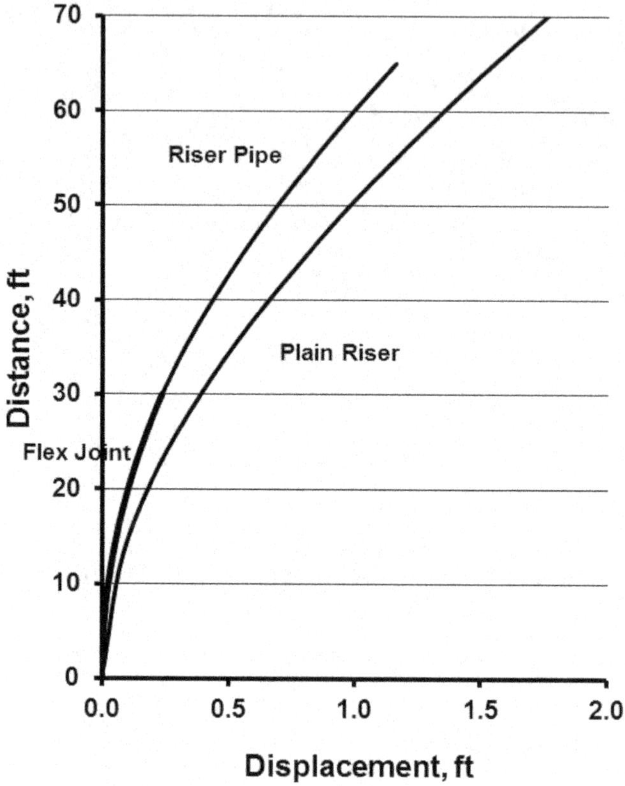

Figure 4-14. Deflections with and without tapered flex joint.

Figure 4-14 shows the predicted displacement over the 30 ft tapered flex joint and a portion of the adjoining riser pipe. The remaining part of the riser bends slightly back to the surface and matches the rig offset, $\delta = 50$ ft. The constants linking the two solutions are

$C_1 = 0.002463368$
$C_2 = 8.863968561$
$C_3 = 0.057714965$
$C_4 = -4.050889538$
$a_0 = 0.080871582$
$a_1 = -12.15691614$
$a_2 = 1143.526858$
$a_3 = -7931.035597$

The maximum bending stress in the flex joint is determined from

$$\sigma_b = E\frac{c}{\rho} = Ec\frac{d^2y_1}{dx^2}(0) \tag{4-124}$$

The curvature at $x = 0$ is determined as follows

$$\frac{dy_1^2}{dx^2} = C_1 D_1 (D_1 - 1)(h - x)^{D_1 - 2} + C_2 D_2 (D_2 - 1)(h - x)^{D_2 - 2} \tag{4-125}$$

Letting $x = 0$

$$\frac{dy_1^2}{dx^2}(0) = C_1 D_1 (D_1 - 1)(h)^{D_1 - 2} + C_2 D_2 (D_2 - 1)(h)^{D_2 - 2} \tag{4-126}$$

Using C_1 and C_2, determined earlier, gives the bending stress at the base of the flex joint of

$$(\sigma_b)_{bottom} = 12{,}700 \text{ psi.} \tag{4-127}$$

The corresponding axial stress is $(\sigma_a)_{bottom} = 1{,}226$ psi. The bending stress at the top of the tapered flex joint is $(\sigma_b)_{top} = 17{,}900$ psi, which is the same as the bending stress at the lower end of the riser pipe.

The solution without a tapered flex joint was given earlier (Fig. 4-6) and is also included in Fig. 4-14 for comparison. As shown earlier, the bending stress at the base of a uniform riser pipe (without a tapered flex joint) is 25,759 psi, which is substantially larger than the tapered flex joint stresses.

4.1.7 Broader Applications of the Closed Form Solution

The accuracy of the power series solution depends on the accuracy of the coefficients in the series. The series does converge rather quickly (usually less than 100 terms). The coefficients for $a_n \geq 0$ can be expressed in terms of the first four. As explained earlier, a_0, a_1, a_2, a_3, are first determined from boundary conditions. The remaining a_n's are determined using the recurrence formula. The algebraic calculations required to obtain a_0, a_1, a_2, a_3 depend on the $\sum F_n, \sum G_n, \sum H_n, \sum I_n$ values. Their precision is critical to the determination of the a's. Even with double precision, the closed form series approach is limited to water depths of 1500 – 2000 ft for typical riser sizes.

4.1.7.1 Combining Two or Multiple Sections

This numerical problem can be overcome to some extent by combining solutions to adjoining riser sections. For example, consider the riser represented by the sketch in Fig. 4-15 showing two sections, y_1 and y_2. Each section has reduced numerical values of α and β. The lower section is referenced from a global xy coordinate system. The deflection of the upper section is referenced to local xy_2 coordinate system.

The total solution is determined by matching boundary conditions at their common point. This leads to eight algebraic equations.

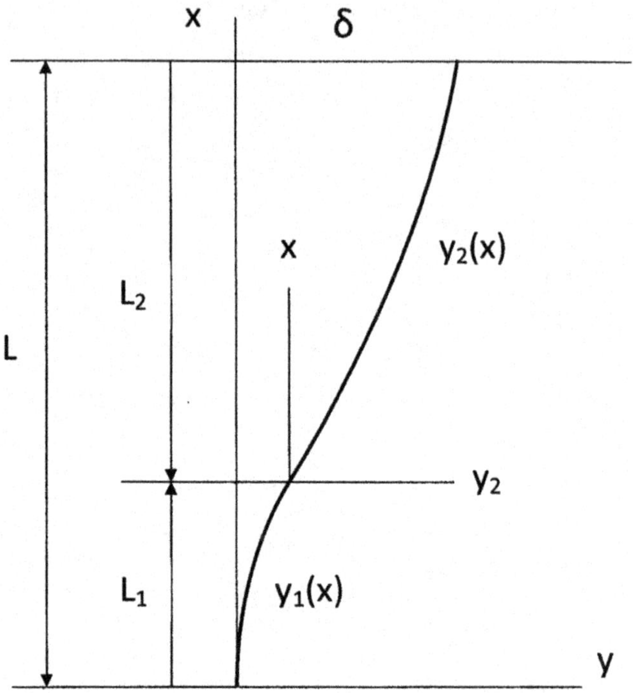

Figure 4 -15. Local/global coordinate systems.

The x coordinate will be replaced by the dimensionless coordinate, $\varsigma = \frac{x}{L}$. Here L is the total length of the riser or water depth. The location of the connection point between the two section is identified by $\varsigma_1 = \frac{L_1}{L}$ while the location of the top with respect to y_2x reference frame is identified by $\varsigma_2 = \frac{L_2}{L}$.

The boundary conditions are

$$y_1(0) = 0 \qquad y_1(0) = 0 \tag{4-128a}$$

$$\frac{d^2y_1}{dx^2}(0) = 0 \qquad \frac{d^2y_1}{d\varsigma^2}(0) = 0 \tag{4-128b}$$

$$y_2(0) = 0 \qquad y_2(0) = 0 \tag{4-128c}$$

$$\frac{dy_1}{dx}(L_1) = \frac{dy_2}{dx}(0) \qquad \frac{dy_1}{d\varsigma}(\varsigma_1) = \frac{dy_2}{d\varsigma}(0) \tag{4-128d}$$

$$\frac{d^2y_1}{dx^2}(L_1) = \frac{d^2y_2}{dx^2}(0) \qquad \frac{d^2y_1}{d\varsigma^2}(\varsigma_1) = \frac{d^2y_2}{d\varsigma^2}(0) \tag{4-128e}$$

$$\frac{d^3y_1}{dx^3}(L_1) = \frac{d^3y_2}{dx^3}(0) \qquad \frac{d^3y_1}{d\varsigma^3}(\varsigma_1) = \frac{d^3y_2}{d\varsigma^3}(0) \tag{4-128f}$$

$$\frac{d^2y_2}{dx^2}(L_2) = 0 \qquad \frac{d^2y_2}{d\varsigma^2}(\varsigma_2) = 0 \tag{4-128g}$$

$$y_2(L_2) = \delta - y_1(L_1) \qquad y_2(\varsigma_2) + y_1(\varsigma_1) = \delta \tag{4-128h}$$

There are eight (8) unknowns, $a_{10}a_{11}a_{12}a_{13}$ and $a_{20}a_{21}a_{22}a_{23}$. The mathematical expressions for $y_1(x)$ and $y_2(x)$ along with the boundary conditions, produce eight (8) linear algebraic equations.

Recall the differential equation of bending accounting for variable tension is

$$EI\frac{d^4y}{dx^4} - \frac{d}{dx}\left\{\left[(F_B + wx) + (L - x)(A_0\gamma_0 - A_i\gamma_i)\right]\frac{dy}{dx}\right\} = 0 \tag{4-129}$$

or

$$EI\frac{d^4y}{dx^4} - \frac{d}{dx}\left\{\left[(F_B + LA_0\gamma_0 - LA_i\gamma_i) + x(w - A_0\gamma_0 + A_i\gamma_i)\right]\frac{dy}{dx}\right\} = 0 \tag{4-130}$$

This equation will now be adapted to Sections #1 and #2.

Section #1 - $y_1(\varsigma)$

$$EI\frac{d^4y_1}{dx^4} - \frac{d}{dx}\left\{\left[(F_B + LA_0\gamma_0 - LA_i\gamma_i) + x(w - A_0\gamma_0 + A_i\gamma_i)\right]\frac{dy_1}{dx}\right\} = 0 \tag{4-131}$$

By letting $\varsigma = \frac{x}{L}$, Eq. (4-131) is rewritten as

$$\frac{d^4y_1}{d\zeta^4} - \frac{d}{d\zeta}\left[(\beta_1 + \alpha_1\zeta)\frac{dy_1}{d\zeta}\right] = 0 \tag{4-132}$$

where

$$\alpha_1 = \frac{(w - A_e\gamma_w + A_i\gamma_m)L^3}{EI}$$

$$\beta_1 = \frac{(F_B + LA_e\gamma_w - LA_i\gamma_m)L^2}{EI}$$

Equation (4-132) now becomes

$$\frac{d^4y_1}{d\varsigma^4} - \alpha_1\zeta\frac{d^2y_1}{d\varsigma^2} - \beta_1\frac{d^2y_1}{d\varsigma^2} - \alpha_1\frac{dy_1}{d\varsigma} = 0,\ 0 \le \varsigma \le \varsigma_1 \tag{4-133}$$

Its solution is of the form

$$y_1(\zeta) = \sum_{n=0}^{\infty} a_n \zeta^n \tag{4-134}$$

Substituting Eq. (4-134) and its derivatives into Eq. (4-133) gives a power series in ζ whose sum is equal to zero. Each coefficient must be equal to zero to satisfy Eq. (4-133) giving the following recurrence equation.

$$a_n = \frac{\beta_1}{n(n-1)} a_{n-2} + \frac{\alpha_1 (n-3)}{n(n-1)(n-2)} a_{n-3} \tag{4-135}$$

The solution continues as before.

$$y_1(\zeta) = a_{10} + a_{11}\zeta + a_{12}\zeta^2 + a_{13}\zeta^3 + \left(\sum_{n=4}^{\infty} F_{1n}\zeta^n\right) a_{10} + \left(\sum_{n=4}^{\infty} G_{1n}\zeta^n\right) a_{11} + \left(\sum_{n=4}^{\infty} H_{1n}\zeta^n\right) a_{12} + \left(\sum_{n=4}^{\infty} I_{1n}\zeta^n\right) a_{13} \tag{4-136}$$

$$\frac{dy_1}{d\varsigma} = a_{11} + 2a_{12}\zeta + 3a_{13}\zeta^2 + \left(\sum_{n=4}^{\infty} nF_{1n}\zeta^{n-1}\right) a_{10} + \left(\sum_{n=4}^{\infty} nG_{1n}\zeta^{n-1}\right) a_{11} + \left(\sum_{n=4}^{\infty} nH_{1n}\zeta^{n-1}\right) a_{12} + \left(\sum_{n=4}^{\infty} nI_{1n}\zeta^{n-1}\right) a_{13} \tag{4-137}$$

$$\frac{d^2y_1}{d\varsigma^2} = 2a_{12} + 6a_{13}\zeta + \left(\sum_{n=4}^{\infty} n(n-1)F_{1n}\zeta^{n-2}\right) a_{10} + \left(\sum_{n=4}^{\infty} n(n-1)G_{1n}\zeta^{n-2}\right) a_{11} + \left(\sum_{n=4}^{\infty} n(n-1)H_{1n}\zeta^{n-2}\right) a_{12} + \left(\sum_{n=4}^{\infty} n(n-1)I_{1n}\zeta^{n-2}\right) a_{13} \tag{4-138}$$

$$\frac{d^3y_1}{d\varsigma^3} = 6a_{13} + \left(\sum_{n=4}^{\infty} n(n-1)(n-2)F_{1n}\zeta^{n-3}\right) a_{10} + \left(\sum_{n=4}^{\infty} n(n-1)(n-2)G_{1n}\zeta^{n-3}\right) a_{11} + \left(\sum_{n=4}^{\infty} n(n-1)(n-2)H_{1n}\zeta^{n-3}\right) a_{12} + \left(\sum_{n=4}^{\infty} n(n-1)(n-2)I_{1n}\zeta^{n-3}\right) a_{13} \tag{4-139}$$

Section # 2 - $y_2(\varsigma)$

$$EI\frac{d^4y_2}{dx^4} - \frac{d}{dx}\left\{\left[(F_{B2} + L_2A_0\gamma_0 - L_2A_i\gamma_i) + x(w - A_0\gamma_0 + A_i\gamma_i)\right]\frac{dy_2}{dx}\right\} = 0 \tag{4-140}$$

where $F_{B2} = F_B + wL_1$ and $L_2 = L - L_1$. Applying $\varsigma = \dfrac{x}{L}$, Eq. (4-140) becomes

$$\frac{d^4y_2}{d\zeta^4}-\frac{d}{d\zeta}\left[(\beta_2+\alpha_2\zeta)\frac{dy_2}{d\zeta}\right]=0,\ 0\le\varsigma\le\varsigma_2 \tag{4-141}$$

where $\varsigma_2=\frac{L_2}{L}$ and

$$\alpha_2=\frac{(w-A_o\gamma_w+A_i\gamma_m)L^3}{EI}$$

$$\beta_2=\frac{(F_{B2}+L_2A_o\gamma_w-L_2A_i\gamma_m)L^2}{EI}$$

Notice that L_2 is inside the bracket while L is outside, a slight variation from the previous formula. This comes about because $dx = Ld\varsigma$. The solution proceeds as before.

$$y_2(\zeta)=a_{20}+a_{21}\zeta+a_{22}\zeta^2+a_{23}\zeta^3+\left(\sum_{n=4}^{\infty}F_{2n}\zeta^n\right)a_{20}+\left(\sum_{n=4}^{\infty}G_{2n}\zeta^n\right)a_{21}+\left(\sum_{n=4}^{\infty}H_{2n}\zeta^n\right)a_{22}+\left(\sum_{n=4}^{\infty}I_{2n}\zeta^n\right)a_{23} \tag{4-142}$$

$$\frac{dy_2}{d\varsigma}=a_{21}+2a_{22}\zeta+3a_{23}\zeta^2+\left(\sum_{n=4}^{\infty}nF_{2n}\zeta^{n-1}\right)a_{20}+\left(\sum_{n=4}^{\infty}nG_{2n}\zeta^{n-1}\right)a_{21}+\left(\sum_{n=4}^{\infty}nH_{2n}\zeta^{n-1}\right)a_{22}+\left(\sum_{n=4}^{\infty}nI_{2n}\zeta^{n-1}\right)a_{23} \tag{4-143}$$

$$\frac{d^2y_2}{d\varsigma^2}=2a_{22}+6a_{23}\zeta+\left(\sum_{n=4}^{\infty}n(n-1)F_{2n}\zeta^{n-2}\right)a_{20}+\left(\sum_{n=4}^{\infty}n(n-1)G_{2n}\zeta^{n-2}\right)a_{21}+\left(\sum_{n=4}^{\infty}n(n-1)H_{2n}\zeta^{n-2}\right)a_{22}+\left(\sum_{n=4}^{\infty}n(n-1)I_{2n}\zeta^{n-2}\right)a_{23} \tag{4-144}$$

$$\frac{d^3y_2}{d\varsigma^3}=6a_{23}+\left(\sum_{n=4}^{\infty}n(n-1)(n-2)F_{2n}\zeta^{n-3}\right)a_{20}+\left(\sum_{n=4}^{\infty}n(n-1)(n-2)G_{2n}\zeta^{n-3}\right)a_{21}+\left(\sum_{n=4}^{\infty}n(n-1)(n-2)H_{2n}\zeta^{n-3}\right)a_{22}+\left(\sum_{n=4}^{\infty}n(n-1)(n-2)I_{2n}\zeta^{n-3}\right)a_{23} \tag{4-145}$$

The values of F_n, G_n, H_n, I_n are different for both sections, because α and β are different. Applying Eqs. (4-136) through (4-145) to the boundary conditions gives

$$\begin{bmatrix} A_{11} & A_{12} & A_{13} & A_{14} & A_{15} & A_{16} & A_{17} & A_{18} \\ A_{21} & A_{22} & A_{23} & & & & & \\ A_{31} & & & & & & & \\ A_{41} & & & & & & & \\ A_{51} & & & & etc & & & \\ A_{61} & & & & & & & \\ A_{71} & & & & & & & \\ A_{81} & & & & & & & A_{88} \end{bmatrix} \begin{Bmatrix} a_{10} \\ a_{11} \\ a_{12} \\ a_{13} \\ a_{20} \\ a_{21} \\ a_{22} \\ a_{23} \end{Bmatrix} = \begin{Bmatrix} 0 \\ 0 \\ 0 \\ 0 \\ 0 \\ 0 \\ 0 \\ \delta \end{Bmatrix} \tag{4-146}$$

Boundary conditions 1, 2, 3 give $a_{10} = 0$, $a_{12} = 0$, and $a_{20} = 0$. The remaining boundary conditions give

$$a_{11} + 3a_{13}\varsigma_1^2 + \left(\sum_4 nG_{1n}\varsigma_1^{n-1}\right)a_{11} + \left(\sum_4 nI_{1n}\varsigma_1^{n-1}\right)a_{13} - a_{21} = 0$$

$$6a_{13}\varsigma_1 + \left(\sum_4 n(n-1)G_{1n}\varsigma_1^{n-2}\right)a_{11} + \left(\sum_4 n(n-1)I_{1n}\varsigma_1^{n-2}\right)a_{13} - 2a_{23} = 0$$

$$6a_{13} + \left(\sum_4 n(n-1)(n-2)G_{1n}\varsigma_1^{n-3}\right)a_{11} + \left(\sum_4 n(n-1)(n-2)I_{1n}\varsigma_1^{n-3}\right)a_{13} - 6a_{23} = 0$$

$$2a_{22} + 6a_{23}\varsigma_2 + \left(\sum_4 n(n-1)G_{2n}\varsigma_2^{n-2}\right)a_{21} + \left(\sum_4 n(n-1)H_{2n}\varsigma_2^{n-2}\right)a_{22} + \left(\sum_4 n(n-1)I_{2n}\varsigma_2^{n-2}\right)a_{23} = 0$$

$$a_{21}\varsigma_2 + a_{22}\varsigma_2^2 + a_{23}\varsigma_2^3 + \left(\sum_4 G_{2n}\varsigma_2^n\right)a_{21} + \left(\sum_4 H_{2n}\varsigma_2^n\right)a_{22} + \left(\sum_4 I_{2n}\varsigma_2^n\right)a_{23} + a_{11}\varsigma_1 + a_{13}\varsigma_1^3 + \left(\sum_4 G_{1n}\varsigma_1^n\right)a_{11} + \left(\sum_4 I_{1n}\varsigma_1^n\right)a_{13} = \delta$$

These equations were solved together to obtain the remaining five unknown constants. This method was used to determine riser deflections for the following input data;

L_1 =	1000
L =	2000 ft
w =	135 lb/ft
γ_o =	64 lb/ft^3

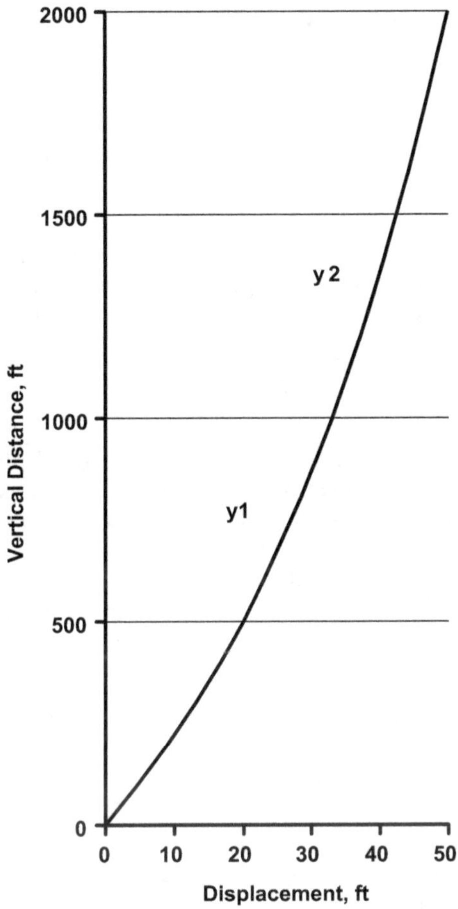

Figure 4-16. Riser displacement represented by two sections.

mud wt	12 ppg
riser OD	22 in
riser ID	20.5 in
A_o	2.6398 ft^2
A_i	2.2921 ft^2
γ_i	89.76 lb/ft^3
E =	2.90E + 07 psi
I =	2829.7 in^4
F_B =	200,000 lb (plus is tension)

The results are shown in Fig. 4-16. Note that F_B is well above $F_{B,cr}$ as shown in Fig. 4-1.

The remaining coefficients are

$a_{11} =$ 95.687
$a_{13} =$ −1178.429
$a_{21} =$ 43.998
$a_{22} =$ −25.445
$a_{23} =$ 19.699

Building riser deflections in segments offers an interesting way to handle ocean currents that may vary with depth. It is also a useful means of analyzing risers containing buoys.

4.1.7.2 Intermediate Buoys in Riser Pipe

Experience has shown that support buoys are needed along risers that are greater than 2000 ft. The inclusion of buoys into the analysis will now be considered. In this case the buoy is considered as a discrete buoy even though riser buoys do occupy some length (Fig. 4-17). Buoy length can be factored into an analysis with the help of a freebody diagram.

Fluid force on the buoy is represented by the symbol Q (lbs). This force is much like a concentrated force on a classical short beam. The change in shear has to be considered. In addition there is a change in tension because of the buoy.

The connecting boundary conditions for the top and bottom portions of the riser are given below.

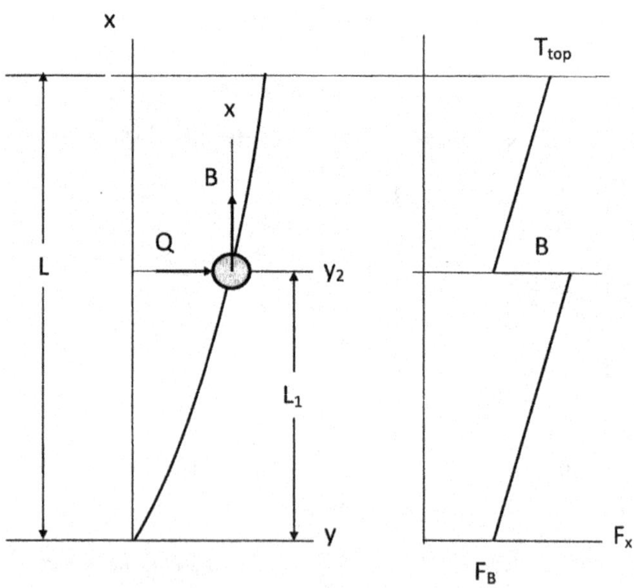

Figure 4-17. Intermediate Buoy support.

The set of matching boundary conditions for a riser with one buoy and two riser sections becomes

$$y_1(0)=0 \qquad y_1(0)=0 \tag{4-147a}$$

$$\frac{d^2y_1}{dx^2}(0)=0 \qquad \frac{d^2y_1}{d\varsigma^2}(0)=0 \tag{4-147b}$$

$$y_2(0)=0 \qquad y_2(0)=0 \tag{4-147c}$$

$$\frac{dy_1}{dx}(L_1)=\frac{dy_2}{dx}(0) \qquad \frac{dy_1}{d\varsigma}(\varsigma_1)=\frac{dy_2}{d\varsigma}(0) \tag{4-147d}$$

$$\frac{d^2y_1}{dx^2}(L_1)=\frac{d^2y_2}{dx^2}(0) \qquad \frac{d^2y_1}{d\varsigma^2}(\varsigma_1)=\frac{d^2y_2}{d\varsigma^2}(0) \tag{4-147e}$$

$$EI\frac{d^3y_1}{dx^3}(L_1)-T_1\frac{dy_1}{dx}(L_1)+Q=EI\frac{d^3y_2}{dx^3}(0)-T_2\frac{dy_2}{dx}(0) \tag{4-147f}$$

$$\frac{d^3y_1}{d\varsigma^3}(\varsigma_1)-\frac{T_1L^2}{EI}\frac{dy_1}{d\varsigma}(\varsigma_1)+\frac{QL^3}{EI}=\frac{d^3y_2}{d\varsigma^3}(0)-\frac{T_2L^2}{EI}\frac{dy_2}{d\varsigma}(0)$$

$$\frac{d^2y_2}{dx^2}(L_2)=0 \qquad \frac{d^2y_2}{d\varsigma^2}(\varsigma_2)=0 \tag{4-147g}$$

$$y_2(L_2)=\delta-y_1(L_1) \qquad y_2(\varsigma_2)+y_1(\varsigma_1)=\delta \tag{4-147h}$$

where (see Fig. 4-18)

$$T_2=T_1-B \tag{4-148}$$

and

$$V_2=V_1+Q \tag{4-149}$$

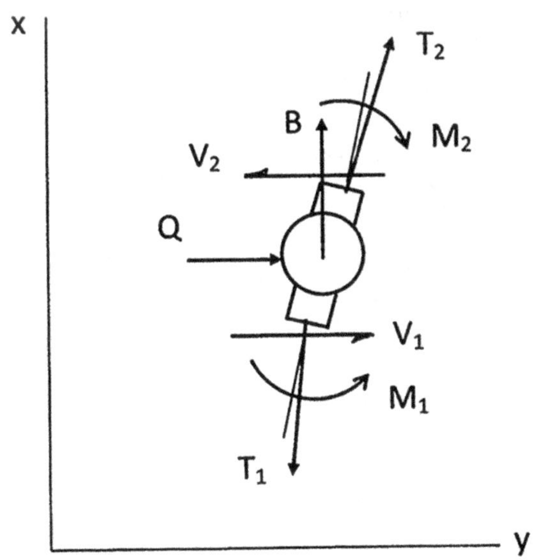

Figure 4-18. Freebody diagram of Buoy.

Slopes at the buoy come into Eq. (4-147f) with or without lateral force, Q. because of ΔT across the buoy. The shear forces in Fig. 4-18 are shown in their plus direction. In both cases

$$V = EI\frac{d^3y}{dx^3} - T\frac{dy}{dx}$$

Note that

$$\alpha_1 = \frac{(w - A_o\gamma_w + A_i\gamma_m)L^3}{EI}$$

$$\beta_1 = \frac{(F_B + LA_o\gamma_w - LA_i\gamma_m)L^2}{EI}$$

$$\alpha_2 = \frac{(w - A_o\gamma_w + A_i\gamma_m)L^3}{EI}$$

$$\beta_2 = \frac{(F_{B2} + L_2A_o\gamma_w - L_2A_i\gamma_m)L^2}{EI}$$

$$F_{B2} = F_B + wL_1 - B$$

As before, there are eight (8) unknowns, $(a_0a_1a_2a_3)_1$ and $(a_0a_1a_2a_3)_2$. The solutions to y_1 and y_2 can be expressed in terms of these constants generating nine (8) linear algebraic equations yielding these constants.

Boundary conditions 1, 2, 3 give $a_{10} = 0$, $a_{12} = 0$, and $a_{20} = 0$. The remaining boundary conditions give

$$a_{11} + 3a_{13}\varsigma_1^2 + \left(\sum_4 nG_{1n}\varsigma_1^{n-1}\right)a_{11} + \left(\sum_4 nI_{1n}\varsigma_1^{n-1}\right)a_{13} - a_{21} = 0 \quad \text{(4-150)}$$

$$6a_{13}\varsigma_1 + \left(\sum_4 n(n-1)G_{1n}\varsigma_1^{n-2}\right)a_{11} + \left(\sum_4 n(n-1)I_{1n}\varsigma_1^{n-2}\right)a_{13} - 2a_{23} = 0 \quad \text{(4-151)}$$

$$\begin{aligned} 6a_{13} &+ \left(\sum_4 n(n-1)(n-2)G_{1n}\varsigma_1^{n-3}\right)a_{11} + \left(\sum_4 n(n-1)(n-2)I_{1n}\varsigma_1^{n-3}\right)a_{13} \\ &- \frac{T_1L^2}{EI}\left[a_{11} + 3a_{13}\varsigma_1^2 + \left(\sum_4 nG_{1n}\varsigma_1^{n-1}\right)a_{11} + \left(\sum nI_{1n}\varsigma_1^{n-1}\right)a_{13}\right] \\ &+ \frac{QL^3}{EI} = 6a_{23} - \frac{T_2L^2}{EI}a_{21} \end{aligned} \quad \text{(4-152)}$$

$$2a_{22} + 6a_{23}\varsigma_2 + \left(\sum_4 n(n-1)G_{2n}\varsigma_2^{n-2}\right)a_{21} + \left(\sum_4 n(n-1)H_{2n}\varsigma_2^{n-2}\right)a_{22}$$
$$+ \left(\sum_4 n(n-1)I_{2n}\varsigma_2^{n-2}\right)a_{23} = 0 \quad (4\text{-}153)$$

$$a_{21}\varsigma_2 + a_{22}\varsigma_2^2 + a_{23}\varsigma_2^3 + \left(\sum_4 G_{2n}\varsigma_2^n\right)a_{21} + \left(\sum_4 H_{2n}\varsigma_2^n\right)a_{22} + \left(\sum_4 I_{2n}\varsigma_2^n\right)a_{23}$$
$$+ a_{11}\varsigma_1 + a_{13}\varsigma_1^3 + \left(\sum_4 G_{1n}\varsigma_1^n\right)a_{11} + \left(\sum_4 I_{1n}\varsigma_1^n\right)a_{13} = \delta \quad (4\text{-}154)$$

Solutions give displacements for both sections, above and below the buoy. Tension distributions are

$$T(x) = F_B + wx \text{ (Section \#1)} \quad (4\text{-}155)$$

$$T(x) = T_2 + w(x - L_1) \text{ (Section \#2)} \quad (4\text{-}156)$$

Consider the riser example of Fig. 4-15 but with the addition of a single buoy located at $L_1 = 1000$ ft. Other locations can be selected. This location is an arbitrary choice. The effect of the buoy on lateral displacements is hardly noticed unless extreme buoy forces are chosen. The effect of the buoy on bending moment is apparent in Fig. 4-19. Input data is the same as used in the previously example (Fig. 4-16) with additional values of buoyancy (B) and side force (Q) as illustrated in Fig. 4-18.

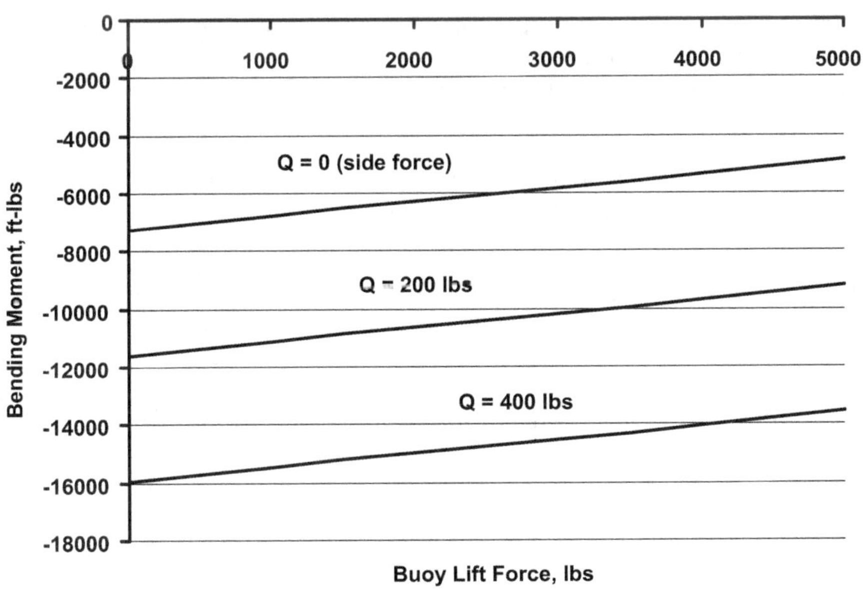

Figure 4-19. Bending moments at Buoy.

For the sake of simplicity, the buoy was assumed to be a discrete buoy; however, the freebody diagram of Fig. 4-18 can be expanded to allow for buoy length and distributed side load effects. Staying with the discrete buoy assumption

$$M_1 = M_2 = \frac{EI}{L^2}\frac{d^2 y_2}{d\varsigma^2}(0) \tag{4-157}$$

Dropping the subscript,

$$M = \frac{EI}{L^2} 2a_{22} \left| \frac{1\ \text{ft}^2}{144\ \text{in}^2} \right| \text{ft-lb} \tag{4-158}$$

where a_{22} has units of ft.

The corresponding bending stress for each case is nominal (less than 1000 psi). Other applications include stepwise current with depth.

4.1.8 Method of Segments

The two previous examples, using two riser segments, suggest that this method can be expanded to include several segments. There are several advantages to this approach:

1. Length (depth) capability of the closed form solution is increased.
2. Each segment allows modeling flexibility such as: changes in EI, current loading and multiple buoy placements.
3. The number of algebraic equations and unknowns is greatly reduced. The number of arbitrary constants is 4 x (number of segments), a significant reduction from finite element and finite different methods. For example 10 segments would involve a total of 40 arbitrary constants.
4. The solution to each segment automatically includes the effects of linear changes in effective tension over the segment.

Consider the drawing shown in Fig. 4-20 illustrating the ith segment. The lower boundary of the segment is identified by point i while the upper boundary is identified by i +1. The general differential equation of bending

$$EI\frac{d^4 y_i}{dx^4} - \frac{d}{dx}\left\{\left[(F_{Bi} + L_{2i}A_0\gamma_0 - L_{2i}A_i\gamma_i) + x(w - A_0\gamma_0 + A_i\gamma_i)\right]\frac{dy_i}{dx}\right\} = q(x) \tag{4-159}$$

where $F_{Bi} = F_B + wL_i$ and $L_{2i} = L - L_i$. Applying $\varsigma = \frac{x}{L}$ to Eq. (4-159) gives

$$\frac{d^4 y_i}{d\zeta^4} - \frac{d}{d\zeta}\left[(\beta_i + \alpha_i\zeta)\frac{dy_i}{d\zeta}\right] = f(x), \quad 0 \le \varsigma \le \varsigma_i \tag{4-160}$$

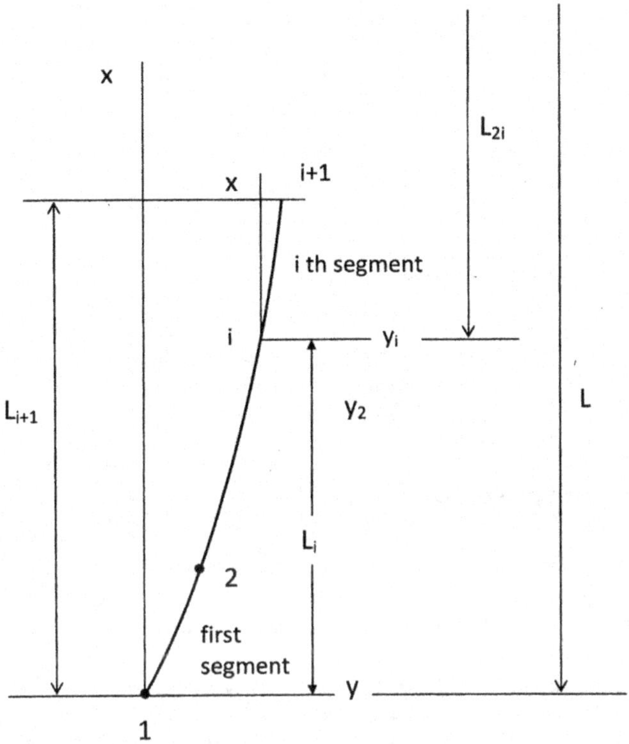

Figure 4-20. Local coordinate system for the method of segments.

where $\varsigma_i = \dfrac{L_{i+1} - L_i}{L}$ and

$$\alpha_i = \frac{(w - A_o\gamma_w + A_i\gamma_m)L^3}{EI} \quad i\text{th segment}$$

$$\beta_i = \frac{(F_{Bi} + L_{2i}A_o\gamma_w - L_{2i}A_i\gamma_m)L^2}{EI} \quad i\text{th segment}$$

Notice that L_{2i} is inside the bracket while L is outside, a slight variation from the previous formula. This comes about because $dx = Ld\varsigma$. The closed form solution for each segment is still

$$y_i(\varsigma) = \sum_{n=0}^{\infty} a_n \zeta^n,\ 0 \le \zeta \le \zeta_i \tag{4-161}$$

By definition, $\varsigma_1 = \dfrac{L_1}{L}$. Boundary conditions for each segment are

$$y_{i-1}(\varsigma_{2(i-1)}) = y_i(0) \tag{4-162a}$$

$$\frac{dy_{i-1}}{d\varsigma}(\varsigma_{2(i-1)}) = \frac{dy_i}{d\varsigma}(0) \tag{4-162b}$$

$$\frac{d^2y_{i-1}}{d\varsigma^2}(\varsigma_{2(i-1)}) = \frac{d^2y_i}{d\varsigma^2}(0) \tag{4-162c}$$

$$\frac{d^3y_{i-1}}{d\varsigma^3}(\varsigma_{2(i-1)}) = \frac{d^3y_i}{d\varsigma^3}(0) \tag{4-162d}$$

These equations satisfy boundary conditions at the interface of each segment, except at buoys. When buoys are located between segments, boundary conditions are modified as outlined in the previous section.

4.1.9 Ultra Deep Risers

The industry is currently operating in water depths near (and even greater than) 5,000 ft. This technology relies heavily on computer models to predict static and dynamic behavior of risers. Experience and team work are huge factors in implementing this technology.

This section addresses the application of closed form solutions to water depths of 5,000 ft and beyond. Structural connections, such as taper flex joints and flex ball joints, at the mudline and surface are very critical. Modeling riser stiffness near these connections is important too. However, much of the riser pipe can be modeled as a steel catenary having zero stiffness (EI = 0).

From a historical perspective, two major events (the Mohole project, 1958–1966 and the Jennifer Project, 1968–1974) preceded current ultra deep drilling and production by several years.

The purpose of the Mohole project was to drill through the earth's crust to the Mohorovicic Discontinuity. The thinnest crust is located offshore in regions of great water depths. The first phase consisted of experimental drilling and coring off Guadalupe, Mexico in 11,700 ft of water. Drilling and coring reached 601 ft beneath the ocean floor going beyond the first layer of the earth's crust at that location. It was significant that drilling took place from an untethered ship at this water depth [5,6].

A few years later, the CIA initiated the Jennifer Project [7-9] to retrieve a Soviet Submarine that sank in 17,000 ft of water. The Hughes Glomar Explorer was built between 1973 and 1974. Heavy lift-pipe was central to the recovery of the submarine. The design and manufacturing relied on both petroleum and military technology. The fabrication of each pipe section followed specifications for large bore battery gun barrels. Large upsets were forged at both ends for tool machining. Tool joint threads were modified buttress threads with a steep taper for quick make-up. The pull specification was 24,000,000 lbs. Six pipe sizes ranged from 15.5 to 12.75 in. (OD) each with a 6-in. bore. Three tool joint sizes ranged from 28 to 25.5 in. The dry weight of the entire 17,000 ft lift pipe was 4,000 tons.

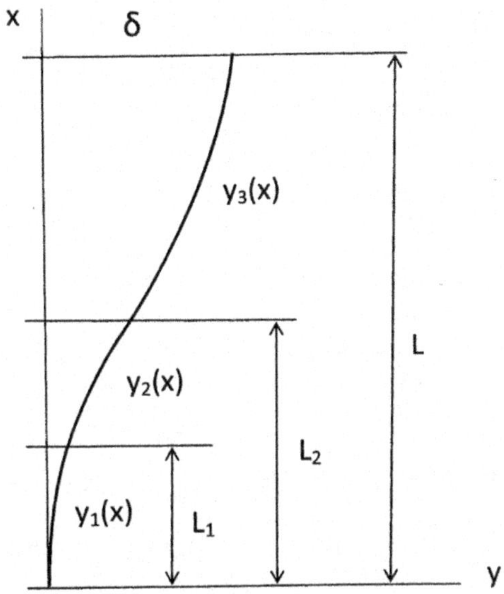

Figure 4-21. Riser with multiple sections.

At this point in time, Semi Submersibles were already being used to drill within the continental shelf. Much of the technology of handling marine risers through a moon pool using heave compensators was already in use.

4.1.9.1 Tapered Flex Joint, Short Riser Pipe, Steel Catenary

Solutions given earlier are combined to predict static deflections and stresses in very long risers. The riser configuration under consideration is (Fig. 4-21) is divided into three sections.

y_1 — tapered flex joint (or flex/ball joint)
y_2 — short elastic riser pipe ($EI \neq 0$)
y_3 — steel catenary approximation ($EI = 0$)

The solution to each section is determined using the individual solutions discussed earlier. The total solution is established by matching common boundary conditions.

4.1.9.2 Tapered Flex Joint Section

The differential equation of bending for the tapered flex joint is

$$\frac{d^2}{dx^2}\left[EI(x)\frac{d^2 y_1}{dx^2}\right] - F_B\frac{d^2 y_1}{dx^2} = 0 \qquad (4\text{-}163)$$

where F_B is the internal force at the bottom of the tapered flex joint. This force is assumed to be constant throughout the flex joint. Therefore, $F_{B2} \approx F_B$. The hydrostatic fluid terms have been dropped for simplicity and because of their negligible effect. As explained earlier, the solution to Eq. (4-163) is

$$y_1(x) = C_1 (h - x)^{D_1} + C_2 (h - x)^{D_2} + C_3 x + C_4 \tag{4-164}$$

The four constants of integration are determined from boundary conditions.

4.1.9.3 Short Riser Pipe Section

The distance between the top of the flex joint and the inflection point is anticipated to be short (~ 100 ft) so this section can be analyzed by short beam theory. Therefore the differential equation of bending over section 2 is

$$EI \frac{d^4 y_2}{dx^4} - T_{\text{eff}} \frac{d^2 y_2}{dx^2} = 0 \tag{4-165}$$

where $T_{\text{eff}} = (T_{\text{eff}})_{\text{average}}$ and

$$(T_{\text{eff}})_{\text{average}} = b - L_1 (A_0 \gamma_0 - A_i \gamma_i) + \frac{1}{2} \frac{L_2 - L_1}{L} a \tag{4-166}$$

The total solution to Eq. (4-165) is

$$y_2(x) = C_5 x + C_6 + C_7 \sin \beta x + C_8 \cos \beta x \tag{4-167}$$

where

$$\beta^2 = \frac{T_{eff}}{EI}$$

The distance to the point of inflection, L_2, is initially unknown but will be determined by iteration.

4.1.9.4 Steel Catenary Section

The steel catenary approximation of long suspended tubulars is an attractive model since catenary theory is simple and gives reasonable engineering results for cases where the effects of bending stiffness are relatively small.

When marine risers are very long they behave somewhat like cables suspended between a top support and a bottom supports. While this is not exactly true much can be learned about the behavior of marine risers in very deep water. In this case beam stiffness (EI) plays a minor role in how risers move from side to side. Tension and cable weight dictate lateral motion. Plunkett [10] shows that relatively small stiffness effects only areas near the end supports and the deviation from the catenary model can be found as a rapidly converging series.

The use of catenary theory to approximate elastically flexible riser pipe depends on side load distribution and anticipated curvature in the riser. For example vortex induced vibration analysis may require the effects of bending stiffness, especially when higher modes of vibration are involved.

The differential equation with the stiffness term is

$$EI\frac{d^4y_3}{dx^4} - \frac{d}{dx}\left[(b + ax)\frac{dy_3}{dx}\right] = 0 \tag{4-168}$$

where

$$a = w - A_0\gamma_0 + A_i\gamma_i$$

$$b = F_B + LA_0\gamma_0 - LA_i\gamma_i$$

Eliminating the stiffness term gives the differential equation for steel catenaries with hydrostatic terms included.

$$\frac{d}{dx}\left[(b + ax)\frac{dy_3}{dx}\right] = 0 \tag{4-169}$$

where

$$a = w - A_0\gamma_0 + A_i\gamma_i$$

$$b = F_{B3} + (L - L_2)(A_0\gamma_0 - A_i\gamma_i)$$

The solution to Eq. (4-169) is

$$y_3(x) = C_9\ell n(b + ax) + C_{10} \tag{4-170}$$

4.1.9.5 Boundary Conditions

For the sake of simplicity, assume there are no buoys. Since there are 10 constants of integration, ten boundary conditions are needed.

$$y_1(0) = 0 \tag{4-171a}$$

$$\frac{dy_1}{dx}(0) = 0 \tag{4-171b}$$

$$y_1(L_1) = y_2(L_1) \tag{4-171c}$$

$$\frac{dy_1}{dx}(L_1) = \frac{dy_2}{dx}(L_1) \tag{4-171d}$$

$$\frac{d^2y_1}{dx^2}(L_1) = \frac{d^2y_2}{dx^2}(L_1) \tag{4-171e}$$

$$\frac{d^3y_1}{dx^3}(L_1) = \frac{d^3y_2}{dx^3}(L_1) \tag{4-171f}$$

$$y_2(L_2) = y_3(L_2) \tag{4-171g}$$

$$\frac{dy_2}{dx}(L_2) = \frac{dy_3}{dx}(L_2) \tag{4-171h}$$

$$\frac{d^3y_2}{dx^3}(L_2) = 0 \tag{4-171i}$$

$$y_3(L) = \delta \tag{4-171j}$$

L_2 is chosen to coincide with zero moment in the short pipe, i.e., $\frac{d^2y_2}{dx^2}(L_2) = 0$. Initially, L_2 is unknown but can be determined by trial. Note that while the curvature in the short pipe is zero at L_2, the curvature in the steel catenary riser is not. This discontinuity is discussed in Appendix 4B.

The third derivative (Eq. (4-171f)) at L_1 is justified as follows recalling that

$$V = \frac{dM}{dx} - F\frac{dy}{dx}. \tag{4-172}$$

Since $V_1(L_1) = V_2(L_1)$ and the slopes and force, F, match

$$\frac{d^3y_1}{dx^3}(L_1) = \frac{d^3y_2}{dx^3}(L_1). \tag{4-173}$$

The third derivative at L_2 (Eq. (4-171g)) is justified as follows;

$$V_2(L_2) = V_3(L_2) \tag{4-174}$$

$$\left(\frac{dM}{dx} - F\frac{dy}{dx}\right)_2 = \left(-F\frac{dy}{dx}\right)_3 \tag{4-175}$$

The minus sign on the right side is consistent with the sign convention for shear. Matching the shear in the short pipe with the shear in the catenary at L_2 gives

$$\frac{d^3y_2}{dx^3}(L_2) = 0 \tag{4-176}$$

The boundary conditions (Eq. (4-171)) produce a matrix equation, which yields the unknown constants. The solution is complete, giving deflections and curvatures in each of the three sections.

Consider the following riser case in 6000 ft of water.

Dry weight per unit length, lb/ft	142.8
Buoyancy factor	0.817
Buoyed weight, lb/ft	116.6676
Vertical force (F_B)or pull on stress joint, lb	300,000
Top lateral position, ft	400
Water depth, ft	6000

Assuming the distance from the flex joint base to the point of inflection is 150 and offset at x = 150 ft is 10 ft then the length of the catenary is 5850 ft and offset is 390 ft relative to the point of inflection.

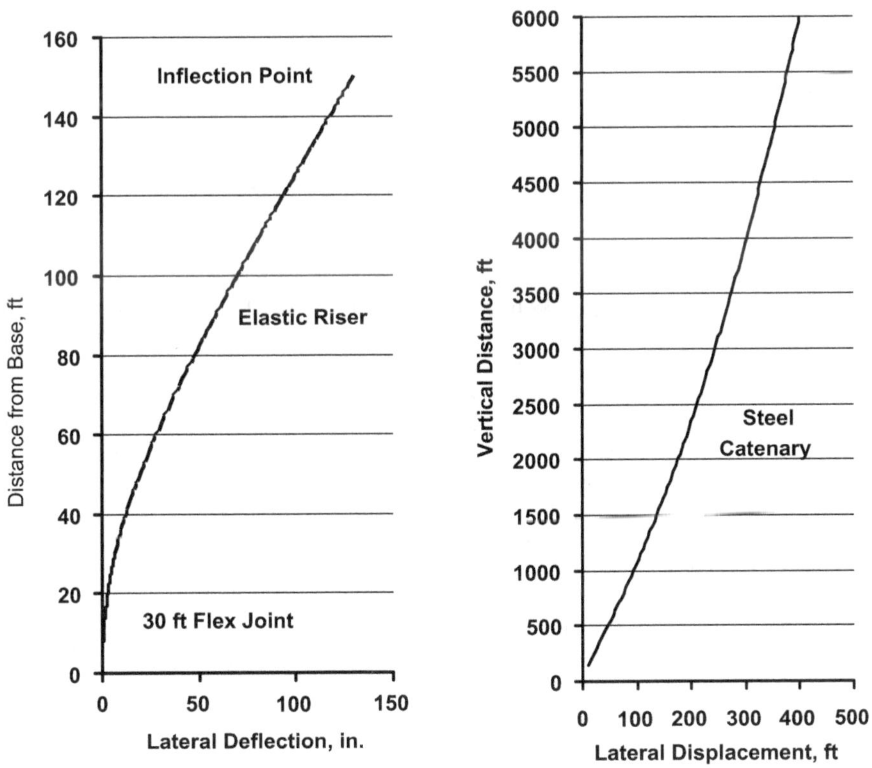

Figure 4-22. Lower end deflection of composite riser with tapered flex joint, short riser pipe and steel catenary.

Noting that the slope of the catenary at the bottom end is

$$\tan\theta = \frac{dy}{dx}(0) = \frac{V}{F_B}$$

Starting with Eq. (4-169), integrating and using boundary conditions of $y(0) = 0$ and $y(L) = 390$ gives the shear at the bottom of the catenary of

$$V = F_B \frac{\delta a}{b} \frac{1}{\ln\left(1 + \frac{aL}{b}\right)}$$

$$V = 32{,}794 \text{ lbs}$$

With $F_B = 300{,}000$ lbs and the shear force (32,794 lbs), the deflection of the tapered flex joint and short beam was determined. The deflections for the tapered flex joint, $y_1(x)$ and the short beam, $y_2(x)$ are shown in Fig. 4-22.

The tiny gap at the lower end of the steel catenary deflection is filled in with the displacement shown in the left plot. The slope of the upper end of the short riser pipe is $\frac{dy_2}{dx}(L_2) = 0.1086$ radians. The slope of the steel catenary at its lower end is $\frac{V}{F_B} = \frac{32{,}794}{300{,}000} = 0.1098$ radians. So the displacements and slopes match. Note also the rather flat shape of y_2 near the inflection point meaning its third derivative is zero as assumed in the boundary conditions.

4.2 DYNAMIC ANALYSIS—NATURAL FREQUENCIES

The accuracy of dynamic response predictions of marine risers depends on oceanographic data, forces produced by fluid motion, definition of rig response to environmental loading, assumptions in modeling the riser systems, and mathematical solution of the differential equations of motion. Experimental studies by Sarpkaya [11-13] give insight into fluid force magnitude produced by unsteady flow past tubulars and vortex shedding.

Response predictions depend on basic understanding of how fluid forces are applied to these pipes and a reasonable mathematical model to predict response. It is generally accepted that riser response can be view as having: 1) motion inline with ocean current and 2) motion transverse to ocean currents caused by vortex shedding. The actual motion is some combination of the two.

Modal analysis is useful for predicting riser response to these loads. Natural modes and frequencies are central to this method. Fluid forces are modeled by lift and drag force components and riser response to these forces are determined in terms of its natural modes. Methods of predicting lateral modes and their natural frequencies will be discussed first. This will be followed by a discussion of in-line vibrations and vortex induced vibrations (VIV).

4.2.1 Differential Equation of Motion

The mathematics for obtaining natural frequencies of lateral modes was explained in Chapter 2 (also see Dareing and Huang [14]). It is important to note that fluid densities inside and out are usually quite different. The freebody diagram shown in Fig. 2-12 still applies.

The differential equation of free vibration for marine risers is

$$EI\frac{\partial^4 y}{\partial x^4} - \frac{\partial}{\partial x}\left\{\left[(F_B + wx) + (L - x)(A_o\gamma_o - A_i\gamma_i)\right]\frac{\partial y}{\partial x}\right\} + m\frac{\partial^2 y}{\partial t^2} = 0 \quad (4\text{-}177)$$

The solution to Eq. (4-177) is assumed to be

$$y(x,t) = Y(x)\sin\omega t$$

After substitution and collecting terms, Eq. (4-177) becomes

$$\frac{d^4Y}{d\zeta^4} - \frac{d}{d\zeta}\left[(\beta + \alpha\zeta)\frac{dY}{d\zeta}\right] - \lambda^4 Y = 0 \quad (4\text{-}178)$$

where

$$\alpha = \frac{(w - A_o\gamma_w + A_i\gamma_m)L^3}{EI}$$

$$\beta = \frac{(F_B + LA_o\gamma_w - LA_i\gamma_m)L^2}{EI}$$

$$\lambda^4 = \frac{m\omega^2 L^4}{EI}$$

$$\zeta = \frac{x}{L}$$

Each parameter is dimensionless and $Y(x)$ is a function only of x.

4.2.1.1 Ball Joint Attachment

As an example consider a riser structurally pin-connected ($K_\phi = 0$) at the BOP stack. The set of boundary conditions are

$$Y(0) = 0 \quad (4\text{-}179a)$$

$$\frac{d^2Y}{d\varsigma^2}(0) = 0 \quad (4\text{-}179b)$$

$$Y(1) = 0 \quad (4\text{-}179c)$$

$$\frac{d^2Y}{d\varsigma^2}(1) = 0 \quad (4\text{-}179d)$$

If the bottom ball joint has built in rotational (bending) stiffness, i.e., there is an elastic resistance to rotation, the second boundary condition (Eq. (4-179b)) is expressed as

$$EI\frac{d^2y}{dx^2}(0) = K_\phi \frac{dy}{dx}(0) \tag{4-180}$$

The solution to Eq. (4-178) is represented by the power series

$$Y(\zeta) = \sum_{n=0}^{\infty} a_n \zeta^n \tag{4-181}$$

Following the procedure given in Chapter 2, the coefficients can be related by the recurrence formula

$$a_n = \frac{\beta}{n(n-1)}a_{n-2} + \frac{\alpha(n-3)}{n(n-1)(n-2)}a_{n-3} + \frac{\lambda^4}{n(n-1)(n-2)(n-3)}a_{n-4} \tag{4-182}$$

And related to a_0, a_1, a_2, a_3 by

$$a_n = F_n a_0 + G_n a_1 + H_n a_2 + I_n a_3 \tag{4-183}$$

Using Eq. (4-181) and the first two boundary conditions, we find that $a_0 = 0$ and $a_2 = 0$. The third and fourth boundary conditions give

$$\begin{aligned} a_1 + \sum_{n=3}^{\infty} a_n &= 0 \\ \sum_{n=3}^{\infty} n(n-1)a_n &= 0 \end{aligned} \tag{4-184}$$

Substituting for a_n, gives

$$\left[1 + \sum_{n=3}^{\infty} G_n\right] a_1 + \left[\sum_{n=3}^{\infty} I_n\right] a_3 = 0 \tag{4-185a}$$

$$\left[\sum_{n=3}^{\infty} n(n-1)G_n\right] a_1 + \left[\sum_{n=3}^{\infty} n(n-1)I_n\right] a_3 = 0 \tag{4-185b}$$

For non-trivial solutions, the determinant, D, of the coefficient matrix must be zero or

$$D\,(\alpha, \beta, \lambda) = 0 \tag{4-186}$$

Table 4-4. Natural frequencies of first five lateral modes of vibration [14].

α	β	λ_1	λ_2	λ_3	λ_4	λ_5
0	0	3.142	6.283	9.425	12.566	15.708
	100	5.738	8.614	11.380	14.206	17.103
	200	6.746	9.861	12.656	15.419	18.221
	300	7.437	10.760	13.632	16.398	19.165
	400	7.975	11.477	14.435	17.228	19.987
50	0	4.234	7.083	10.020	13.034	16.090
	100	6.029	8.969	11.735	14.536	17.401
	200	6.934	10.105	12.919	15.679	18.470
	300	7.580	10.950	13.845	16.616	19.379
	400	8.092	11.635	14.615	17.417	20.176
100	0	4.778	7.649	10.513	13.449	16.443
	100	6.268	9.275	12.053	14.841	17.683
	200	7.101	10.325	13.161	15.924	18.706
	300	7.711	11.126	14.044	16.823	19.585
	400	8.202	11.784	14.786	17.598	20.359
150	0	5.162	8.093	10.934	13.824	16.772
	100	6.473	9.544	12.342	15.124	17.949
	200	7.251	10.528	13.387	16.156	18.932
	300	7.832	11.291	14.232	17.021	19.783
	400	8.305	11.925	14.948	17.771	20.536
200	0	5.465	8.462	11.303	14.165	17.079
	100	6.654	9.786	12.607	15.388	18.202
	200	7.388	10.715	13.598	16.375	19.149
	300	7.945	11.447	14.410	17.210	19.973
	400	8.402	12.058	15.103	17.938	20.707
250	0	5.719	8.781	11.633	14.479	17.368
	100	6.816	10.006	12.851	15.636	18.442
	200	7.515	10.889	13.797	16.583	19.356
	300	8.051	11.593	14.580	17.390	20.157
	400	8.494	12.186	15.252	18.099	20.872
300	0	5.938	9.062	11.931	14.770	17.641
	100	6.964	10.209	13.079	15.871	18.671
	200	7.633	11.052	13.985	16.781	19.555
	300	8.151	11.732	14.741	17.564	20.334
	400	8.581	12.308	15.395	18.254	21.032

For given values of α and β, the solution to Eq. (4-186) yield the values of λ. Eigenvalues, λ, for the first five modes are given in Table 4-4.

Assume the following conditions for a 500-ft, 24 × 5/8 in. riser.

$$w = 214 \text{ lb/ft}$$
$$A_e = 3.14 \text{ ft}^2$$
$$A_i = 2.99 \text{ ft}^2$$
$$\gamma_w = 64.8 \text{ lb/ft}^3$$
$$\gamma_m = 85 \text{ lb/ft}^3$$
$$I = 3{,}136.9 \text{ in}^4$$
$$E = 30(10^6) \text{ psi}$$
$$F_B = 286{,}000 \text{ lb}$$
$$m = 20.8 \text{ slugs/ft}$$

Corresponding dimensionless numbers

$$\alpha = 50$$
$$\beta = 100$$
$$\lambda_1 = 6.029$$

The natural circular frequency of the first mode is $\omega_1 = 0.815$ rad/sec. The period of oscillation of the first mode, then, is 7.71 seconds.

4.2.1.2 Structurally Fixed at BOP

If the bottom end of the riser structurally fixed to the BOP stack, Eq. (4-178) still applies but the boundary conditions become

$$Y(0) = 0 \tag{4-187a}$$
$$\frac{dY}{d\varsigma}(0) = 0 \tag{4-187b}$$
$$Y(1) = 0 \tag{4-187c}$$
$$\frac{d^2Y}{d\varsigma^2}(1) = 0 \tag{4-187d}$$

Using Eq. (4-181) and the first two boundary conditions gives $a_0 = a_1 = 0$. The second two give

$$\left[\sum_{n=3}^{\infty} H_n\right] a_2 + \left[\sum_{n=3}^{\infty} I_n\right] a_3 = 0 \tag{4-188a}$$

$$\left[\sum_{n=3}^{\infty} n(n-1)H_n\right] a_2 + \left[\sum_{n=3}^{\infty} n(n-1)I_n\right] a_3 = 0 \tag{4-188b}$$

Expanding the coefficient matrix gives the characteristic equation. Its solution gives the natural frequencies and mode shapes for a fixed boundary condition at the lower end.

4.2.2 Natural Frequencies Based on Steel Catenary Model

The analysis giving natural frequencies for steel catenaries modeled for ultra deep waters is explained in Chapter 2. Consider the following example for a 5000-ft riser.

$$D_0 = 20 \text{ in}$$
$$D_i = 19 \text{ in}$$
$$L = 5000 \text{ ft}$$
$$\gamma_0 = 65 \text{ lb/ft}^3$$
$$\gamma_i = 12 \text{ ppg}$$
$$F_B = 300{,}000 \text{ lb}$$

$$a = 695{,}768$$
$$b = 125{,}375$$

The eigenvalues for the first five (5) modes are given in Table 4-5. The eigenvalue is related to natural frequency by

$$\Psi^2 = m(\omega L)^2$$

These number indicate that the higher modes are most likely involved in inline motion and VIV response.

4.2.3 Alternate Methods for Determining Natural Frequencies

4.2.3.1 *Uniform Tension*

According to Timoshenko [15, p. 376] natural frequencies for a simply supported beam under an axial tension load can be determined by

Table 4-5. Natural frequencies of first five modes for a 5000 ft riser (based on Eq. (2-148)).

Mode	Ψ	ω (rad/sec)	f cps	T (sec)
1	2363	0.173	0.0275	36.4
2	5426	0.397	0.063	15.8
3	8505	0.622	0.099	10.1
4	11,585	0.847	0.135	7.47
5	14,665	1.072	0.171	5.56

$$p_i = \frac{ai^2\pi^2}{L^2}\sqrt{1+\frac{SL^2}{i^2EI\pi^2}}, \quad i = 1,2,3,\ldots \tag{4-189}$$

where

$a^2 = \dfrac{EI}{m}$

m — mass per unit length
S — axial tension force in beam

If the beam is very flexible (wire) the second term is very large in comparison with unity and the natural frequency can be approximated by

$$p_i = \frac{i\pi}{L}\sqrt{\frac{S}{m}} \tag{4-190}$$

which predicts natural frequencies of a tight wire.

Equation (4-189) converts for use in predicting natural frequencies of marine risers by assuming uniform tension equal to bottom tension plus half of the weight of the riser [14].

$$\lambda_i = i\pi\sqrt[4]{1+\frac{\beta+0.5\alpha}{i^2\pi^2}} \tag{4-191}$$

Table 4-6. Natural Frequencies Computed with Eq. (4-191).

	λ_1	λ_2	λ_3	λ_4	λ_5
$\alpha = 10$					
$\beta = 10$	3.958	6.810	9.799	12.855	15.941
100	5.803	8.690	11.455	14.274	17.164
1000	10.004	14.250	17.655	20.701	23.574
$\alpha = 100$					
$\beta = 10$	5.244	7.916	10.723	13.620	16.586
100	6.303	9.299	12.068	14.850	17.688
1000	10.113	14.401	17.834	20.898	23.783
$\alpha = 1000$					
$\beta = 10$	8.463	12.136	15.187	18.021	20.787
100	8.808	12.605	15.728	18.600	21.379
1000	11.049	15.701	19.382	22.620	25.622

Table 4-7. Natural Frequencies based on Eq. (4-191).

Mode	λ	ω rad/sec	f cps
1	9.5075	0.4193	0.0667
2	13.5642	0.8535	0.1358
3	16.8468	1.3166	0.2095
4	19.8133	1.8211	0.2898
5	22.6397	2.3777	0.3784

Equation (4-191) gives fairly accurate predictions of natural frequencies. The greatest error occurs when β (or F_B) is near zero. Table 4-6 gives values of λ_i based on Eq. (4-191). Comparison of the numbers in Table 4-6 with those in Table 4-4 shows the values are reasonably close.

For example consider the following:

L = 1500 ft
Riser size — 22 in. OD × 20.75 in. ID
w — 142.8 lb/ft (based on nominal cross section dimensions)
Mud weight — 12 ppg (89.7 lb/ft³)
Sea water — 8.56 ppg (64 lb/ft³)
E — 29 $(10)^6$ psi
I — 2399 in.⁴ (based on nominal cross section dimensions)
F_B — 100,000 lbs
Flex/ball joint ($K_\phi = 0$)

Using these numbers $\alpha = 1290$ and $\beta = 173$.

Natural frequencies for this example are displayed in Table 4-7. They are calculated using Eq. (4-191) and $\lambda^4 = \dfrac{m\omega^2 L^4}{EI}$.

4.2.3.2 *Iteration on Lateral Mode*

Graves and Dareing [16] give a method for approximating natural frequencies of marine risers in deep water. The method is direct and easy to use. It is especially useful for determining the natural frequency of the higher modes. The approach is based on dividing the riser into an number of half sine waves depending on the desired mode.

To determine the natural frequency of the nth mode, the riser length is divided into n half wave sections. Each section will be treated as a pinned-pinned freely vibrating pipe eventually having the same frequency as determined by

$$f_n = f_i \approx \frac{1}{2\pi}\sqrt{\left(\frac{\pi}{\ell_i}\right)^4 \frac{EI}{\rho} + \left(\frac{\pi}{\ell_i}\right)^2 \frac{\overline{T}_i}{\rho}} \tag{4-192}$$

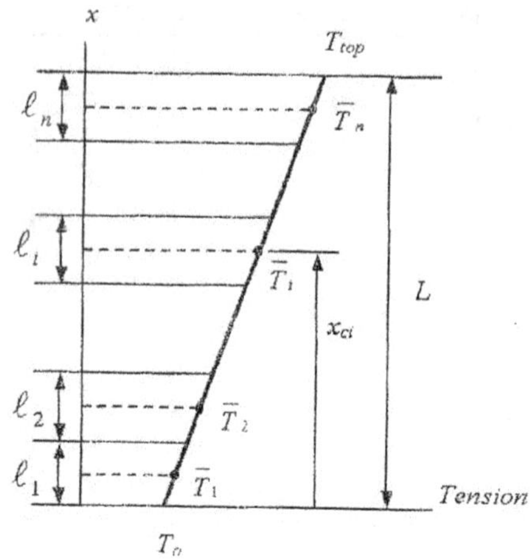

Figure 4-23. Diagram of average tension in each segment.

$\overline{T}_i$ is the average effective tension over the ith section (see Fig 4-23). However, initially, this frequency is unknown but will be determined by making adjustments to $\overline{T}_i$ and ℓ_i in an iterative process.

For example, assume the natural frequency of the third mode ($n = 3$) is desired.

$$\ell_1 = \frac{L}{\left(1 + \frac{\ell_2}{\ell_1} + \frac{\ell_3}{\ell_1}\right)} \tag{4-193}$$

Similar expressions can be written for ℓ_2 and ℓ_3. Since the frequencies of each of the three sections are equal, according to Eq. (4-192)

$$\frac{\ell_2}{\ell_1} = \left[\frac{\ell_2^2 \overline{T}_2 + \pi^2 EI}{\ell_1^2 \overline{T}_1 + \pi^2 EI}\right]^{\frac{1}{4}} \tag{4-194}$$

and

$$\frac{\ell_3}{\ell_1} = \left[\frac{\ell_3^2 \overline{T}_3 + \pi^2 EI}{\ell_1^2 \overline{T}_1 + \pi^2 EI}\right]^{\frac{1}{4}} \tag{4-195}$$

New values of the length ratios are calculated by Eqs. (4-194) and (4-195) and substituted back into Eq. (4-193) to obtain revised half wave lengths. The new lengths and new effective tensions are substituted into Eq. (4-192) to check frequency values. New values of effective tension are calculated and both new lengths and new effective tensions are put into Eqs. (4-194) and (4-195). The process is repeated until the frequencies (using Eq. (4-192)) of successive iterations converge within acceptable limits.

Table 4-8. Comparison of numerical and analytical results with bending stiffness and hydrostatic effects [16].

Mode	Numerical f_n (Hz)	Analytical f_n (Hz)	% Error
1	0.1309	0.1297	0.8905
2	0.2870	0.2871	0.4776
3	0.4905	0.4914	0.3435
4	0.7525	0.7540	0.2520
5	1.079	1.081	0.1944

The results using this method were compared with results from the closed form solution explained previously. The input to both methods is given below.

$$L = 500 \text{ ft}$$
$$D_0 = 24 \text{ in.}$$
$$D_i = 22.75 \text{ in.}$$
$$\rho = 20.8 \text{ slugs/ft}$$
$$w = 214 \text{ lb/ft}$$
$$T_0 = 286{,}000 \text{ lb}$$
$$E = 30 \times 10^6 \text{ lb/in}^2$$
$$\gamma_0 = 64.8 \text{ lb/ft}^3$$
$$\gamma_i = 85 \text{ lb/ft}^3$$

A numerical comparison of the two methods is given in Table 4-8. The percent error is very small for the numerical method and the percent error decreases for increasing mode number.

4.3 DYNAMIC ANALYSIS—FORCED VIBRATION

The dynamic response of marine risers to ocean currents and waves is complicated by fluid forces moving against these long pipes in different directions. Recent research has addressed out of plane response to these environmental forces. In many cases, however, currents and wave advance against risers from the same direction so in-plane (or in-line) analyses gives reasonable engineering results.

There are many good software packages available to make these calculations. Approaches are typically based on finite element or finite difference methods [17-23]. Design variations along risers are somewhat easer to model with the finite element method and for this reason several commercial computer programs are based on the method. Modal analysis offers a different approach and will be applied in the following discussion.

4.3.1 In-Line Vibration Caused by Ocean Currents and Waves

The modal analysis method was used by Dareing and Huang [24] to determine the dynamic response of marine risers to rig offset and surge as well as ocean current.

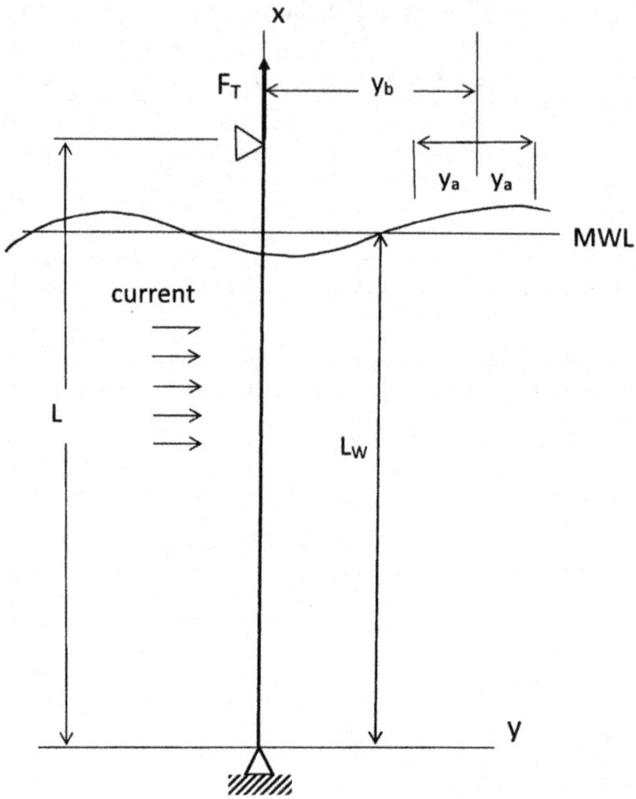

Figure 4-24. Parameters used in the equation of motion.

The modal analysis approach is a useful engineering tool because it gives time domain and frequency domain solutions directly from eigenvalues: classical equations of mechanical vibrations solutions for both are well documented.

Figure 4-24 shows the physical arrangement of the riser modal. Fluid loading and riser response are assumed to be in the same vertical plane. Fluid forces on the riser are modeled by the Morison equation.

In applying the modal method, the nonlinear fluid drag term in the governing differential equation of motion is linearized by the use of an equivalent damping coefficient, which is adjusted during the successive calculations in an iterative procedure. The calculations converge quickly to an acceptable damping coefficient when the static displacement distribution, corresponding to the rig surge amplitude, is assumed initially as the dynamic displacement distribution.

If this assumed that initial dynamic displacement distribution is smaller than the actual dynamic displacement distribution, the first set of displacement calculations produces a low damping coefficient along with a large dynamic displacement distribution. The next set of displacement calculations produces a high damping coefficient along with a small displacement distribution.

Damping therefore encourages fast convergence. In order to uncouple the vibration modes, this damping coefficient is assumed to be constant along the riser.

The marine riser is modeled as a vertical beam in tension pinned at the bottom and top ends (see Fig. 4-24). For the sake of simplicity, the riser is represented by a single uniform tubular containing a fluid, such as drill mud, have a density different from the density of sea water. Variable tension, due to riser weight, is a unique aspect of the mathematics. The bottom end is stationary and pinned. The top end is simply supported but time dependent to account for surge and sway of the support vessel. Two sources of motion to the riser are 1) in-line movement of the vessel and 2) direct in-line loading due to water particle motion.

The lateral displacement function, $y(x,t)$, varies with both vertical position, x and time, t. The fluid drag term makes the equation nonlinear while tensile and fluid pressure gradients produce a term in the differential equation having a variable coefficient. These terms are characteristic of the marine riser problem.

4.3.1.1 Formulation of Equation of Motion

The development of the differential equation of motion is given in ref. [24]. The differential equation for this case is

$$m\frac{\partial^2 y}{\partial t^2} - C_m \frac{\gamma_0}{g}\frac{\pi}{4} d_0^2 \left(a_w - \frac{\partial^2 y}{\partial t^2} \right) - C_d \frac{\gamma_0 d_0}{2g} \left| v_c + v_w - \frac{\partial y}{\partial t} \right| \left(v_c + v_w - \frac{\partial y}{\partial t} \right) + EI\frac{\partial^4 y}{\partial x^4}$$
$$- \frac{\partial}{\partial x}\left\{ \left[(F_B + wx) + (L - x)(A_0\gamma_0 - A_i\gamma_i) \right] \frac{\partial y}{\partial x} \right\} = 0 \quad (4\text{-}196)$$

The interaction of surrounding fluid with the riser body are modeled by use of Morison's drag and mass terms [25], where the magnitude of the fluid forces depends on fluid velocity and acceleration relative to the riser pipe.

Rig offset and surge are introduced through the top boundary condition. The boundary conditions for the problem are:

$$\begin{aligned} & y(0,t) = 0 \\ & \frac{\partial^2 y}{\partial x^2}(0,t) = 0 \\ & y(L,t) = y_b + y_a \cos(\omega t - \psi_v) \\ & \frac{\partial^2 y}{\partial x^2}(L,t) = 0 \end{aligned} \quad (4\text{-}197)$$

The solution to Eq. (4-196) is complicated by nonlinearity, due to Morison's forces. This complication is removed by replacing the fluid drag force by: 1) a drag force due to the current alone and 2) a viscous damping force due only to the fluid velocity of the wave relative to the moving riser. The nonlinear drag force is then replaced by

$$C_d \frac{\gamma_0 d_0}{2g} v_c^2 + C_{eq}\left(v_w - \frac{\partial y}{\partial t}\right)$$

The equivalent damping coefficient is determined by equating work done by the drag force along the riser during one period of oscillation to the total work done along the riser by an equivalent viscous damping. Assuming a Stokes wave, and the linearized drag term, the resulting linearized differential equation becomes

$$m'\frac{\partial^2 y}{\partial t^2} + C_{eq}\frac{\partial y}{\partial t} + EI\frac{\partial^4 y}{\partial x^4} - \frac{\partial}{\partial x}\left\{\left[(F_B + wx) + (L-x)(A_0\gamma_0 - A_i\gamma_i)\right]\frac{\partial y}{\partial x}\right\}$$
$$= \left[C_m \frac{\gamma_0}{g}\frac{\pi}{4} d_0^2 a_0(x)\right]\cos\omega t + C_{eq} v_0(x)\sin\omega t + C_d \frac{\gamma_0 d_0}{2g} v_c^2(x) \quad (4\text{-}198)$$

The solution to Eq. (4-198) can be expressed in the following form

$$y(x,t) = y_c(x) + y_{st}(x,t) + y^*(x,t) \quad (4\text{-}199)$$

where

$y_c(x)$ — static displacement due to current loading alone
$y_{st}(x,t)$ — displacement component due to rig offset and is calculated as a static displacement for a given rig offset position even though rig offset is time dependent.
$y^*(x,t)$ — forced vibration response and accounts for inertia effects and time-dependent fluid interaction.

By substituting Eq. (4-199) into Eq. (4-198), the above differential equation can be broken out into separate differential equations each in terms of the dependent variables. Each of these equations will now be considered.

4.3.1.1.1 Current Loading

For the current loading case

$$EI\frac{d^4 y_c}{dx^4} - \frac{d}{dx}\left\{\left[(F_B + wx) + (L-x)(A_0\gamma_0 - A_i\gamma_i)\right]\frac{dy_c}{dx}\right\} = C_d \frac{\gamma_0 d_0}{2g} v_c^2(x) \quad (4\text{-}200)$$

Corresponding boundary conditions are

$$y_c(0) = 0$$
$$y_c(L) = 0$$
$$\frac{d^2 y_c}{dx^2}(0) = 0$$
$$\frac{d^2 y_c}{dx^2}(L) = 0$$

4.3.1.1.2 Rig Offset

The second differential equation considers the response to rig movement as a quasi-static motion.

$$y_{st}(x,t) = \left[y_b + y_a \cos\left(\omega t - \psi_v\right)\right] f(x) \tag{4-201}$$

where $f(x)$ is the static deflection with the top of the riser displaced by one unit. This expression satisfies the top moving boundary condition or rig surge. In this case, the differential equation becomes

$$EI\frac{d^4 f}{dx^4} - \frac{d}{dx}\left\{\left[(F_B + wx) + (L-x)(A_0\gamma_0 - A_i\gamma_i)\right]\frac{df}{dx}\right\} = 0 \tag{4-202}$$

and the corresponding boundary conditions are

$$f(0) = \frac{d^2 f}{dx^2}(0) = \frac{d^2 f}{dx^2}(L) = 0 \tag{4-203}$$

and

$$f(L) = 1 \tag{4-204}$$

Residual terms from the left side of Eq. (4-198) are

$$\frac{\partial y_{st}}{\partial t} = \left[-y_0 \omega \sin(\omega t - \psi_v)\right] f(x) \tag{4-205}$$

and

$$\frac{\partial^2 y_{st}}{\partial t^2} = \left[-y_0 \omega^2 \cos(\omega t - \psi_v)\right] f(x) \tag{4-206}$$

which are incorporated into the remaining differential equation.

4.3.1.1.3 Forced Vibrations

The remaining terms give the real time response. The third differential equation is reduced in form by using dimensionless parameters.

$$m'\frac{\partial^2 y^*}{\partial t^2} + C_{eq}\frac{\partial y^*}{\partial t} + \frac{EI}{L^4}\left[\frac{\partial^4 y^*}{\partial \zeta^4} - (\beta + \alpha\zeta)\frac{\partial^2 y^*}{\partial \zeta^2} - \alpha\frac{\partial y^*}{\partial \zeta}\right] = a(\zeta)\cos\omega t + b(\zeta)\sin\omega t \tag{4-207}$$

Boundary conditions which apply to the solution of Eq. (4-207) are

$$y^*(0,t) = y^*(1,t) = \frac{\partial^2 y^*}{\partial \zeta^2}(0,t) = \frac{\partial^2 y^*}{\partial \zeta^2}(1,t) = 0 \tag{4-208}$$

The solutions to Eq. (4-200) and Eq. (4-202) are determined from static analysis and can be obtained directly from either the finite difference method, finite element numerical method or the closed form solution method given in this chapter. Petty [26] also gives a closed form solution to the static riser problem. The solution to Eq. (4-207) is time dependent and will now be expressed terms of natural vibration modes.

Represent y^* by in infinite series

$$y^*(\zeta,t)=\sum_{i=1}^{\infty}q_i(t)Y_i(\zeta) \tag{4-209}$$

where $q_i(t)$ are unknown functions and $Y_i(\zeta)$ are the mode shapes. After substituting Eq. (4-209) into Eq. (4-207), multiplying through by $Y_i(\varsigma)$, and integrating both sides of the equation

$$\ddot{q}_i+2\mu_i p_i\dot{q}_i+p_i^2q_i=\frac{F_{o,i}}{\rho'}\sin(\omega t+\phi_i) \tag{4-210}$$

The solution to this equation represents the amplitude of each mode participating in the vibration response.

The solution to the problem is now complete by substituting each component into Eq. (4-199).

$$y(\zeta,t)=y_c(\zeta)+y_bf(\zeta)+\left[y_af(\zeta)\cos(\omega t-\psi_v)+\sum_{i=1}^{\infty}q_i(t)Y_i(\zeta)\right] \tag{4-211}$$

Bending moment distribution is determined from the second derivative of the displacement function and is represent by

$$M(\zeta,t)=M_c(\zeta)+M_{st}(\zeta,t)+\sum_{i=1}^{\infty}q_i(t)M_i(\zeta) \tag{4-212}$$

4.3.1.2 *Example Calculation*

An application of these equations is given below. The input data are consistent with conditions in the North Sea at the time ref. [24] was written. Since the approach is based on modal analysis, the first step is to determine the natural modes of the riser. This was done in this case using the solution given by Dareing and Huang [14]. The equivalent damping coefficient was determined by successive iterations stating with the static displacement function and refining C_{eq} until there is no change in the dynamic displacement function.

$L = 575$ ft

$L_w = 500$ ft

$d_0 = 18$ in.OD

$d_i = 17$ inID

$F_B = 50{,}000$ lbs

$y_b = 15$ ft

$y_a = 5$ ft

$H = 25$ ft
$T = 11$ sec
$v_c = 3$ ft/sec
$\psi_v = 90$ deg.
$\gamma_0 = 64.8$ lb/ft^3
$\gamma_i = 85$ lb/ft^3
$C_d = 1$
$C_m = 2$

The corresponding values for α and β are

$\alpha = 97.797$
$\beta = 58.474$

and the shape of the first five modes for this input is shown in Table 4-9.

Note that the actual amplitude of each mode participating in the final solution will not be the same but will depend on the solution to Eq. (4-210). The maximum amplitudes of each mode shown in Table 4-9 have been normalized to one. The corresponding frequencies for each of the first five modes are

$\lambda_1 = 5.779$
$\lambda_2 = 8.692$
$\lambda_3 = 11.472$
$\lambda_4 = 14.297$
$\lambda_5 = 17.187$

where

$$\lambda_i = \sqrt[4]{\frac{\rho' p_i^2 L^4}{EI}} \tag{4-213}$$

The total displacement amplitude is sum of the four solution components (see Fig. 4-25). Corresponding bending moments are shown in Fig. 4-26.

Modal analysis provides a direct approach to the response solution. The method requires the predetermination of natural mode shapes and frequencies. The above example indicates that only a few (5 modes in this case) modes participate in the dynamic response. Linearizing the drag term and assuming damping is constant along the riser uncoupled the modes and greatly simplified the solution.

4.3.2 Vortex-Induced Vibrations

The previous section addressed vibrations in a vertical plane oriented in-line with fluid flow. Current and wave action were assumed to occur in the same direction and this produced static and dynamic motion in one plane. Since particle motion in waves varies with time, in line riser motion also vary with time. The total displacement along riser pipe has static and dynamic components. A static analysis is a good start point for riser design. Dynamic forces and stresses must be considered in light of addition side loading and possible fatigue damage. Fatigue loading caused by wave action should be considered; however, the cyclic frequency of waves is small ($f \approx 0.1$ cps). The largest cyclic bending stresses are near the surface as shown in the previous example.

Vortex induced vibrations (VIV) occur transverse to the direction of current and wave flow as explained in Chapter 1. Ocean currents, uniform with depth, can excite lateral vibration modes over the full length of riser pipe. Also, the frequency of excitation (or vortex shedding) is typically larger than waves passing across a riser pipe. For example, the frequency of vortex shedding on an 18-in. stationary cylinder is $f = 0.29$ cps assuming a current velocity of 2 ft/sec. The frequency of vortex shedding is about 3 times higher than wave frequencies. Another significant difference is the total involvement of participating modes

Table 4–9. Shape of first five modes.

Location (ft)	First mode	Second mode	Third mode	Fourth mode	Fifth mode
575	0	0	0	0	0
550	−0.1107	0.2387	−0.3712	0.496	−0.6162
525	−0.2211	0.4647	−0.6905	0.8621	−0.9747
500	−0.3302	0.6646	−0.9115	1	−0.9236
475	−0.4367	0.8258	−1	0.87	−0.4815
450	−0.5391	0.937	−0.9404	0.5041	0.1664
425	−0.6359	0.9893	−0.7387	−0.0005	0.7435
400	−0.7251	0.9774	−0.4229	−0.5063	1
375	−0.805	0.8998	−0.0397	−0.8723	0.8217
350	−0.8737	0.7595	0.352	−0.9939	0.2835
325	−0.9293	0.5642	0.6898	−0.8335	−0.3802
300	−0.97	0.3259	0.9179	−0.4337	−0.8756
275	−0.994	0.0607	0.9965	0.0925	−0.9801
250	−1	−0.2125	0.9095	0.5927	−0.6437
225	−0.9865	−0.4729	0.6684	0.9188	−0.0155
200	−0.9527	−0.6996	0.3115	0.9716	0.6202
175	−0.8978	−0.873	−0.1014	0.7319	0.9714
150	−0.8218	−0.9769	−0.4982	0.2692	0.8737
125	−0.7251	−1	−0.8073	−0.2769	0.3692
100	−0.6088	−0.9375	−0.9707	−0.7379	−0.3084
75	−0.4748	−0.792	−0.9555	−0.9684	−0.8408
50	−0.3259	−0.574	−0.761	−0.8929	−0.9739
25	−0.166	−0.3017	−0.4214	−0.5323	−0.6413
0	0	0	0	0	0

under VIV conditions. Bending, under uniform flow conditions, is greatest near the lower end because of the shape of lateral modes. Variations in current velocity with depth (shear flow) bring in a challenging aspect to the problem.

4.3.2.1 Self-Exciting Mechanism

Numerous test have been conducted on spring mounted cylinders, which allow transverse movement with respect to current flow [27]. Test data show how the ratio of shedding frequency to natural frequency, $\frac{f_s}{f}$, changes with the reduced velocity parameter, $V_r = \frac{V}{fD}$. Experimental studies [28,29] show that when V_r is less than 5 the vibration responds as a force vibration. Over the range $5 \le V_r \le 7$, the vortex shedding frequency locks into the natural frequency of the spring mounted cylinder. Over the lock-in region, the frequency ratio is one (1) and response amplitude is the greatest. Outside of V_r =7 the system responds as a forced vibration.

Much work has been conducted on VIV [30-32]. Vandiver [33] identifies five problem areas facing VIV:

1. The understanding of the fluid mechanics of the fluid-structure interaction

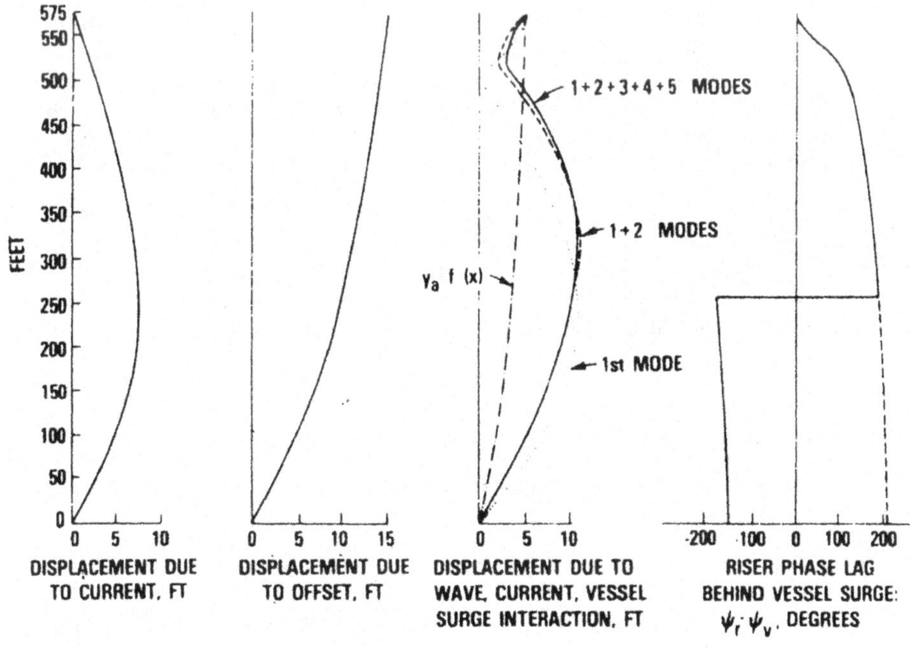

Figure 4-25. Modal participation in total response.

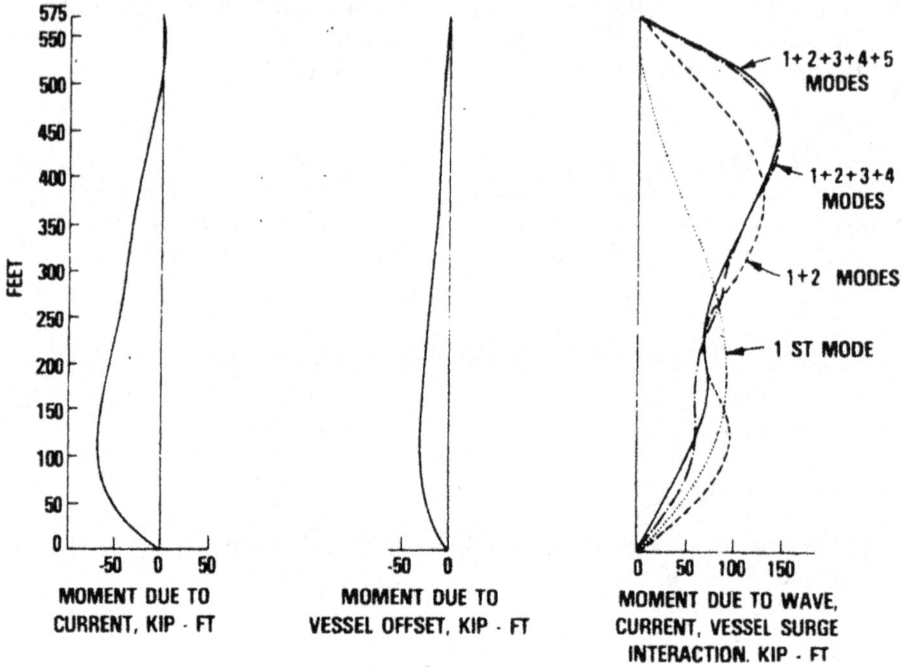

Figure 4-26. Bending moment from each loading.

2. The development of better structural dynamics modeling techniques
3. The acquisition of high quality, full scale, riser response data for the purpose of calibration of response prediction models.
4. The analysis of multi-channel field data for comparison to response prediction models
5. The development of VIV suppression and protection technologies.

Measurements of vortex induced vibration indicate that the formation of vortices in the wake also produce in-line vibrations. However, the magnitude of these vibrations is an order of magnitude smaller than those in the transverse directions [33]. Other causes of in-line motion are wave action and vessel surge.

4.3.2.2 Modal Analysis

Different approaches are being considered for predicting riser response to vortex shedding. Among these approaches are [33]:

- model analysis response methods
- real time response of riser structure to fluid loading models
- interfacing the real time response of riser structure with Navier-Stokes equations

The model analysis can easily accommodate shear flow, or fluid flow that varies with ocean depth. This brings about variations in lift force magnitude and vortex shedding frequency with depth.

The equation of bending of risers in its simplest form is

$$EI\frac{\partial^4 y}{\partial x^4}-\frac{\partial}{\partial x}\left\{\left[(F_B+wx)+(L-x)(A_o\gamma_o-A_i\gamma_i\right]\frac{\partial y}{\partial x}\right\}+m\frac{\partial^2 y}{\partial t^2}+c\frac{\partial y}{\partial t}=F(x,t) \tag{4-214}$$

where damping has been linearized and the mass term is the total effective mass.

The term on the right side represents lift force caused by vortex shedding.

Natural modes and frequencies are determined from

$$EI\frac{\partial^4 y}{\partial x^4}-\frac{\partial}{\partial x}\left\{\left[(F_B+wx)+(L-x)(A_o\gamma_o-A_i\gamma_i\right]\frac{\partial y}{\partial x}\right\}+m\frac{\partial^2 y}{\partial t^2}=0 \tag{4-215}$$

Approximate methods discussed earlier may be useful for obtaining mode shapes and frequencies. The mode shape of long risers is important here as the greatest bending may be near the lower end depending on the current profile (see Fig. 4-27).

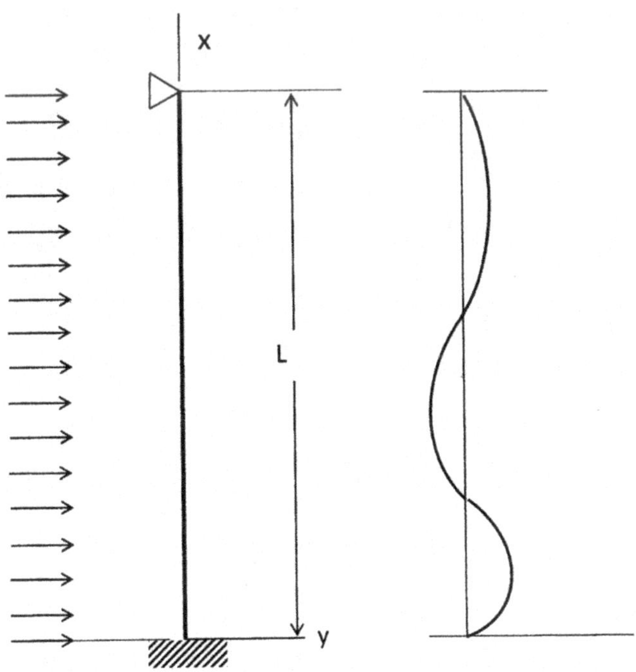

Figure 4-27. Marine riser in a uniform current stream.

By assuming uniform current, the forcing function associated with vortex shedding can be expressed as

$$F(z,t) = F_0 \cos \omega t \tag{4-216}$$

The modal analysis method has basically three parts:

1. Establish the natural vibration modes of the structure and corresponding natural frequencies. Mode shapes and frequencies are established based on free non-damped.
2. Establish modal mass, stiffness, damping and force. It is usually assumed that modal damping is uncoupled even though this may not be exactly true.
3. Solve the time dependant equations. Recognize that the vibration response is the sum of modal responses. In most cases, only the first four or five (or less) need to be considered.

The solution is assumed to be of the form

$$Y(\zeta,t) = \sum_{1}^{n} X_i(\zeta) q_i(t) \tag{4-217}$$

The problem is one of determining modal mass, modal damping, modal stiffness, and modal force. Modal stiffness is somewhat involved because of the variable coefficient. The method of modal analysis is well documented [15].

Appendix 4A

CABLES SUSPENDED IN AIR

The differential equation for the deflection for steel catenaries suspended in air can be extracted from the bending equation developed for elastically flexible pipes. For the sake of simplicity, hydrostatic effects are omitted.

$$EI\frac{d^4y}{dx^4} - \frac{d}{dx}\left[(wx + F_B)\frac{dy}{dx}\right] = 0 \tag{A-1}$$

Dropping the bending term leaves

$$\frac{d}{dx}\left[(wx + F_B)\frac{dy}{dx}\right] = 0 \tag{A-2}$$

which is the catenary equation for small lateral displacements, a condition imposed on Eq. (A-1).

TOP SUPPORT OFFSET FROM VERTICAL AXIS

First consider cable deflection when top support is offset from the bottom support by δ. The direct integration of Eq. (A-2) gives

$$(wx + F_B)\frac{dy}{dx} = C_1 \tag{A-3}$$

With reference to Fig. A4-1

$$\left.\frac{dy}{dx}\right|_{x=0} = \frac{H_0}{F_B} \tag{A-4}$$

The boundary condition at $x = 0$ gives $C_1 = H_0$.

$$\frac{dy}{dx} = \frac{H_0}{wx + F_B} \tag{A-5}$$

The second integration gives

$$y = \frac{H_0}{w}\ell n(wx + F_B) + C_2 \tag{A-6}$$

Using the top boundary, $y(L) = \delta$,

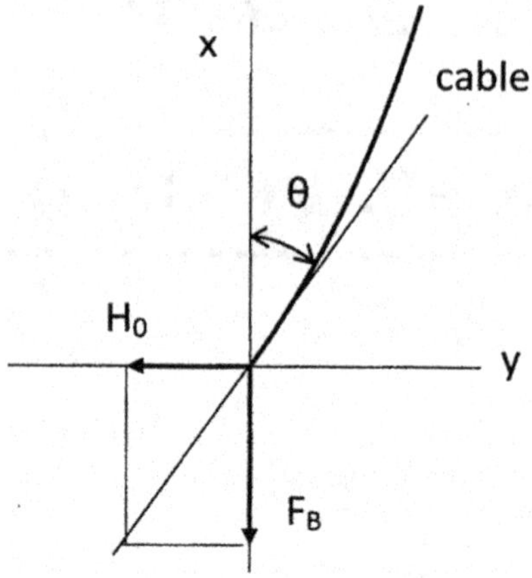

Figure 4A-1. Bottom end of cable.

$$y = \delta + \frac{H_0}{w} \ell n \left(\frac{wx + F_B}{wL + F_B} \right) \tag{A-7}$$

Using the condition that at $x = 0$, $y = 0$, we obtain

$$H_0 = -\frac{\delta w}{\ell n \left(\frac{F_B}{wL + F_B} \right)} \tag{A-8}$$

From this equation, H_0 can be determined for a given rig offset, δ, and bottom tension (F_B),

Curvature along the deflected catenary is expressed by

$$\frac{d^2y}{dx^2} = -\frac{w}{(wx + F_B)} \frac{dy}{dx} \tag{A-9}$$

or by substitution

$$\frac{1}{\rho} = \frac{d^2y}{dx^2} = \frac{-wH_0}{(wx + F_B)^2} \tag{A-10}$$

where H_0 is given in terms of F_B as defined by Eq. A-8.

Tension effects maximum curvature but there is a trade-off between reduced curvature and cost. It is tempting to use the catenary deflection function to approximate deflection over portions of risers that have relative low curvature. Caution is advised. At the lower point, curvature in the catenary is a maximum even though there is zero bending.

Eq. (A-7) was used to generate the deflection curve shown in Fig. 4A-2. Input data used are:

- Vertical height = 5000 ft
- Top Offset = 10 ft
- Cable weight = 67 lbs/ft
- Bottom Pull = 50,000 and 100,000 lbs
- Corresponding H_0 = 328 and 456 lbs respectively

Top Tensions for the two cases are shown in the figure.

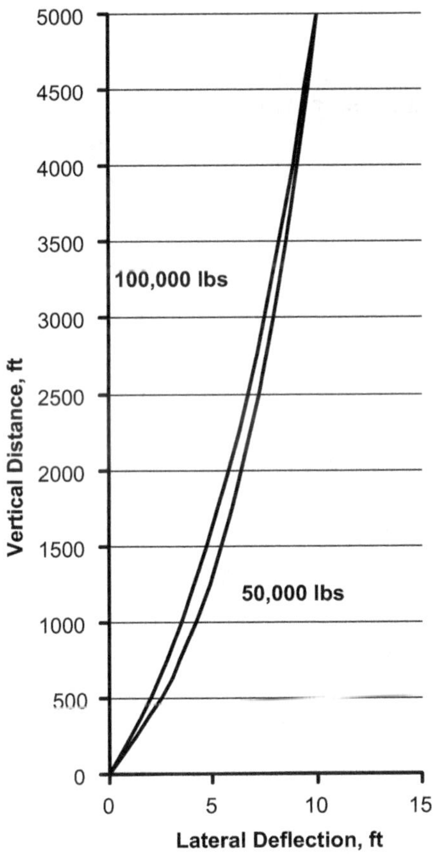

Figure 4A-2. Deflection of cable with top offset.

Appendix 4B

COMPARISON OF STEEL CATENARY APPROXIMATION WITH ELASTIC RISER

Fluid densities can be incorporated into the steel catenary formulation by simplifying the general equation of bending. Starting with

$$EI\frac{d^4y}{dx^4}-\frac{d}{dx}\left\{[(F_B+wx)+(L-x)(A_o\gamma_o-A_i\gamma_i)]\frac{dy}{dx}\right\}=0 \tag{B-1}$$

and dropping the first term gives the steel catenary equation.

$$\frac{d}{dx}\left[(b+ax)\frac{dy}{dx}\right]=0 \tag{B-2}$$

where

$$a=w-A_0\gamma_0+A_i\gamma_i$$

$$b=F_B+LA_0\gamma_0-LA_i\gamma_i$$

Boundary conditions are chosen as

$$y(0)=0$$
$$y(L)=\delta^{\cdot}$$

From the first integration

$$(b+ax)\frac{dy}{dx}=b\frac{H_0}{F_B} \tag{B-3}$$

The second integration gives

$$y(x)=\delta+\frac{b}{a}\frac{H_0}{F_B}\ell n\left(\frac{b+ax}{b+aL}\right) \tag{B-4}$$

By setting $y(0) = 0$

$$\frac{H_0}{F_B}=-\delta\frac{a}{b}\frac{1}{\ell n\left(\dfrac{b}{b+aL}\right)} \tag{B-5}$$

Slope at any point, x, the slope of the deflection curve is

$$\frac{dy}{dx} = -\frac{H_0}{F_B}\frac{b}{b+ax} \tag{B-6}$$

and local curvature is

$$\frac{d^2y}{dx^2} = -\frac{H_0}{F_B}\frac{ab}{(b+ax)^2} \tag{B-7}$$

ELASTIC MODEL

The solution to Eq. (B-1) was given earlier for the following input conditions

$L =$	1500 Ft
$w =$	135 lb/ft
$\gamma_o =$	64 lb/ft^3
mud wt	12 ppg
riser OD	22 In
riser ID	20.5 In
$A_o =$	2.6398 ft^2
$A_i =$	2.2921 ft^2
$\gamma_i =$	89.76 lb/ft^3
$E =$	2.90E + 07 psi
$I =$	2829.7 in^4
$F_B =$	100000 lb
$\delta =$	50 ft
$\alpha =$	1.02E + 03
$\beta =$	1.77E + 02
$H_0 =$	1.00E + 04 lb
$H_0/F_B =$	1.00E − 01

Calculations of deflection, slope, and curvature based on the elastic model are given in Table B-1.

STEEL CATENARY MODEL

Calculations of deflection, slope, and curvature using the catenary model are listed in Table B-2.

The steel catenary model predicts a high bending stress at the lower end, while in reality, there is zero bending stress as indicated by the elastic model.

Table B-1. Elastic model.

Zeta	Displacement ft	Slope rad.	Curvature 1/ft	Radius ft	Bending stress, psi
0	0.00	0.07678	0.00E+00		0.00
0.1	10.80	0.06454	−1.15E−04	8.683E+03	3061.64
0.2	19.25	0.04884	−8.81E−05	1.135E+04	2342.89
0.3	25.71	0.03815	−5.65E−05	1.770E+04	1502.26
0.4	30.88	0.03122	−3.76E−05	2.659E+04	999.88
0.5	35.19	0.02647	−2.68E−05	3.736E+04	711.51
0.6	38.89	0.02299	−2.01E−05	4.980E+04	533.84
0.7	42.13	0.02033	−1.57E−05	6.388E+04	416.16
0.8	45.01	0.01823	−1.25E−05	7.968E+04	333.61
0.9	47.61	0.01653	−1.00E−05	9.980E+04	266.37
1	50.00	0.01548	0.00E+00		0.00

Table B-2. Steel catenary model.

Zeta	Displacement ft	Slope radians	Curvature 1/ft	Radius ft	Bending stress, psi
0	0.00	0.10038	−3.85E−04	2.60E+03	10229.30
0.1	11.89	0.06373	−1.55E−04	6.45E+03	4123.49
0.2	20.04	0.04669	−8.32E−05	1.20E+04	2212.79
0.3	26.25	0.03683	−5.18E−05	1.93E+04	1377.46
0.4	31.26	0.03042	−3.53E−05	2.83E+04	939.25
0.5	35.47	0.02590	−2.56E−05	3.90E+04	681.18
0.6	39.09	0.02256	−1.94E−05	5.15E+04	516.52
0.7	42.27	0.01997	−1.52E−05	6.56E+04	405.07
0.8	45.11	0.01792	−1.23E−05	8.15E+04	326.16
0.9	47.67	0.01625	−1.01E−05	9.91E+04	268.24
1	50.00	0.01487	−8.44E−06	1.18E+05	224.49

REFERENCES

1. American Petroleum Institute Recommended Practice, API RP 16Q.
2. Huang, T. and Dareing, D. W., "Buckling and Lateral Vibration of Drill Pipe," Trans. ASME, *Journal of Engineering for Industry*, Vol. 90, Series B., No. 4, 1968, pp. 613–619.
3. Dareing, D. W.; "Stress Balanced Tapered Columns in Marine Risers", Proceedings of the 37th Petroleum Mechanical Engineering Workshop and Conference (ASME), Dallas, Sept. 13–15, 1981.

4. Gordon, Cormack C. and Don W. Dareing, "Configuring Marine Riser Tapered Stress Joints Used In Top-Supported Applications," *ASME Journal of Energy Resources Technology*, Vol 126, No. 3, Sept 2004.
5. A Bore-Hole to the Earth's Mantle: AMSCO Mohole, NATURE, July 18, 1959.
6. A Hole in the Bottom of the Sea: The Story of the Mohole Project, Willard Bascom, 1961.
7. http://files.asme.org/asmeorg/Communities/History/Landmarks/12708.pdf
8. Tom Dougherty and Jeff Dougherty, "Raising the K-129; a Tale of the Cold War", The Sub Committee Report Magazine, September 2002, Issue 60.
9. Clyde W. Burleson, *The Jennifer Project*, Texas A & M University Press, 1997.
10. Plunkett, R., "Static Bending Stresses in Catenaries and Drill Strings", *Journal of Engineering for Industry*, February 1967.
11. Sarpkaya, T.; "Vortex Shedding and Resistance in Harmonic Flow About Smooth and Rough Circular Cylinders", Proceeding of Behavior of Off-Shore Structures Conference, Trondheim, Norway, Aug. 1976.
12. Sarpkaya, T. (1986) "Force on a circular cylinder in viscous oscillatory flow at low Keulegan-Carpenter numbers", *Journal of Fluid Mechanics*, Vol 165, pp. 61–71.
13. Sarpkaya, T.; "Vortex-Induced Oscillations", *Journal of Applied Mechanics*, June 1979, Vol. 46, pp. 241–258.
14. Dareing, D. W. and Huang, T., "Natural Frequencies of Marine Drilling Risers," *Journal of Petroleum Technology,* July 1976, pp. 813–818.
15. Timoshenko, S.: Vibrations Problems in Engineering, D. Van Nostrand Co., Princeton. N.J. (1956).
16. Graves, Jason and Don W. Dareing; "Direct Method for Determining Natural Frequencies of Marine Risers In Deep Water", ASME Journal of Energy Resources Technology, Vol.126, No. 1, pp. 47–53, March 2004.
17. Morgan, G. W. and Peret, Jack W.: "Applied Mechanics of Marine Riser Systems", a series of articles in *Petroleum Engineer* starting October 1974.
18. Burke, B. G.: "An Analysis of Marine Risers for Deep Water", *Journal of Petroleum Technology*, April 1974.
19. Young, R. D., Fowler, J. R., Fisher, E. A., and Luke, R. R.: "Dynamic Analysis as an Aid to the Design of Marine Risers", ASME, 77-PET-62, 1977.
20. Gardner, T. N. and Kotch, M. A.: "Dynamic Analysis Risers and Caissons by the Element Method", Offshore Technology Conference, OTC 2651, 1976.
21. Burke, B. G.: "An Analysis of Marine Risers for Deep Water", *Journal of Petroleum Technology*, Apr. 1974.
22. Young, R. D., Fowler, J. R., Fisher, E. A., and Luke, R. R.: "Dynamic Analysis as an Aid to the Design of Marine Risers", *ASME*, 77-PET-62, 1977.
23. Gardner, T. N. and Kotch, M. A.: "Dynamic Analysis Risers and Caissons by the Element Method", Offshore Technology Conference, OTC 2651, 1976.
24. Dareing, D. W. and Huang, T., "Marine Riser Vibration Response Determined by Modal Analysis," Trans. ASME, *Journal of Energy Resources Technology*, Vol. 101, Sept. 1979, 159–166.
25. Morison, J. R., O'Brien, M. P., Johnson, J. W. and Schaaf, S. A. (1950), "The Forces Exerted by Surface Waves on Piles", Petroleum Trans., AIME, **189**, 149–157.

26. Petty, T. D., "Analysis of Marine Risers Used with Floating Drilling Vessels", MS Thesis, University of Oklahoma, 1970.
27. J. Kim Vandiver, "Drag Coefficients of Long flexible Cylinders", Offshore Technology Conference paper, OTC 4490, 1983.
28. Stansby, P. K. (1976), "The locking-on of vortex shedding due to the cross-stream vibration of circular cylinders in uniform and shear flow", *Journal of Fluid Mechanics,* 74, 641–665.
29. Sarpkaya, T., (1979), "Vortex-induced oscillations—a selective review", *Journal of Applied Mechanics*, **46**, No. 2.
30. Sarpkaya, T., "Vortex Shedding and Resistance in Harmonic Flow About Smooth and Rough Circular Cylinders", Proceedings of Behavior of Off-Shore Structures Conference, Trondheim, Norway, Aug. 1976.
31. Charles Dalton, Fundamentals of Vortex Induced Vibrations, University of Houston.
32. C. Le Cunff, F. Biolley, E. Fontaine, S. Ethienne and M. L. Facchinetti: "Vortex-Induced Vibrations of Risers: Theoretical, Numerical and Experimental Investigation", Oil & Gas Science and Technology, Vol. 57 (2002), No. 1, pp. 59–69.
33. J. Kim Vandiver; "Research Challenges in the Vortex-Induced Vibration Prediction of Marine Risers", Offshore Technology Conference; OTC 8698 (1998).

Chapter 5

APPLIED DRILLING MECHANICS

It is difficult to discuss drillstring mechanics without including the broader aspects of drilling. The drillstring is central to rotary drilling but it must be viewed as a part of the total drilling system. All subsystems (hoisting, rotary, circulating, well control, and power) must work together. This chapter covers the mechanics of several related topics with emphasis on the role of drillstrings.

5.1 DEVELOPING THE DRILLING PROGRAM

Drilling programs involve drilling teams from both operator and contractor. There is a special process that brings these two teams together to develop and implement a drilling program.

Well planning deals with the selection and application of drilling equipment to drill safely and economically. Safety and cost override everything. Safety includes everything from rig personnel to well control. Cost implies optimum drilling practices, including best rate of penetration, plus lowest footage cost. Drilling economics depends on: skills of rig personnel, logistics, and well planning. Assuming top side operations are conducted efficiently, drilling costs are most affected by the selection and operation of tools used in the bottom hole assembly.

Well planning actually starts with the reservoir development plan, which defines the most economical well spacing and entry point into a reservoir. The plan sets the production tubing size, which in turn sets the casing size and this dictates drill bit size. Well depth and well path affect drill pipe size. Drill bit size changes with each casing interval. The length of the casing depends on formation fracture gradient and pressure gradient of the mud weight.

5.1.1 Operational Requirements

Operational requirements describe what is expected. They are set by management and represent a decision to move ahead with field development. This decision is based an economic study of reservoir performance and production costs in consideration of market demand and timing. This may include expected production rate, well pattern and points of entry into the reservoir, and other relevant information. This declaration initiates and gives a clear direction for the project.

5.1.2 Design Specifications

Design specifications are developed from the project description as stated in the operational requirements. These specifications detail engineering constraints. They specify expected outcomes. Examples of specification categories are

- Performance Requirements
    - Directional well path
    - Fracture gradient
    - Expected loads
- Codes and standards
- Sustainability
- Environmental
- Safety
- Reliability of downhole tools
- Cost considerations
- Well bore life cycle

A comprehensive set of specifications is important and usually becomes part of a drilling contract. It states specifically what is expected and sets the engineering baseline for the drilling program.

5.1.3 Creating Design Alternatives

Well planning, like all projects, is an open ended design activity [1]. There are many possible solutions and each have to be created and evaluated. In creating design alternatives remember that each has to satisfy the specifications, be technically feasible, cost-effective, and safe. Past experience in a given production area is useful in this regard.

5.1.4 Evaluating Alternatives

Drillstring design is typically based on the selection of off-the-shelf components, such as drill pipe, drill collars, drill bits, motors, MWD, etc. For each possible component, there are many brands, sizes, and performance capabilities. The designer must choose among many possible tools to configure drillstrings for each well bore section. In essence, there are many possible drillstring configurations, and the problem is to select the best tool combination to meet drilling specifications. A rationale for making these choices is explained below.

The four basic metrics used in making choices are as follows:

Performance — The capability to achieve needed operational characteristics, plus reliability.
Cost — The estimated cost of the design, including development and manufacturing costs.
Risk — The possibility that performance may not be met because of the design approach, absence of testing, or some specific technical consideration.
Availability — The availability of a design depending upon the stage of development.

Table 5-1. Evaluation scoring matrix.

	Weight	Concept			
	%	A	B	C	D
PERFORMANCE (40)					
Functional requirements	5	0	0	5	0
Reliability/maintenance	10	9	3	9	6
Sustainability	10	8	8	1	8
Compatibility with casing program	15	14	14	10	10
RISK (30)					
Credibility/history of technology	10	10	10	9	7
Experience of personnel	10	10	10	7	7
Accuracy of calculations, testing	10	9	9	8	7
SCHEDULE (20)					
Time of delivery (to market, production site)	10	10	10	9	9
Personnel raining	10	10	10	8	6
COST (10)					
Development	3	3	2	3	1
Implementation	4	3	3	2	2
Environmental protection	3	3	2	2	1
TOTAL SCORE	100%	89%	80%	82%	60%

Each of these metrics can be further expanded into submetrics to detail specific elements that may be important in decision making. Submetrics are given a weighting factor, say, out of 100. Each design alternative is then evaluated against the set of metrics and scored against these weighting factors. The sum of all scores given in each submetric represents a total score which can be compared with the scores given other alternatives. This scoring method can help identify a best choice whether it be drill bits, motors, or rotary steerable tools of total design. Table 5-1 illustrates the scoring method. The scores given the various metrics depends to a large extent on experience.

This numerical evaluation scheme has two key objectives. The first is to have a way to quantify one's judgment against a fixed scale so that each design concept can be rated in the same manner, thus, showing their truest level of merit in comparison with each other. The second objective is to provide a means to examine the rationale of the final scores by looking at the subelements of each concept in order to see the strong and weak features of each. Since all design selections are the result of trade-offs, the scoring system aides in selecting the best drillstring and BHA. Specifics of a drilling program are developed jointly by engineers from both operator and contractor. Communication is critical in implementing the plan.

5.1.5 Drillstring Design

It is helpful in design and project work to view the total project as the sum of subparts or subsystems. In drillstring design, it is useful to view the total system as made up of drill bits, motors, MWD, drill pipe, drill collars, etc. This section gives background on each of these components starting with drill bits.

5.1.5.1 *Drill Bits*

Since the beginning of rotary drilling, numerous drill bits have been developed. The first drill bits were drag bits but these were limited by design and metallurgy to soft formations. Drag bits were followed by roller cone bits having two cones with non-intermeshing teeth. The two cone bits gave low rate of penetration, and it was not until a three roller cone bit with intermeshing teeth was developed that roller cone bits were widely accepted.

There are many different types of roller cone bits and each has unique features and benefits. They can be divided into milled tooth bits and tungsten carbide insert cutters. Bearing performance has been extended over the years through design and seal improvements. PDC and diamond drill bits are also options, and each has been designed for special applications and formations.

5.1.5.2 *Roller Cone Drill Bits*

Three components of these bits are cutters, bearings, and bit body. The cuter are fixed to cones which rotate on bearings relative to the bit body. Cutter elements are either machined directly on the roller cones or are tungsten carbide inserts which are pressed into the cone surfaces. Cutters on milled tooth bits are longer, and the cones are offset to create dragging action, making these bits aggressive in soft formations. Insert bits are designed for hard formations and attack the rock by crushing action rather than chipping and gouging action.

Bearings in these bits have to work under high loads and sometime in an abrasive environment. Bearing cavities in early bit designs were not sealed and bearings wore out within 8 to 10 hours. Advanced designs include dynamic seals and lubricant reservoir, which have greatly extended the life of roller cone bits.

5.1.5.3 *Polycrystalline Diamond Compact (PDC) Bits*

Polycrystalline diamond compact (PDC) bits are drag bits containing multiple cutters made of synthetic diamonds. The size of the synthetic diamonds is about 175 diamonds per carat. These tiny diamonds are bonded to form a disc shape (about the size of a nickel) and backed by a thick layer of tungsten carbide substrate. The synthetic diamond layer is about 0.025 in. thick. The tungsten carbide backing is about 0.115 in. thick. This cutting structure is then bonded onto a metal stud which is pressed into a steel body, which forms the drill bit. When used in tungsten carbide matrix bits, they are bonded onto a short disc which is then bonded into the matrix.

Rock bits of all types fail rock by developing high shear stresses under cutters. PDC cuter develops shear stresses efficiently, provided they can be forced into the rock. The application areas for PDC drill bits are soft to medium hard formations.

5.1.5.4 *Natural Diamond Bits*

Natural diamond drill bits were introduced in the late 1940s. These bits allow exploration of deep reservoirs which usually mean harder and more abrasive formations. As a general rule, the smaller the diamond bit, the better it will perform relative to roller cone bits of the same size. The life of a roller bit bearing reduces directly with bearing size. Also, the rate of penetration with diamond bits is usually higher than the rate of penetration with rock bits in small holes and hard formations. Therefore, diamond bits usually lead to a lower footage cost in deep wells where hole size is smallest. On the other hand, diamond bits do not perform as well as rock bits in larger holes and softer formations.

The flow of drilling mud under a diamond bit is parallel to formation and bit face. Water courses in diamond bits make up part of the "nozzle." The formation makes up the other part. Pressure drop calculations for diamond bits are based on the TFA at the control diameter or circle normally located at the nose of the bit. A diamond bit behaves much like a hydrostatic thrust bearing in this regard. Fluid pressure creates an upward force tending to lift the bit off bottom. This pump-off force has to be overcome by additional bit weight.

Fluid pressure under a diamond bit can create an effective overbalance condition between well bore pressure and formation pressure, making the formation more difficult to drill. An overbalance of 0.5 ppg is common for rock bits. This is equivalent to about 300 psi at 10,000 ft. A pressure drop of 2100 psi under a diamond bit is equivalent to a 3.5-ppg overbalance. The formation cannot distinguish between hydrostatic and dynamic fluid pressure.

5.1.5.5 Cost Analysis

The bit that drills a given formation at the lowest cost per foot is the best choice.

Estimated performance of drill bits is based on a past performance in similar formations. Statistical data is useful for predicting bit performance and for economic comparisons of various bit options. Footage cost predictions are useful in this regard.

Footage cost is calculated using a simple formula.

$$C = \frac{C_{\text{rig}}(t_{\text{T}} + t_{\text{d}}) + C_{\text{bit}})}{D} \tag{5-1}$$

where

C	— footage cost	(\$/ft)
C_{rig}	— rig rate	(\$/hr)
C_{bit}	— bit cost	(\$)
t_T	— trip time	(hr)
t_d	— drilling time	(hr)
D	— footage	(ft)

This formula does not include costs related to hole problems, casing, etc. It is mainly used to evaluate the cost effectiveness of drill bits. Since the average rate of penetration is

$$\text{ROP} = \frac{D}{t_{\text{d}}} \tag{5-2}$$

The drilling cost formula can also be written

$$C = \frac{C_{\text{rig}}(t_{\text{T}} + t_{\text{d}}) + C_{\text{bit}})}{\text{ROP}t_{\text{d}}} \tag{5-3}$$

A useful presentation of the drilling cost equation is the breakeven curve. Rearranging Eq. (5-1).

$$D = \frac{C_{\text{rig}}}{C} t_{\text{d}} + \frac{C_{\text{rig}} t_{\text{t}} + C_{\text{bit}}}{C} \tag{5-4}$$

Equation (5-4) predicts the drilling interval required to achieve footage cost, C, during on bottom drilling time, t_d.

A breakeven curve is developed from Eq. (5-4) as follows.

$$D = K_2 t_{\text{d}} + K_3 \tag{5-5}$$

where

$$K_2 = \frac{C_{\text{rig}}}{C}$$

$$K_3 = \frac{C_{\text{rig}} t_{\text{t}} + C_{\text{bit}}}{C}$$

Rearranging the above equations gives

$$\text{ROP} = \frac{K_2 D}{D - K_3} \tag{5-6}$$

The constants K_2 and K_3 depend on the formation being drilled.

A breakeven curve shows the relation between rate of penetration (ROP) and footage required to achieve a given footage cost (see Fig. 5-1). The ROP curve in Fig. 5-1 is bounded by and K_2 and K_3. K_2 is parallel to the D axis, while K_3 is parallel to the ROP axis.

This cost presentation is useful for comparing economic performance of one drill bit against another. If the footage cost using Bit A is known to be C (\$/ft), and if Bit B can maintain a rate of penetration (ROP) to footage (D_{B}), Bit B leads to a lower footage cost than Bit A. In this example, Bit B breaks even or has the same footage cost as Bit A after drilling (D_{A}) footage.

Downhole drilling motors (PDMs) or turbines are normally charged to a drilling interval only for the time they are on bottom drilling. They are not charged to the interval while tripping. The footage cost equation is modified as

$$C = \frac{C_{\text{rig}}(t_{\text{T}} + t_{\text{d}}) + C_{\text{m}} t_{\text{d}} + C_{\text{bit}}}{D} \tag{5-7}$$

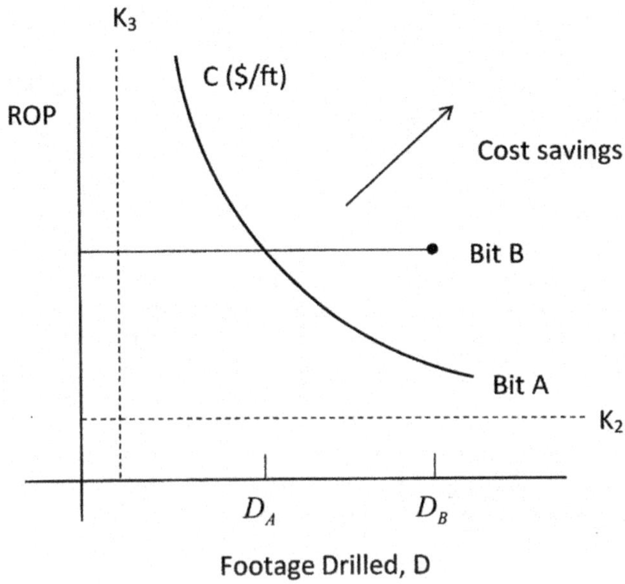

Figure 5-1. Breakeven cost curve.

where C_m is motor rate ($/h). The previous cost diagram still applies where now

$$K_2' = \frac{C_{rig} + C_m}{C} \tag{5-8}$$

K_3 remains the same.

A useful formula compares the ROP required to produce a certain footage cost with a motor (turbine) with the ROP required to produce the same footage cost with conventional rotary.

$$\mathrm{ROP_{motor}} = \frac{K_2' D}{D - K_3} \tag{5-9}$$

$$\mathrm{ROP_{rotary}} = \frac{K_2 D}{D - K_3} \tag{5-10}$$

from which

$$\frac{\mathrm{ROP_{motor}}}{\mathrm{ROP_{rotary}}} = \frac{K_2'}{K_2} \tag{5-11}$$

$$\mathrm{ROP_{motor}} = \left(1 + \frac{C_m}{C_{rig}}\right)\mathrm{ROP_{rotary}} \tag{5-12}$$

Equation (5-12) shows that for the footage cost to be the same for both motor and rotary,

ROP with motor must be $1+\frac{C_{\mathrm{m}}}{C_{\mathrm{rig}}}$ times as much as ROP with conventional rotary.

Experience with downhole motors and turbines suggests that rate of penetration can be increased over conventional rotary drilling by a factor of 1.5 to 1.9 with motors and increased by a factor of 2.0 to 2.9 with turbines.

5.1.5.6 Bottom Hole Assemblies

Bottom hole assemblies include tools such as drill bit, stabilizers, PDM, turbine, MWD, or rotary steerable tools. There may be different tool arrangements for each interval of the well. The reliability and performance of each greatly affect drilling cost.

Bottom hole assemblies (BHA) include all tools from the drill bit to the top stabilizer. Each of the well segments requires a special bottom hole assembly. The location of the casing shoe depends on the formation pressure gradient and drilling mud pressure gradient. This usually means formation changes and thus changes in bottom hole assemblies.

5.1.5.7 Positive Displacement Motors

Motors and turbines may be used for either straight hole or directional drilling but are typically used in directional wells. Motors are used in conjunction with deviation devices which are integral to the motor. Deviation devices that are controlled from the surface are called rotary steerable tools.

Positive displacement drilling motors (PDMs) are a spin-off from the Moineau pump. This pump, sometimes called a progressive cavity pump, is still in use today for transporting food, oil, sewage, and viscous chemicals. Fluid is trapped within cavities (characteristic of the Moineau pump), which progress axially from the intake to the outlet end. Pressure within each cavity may vary as a result of leakage between cavities [2,3].

The centerline distance between rotor and stator is important for the establishment of progressive cavities. If the stator is large and the rotary is small, fluid would simply flow through the void space. Centerline spacing must allow continuous line contact between the rotor and stator in order to create the cavities.

The conversion of a Moineau pump to a downhole drilling motor had two purposes: 1) a positive displacement motor (PDM) located directly above the drill bit could produce more power to the drill bit without rotating the entire drillstring and 2) the PDM could be coupled with a bent sub to force drill bits side ways without the use of whipstocks. Both applications provided huge cost savings. PDMs were introduced when offshore drilling was expanding rapidly during the 1960s.

Positive displacement motors have a direct relationship between output speed and through-put flow rate. This characteristic makes it possible to monitor output speed by monitoring flow rate.

$$N = CQ \tag{5-13}$$

where N is rotational speed (rpm), Q is flow rate through the motor (gpm), and C is a constant depending on geometry of rotor and stator. The constant, C, can also be viewed as

$$C = \frac{n}{q} \tag{5-14}$$

where

n — turns of output shaft (drill bit)
q — corresponding volume thru motor

Since the rotor and stator interact kinematically as a planetary gear train, the number of turns (n) of the output shaft (equal to number of turns of the rotor) per one complete circular travel of the rotor center is

$$n = \frac{a}{b} - 1 \tag{5-15}$$

where

a — number of lobes on the stator
b — number of lobes on the rotor

For example, if a stator has two lobes and the rotor one lobe, $n = 1$ then the output shaft makes one turn for one circular path of the rotor center. Or, if a stator has 10 lobes and the rotor 9, the output rotation of the motor would be 1/9 rotation for each circular path of point O; a much smaller rotation. The travel of point O is counter to the rotation of the rotor.

Consider the volume throughput (q) for one circular movement of the rotor center. Let the two cylinders in Fig. 5-2 represent the pitch cylinders of a rotor

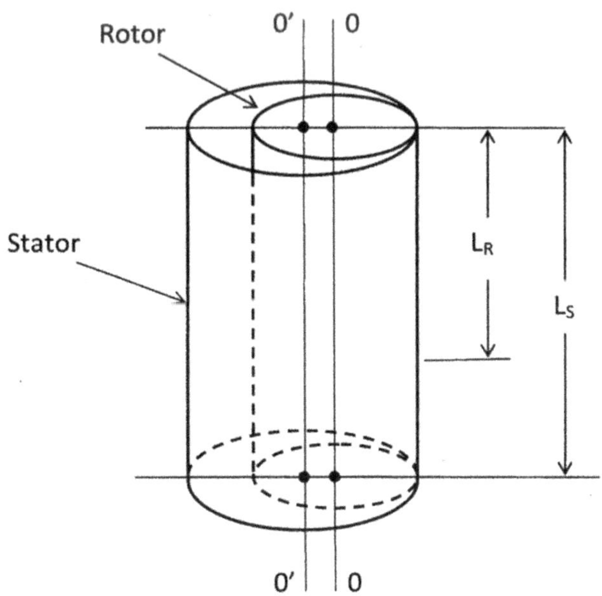

Figure 5-2. Pitch cylinders of rotor and stator.

and stator. Let the length of both pitch cylinders be the length of one stage of the stator, L_s. For the sake of visualization assume:

- center lines O and O' are fixed in space
- both pitch cylinders are torsionally flexible
- top ends of both rotor and stator are fixed
- lower end of the stator makes one complete turn causing the lower end of the rotor to make a/b turns
- rotation of both rotor and stator are linear with distance from the top fixed end

Based on the principle of kinematic inversion, the volume between the rotor and stator over stator length, L_s, is displaced down the motor whether point O is viewed as making a circular path about point O' or both O and O' are viewed as being fixed and the two pitch circles mesh as a gear pair. This volume is

$$q = AL_s \tag{5-16}$$

where

A — cross-sectional area of the space between rotor and stator cavity
L_s — length of one stage as measured on the stator.

Note that this volume is not the same as each isolated cavity entrapped between rotor and stator. It represents the total volume over one stage (L_s).

Combining Eqs. (5-15) and (5-16) with Eq. (5-14) gives

$$C = \frac{n}{q} = \frac{\frac{a}{b} - 1}{AL_s} \tag{5-17}$$

Since $a = b + 1$ for PDMs

$$C = \frac{1}{bAL_s} \tag{5-18a}$$

and since $L_r = \frac{b}{a} L_s$

$$C = \frac{1}{aAL_r} \tag{5-18b}$$

Equation (5-13) then becomes

$$N = \frac{1}{bAL_s} Q \tag{5-19}$$

or

$$N = \frac{1}{aAL_r} Q \tag{5-20}$$

AL_r is the total entrapped volume between rotor and stator over length L_r. However, it is not the cavity volume. In consideration of Eq. (5-19) or Eq. (5-20), the smaller the denominator, the larger the output rotational speed for a given flow rate, Q. Also, the larger the number of lobes, the lower the output speed. That is why multi-lobed motors operate at low speeds and high torque. Note that N and Q are proportional which is characteristic of positive displacement motors and pumps.

The conversion of fluid pressure to mechanical torque is the result of the fluid wedge at the fluid entry or first stage. Fluid density and flow rate have no effect on torque generated by the motor. PDMs are typically made up of several stages and each stage contains an entrapped volume of fluid which travels through the motor. Even though these motors have multiple stages, no work is done by the constant volume cavities and only the top stage converts hydraulic pressure to torque.

Multiple stages, usually about five, perform as dynamic seals so reasonable pressure differentials across these motors can be developed. Performance tests show that even these back-up stages allow a certain amount of fluid leakage. The extra stages do, however, create friction and affect the overall efficiency.

Since the hydraulic horsepower consumed by the motor is equal to the output mechanical output power (assuming no losses)

$$\Delta pQ = TN \tag{5-21}$$

So output torque is

$$T = \frac{Q}{N}\Delta p \tag{5-22}$$

$$T = (bAL_s)\Delta p \tag{5-23}$$

This equation shows that output torque increases with number of lobes on the rotor (and stator). Torque also increases with stage length, which increases the wedging effect of fluid entering the motor.

Typical performance data for PDMs is shown in Fig. 5-3. Pressure differentials for various dynamometer torque settings are recorded at constant flow rates. Applied torque and rotational speeds are recorded. When applied torque is zero, rotational speed is a maximum. Pressure differential is required to overcome internal friction but is lowest under no torque conditions. When torque is increased, rotational speed drops off slightly as a result of fluid leakage through each stage. Higher torque requires higher differential pressures and causes a reduction in rotational speed due to leakage.

When applied torque is taken to the extreme, PDMs stall out. This means that the pressure level is great enough to force fluid through motors without rotation. Fluid leakage is 100%. Even though torque is maximum under stall conditions, power output is zero because the output shaft does not turn. Therefore, operating a PDM near stall conditions produces little or no power to the drill bit. PDMs are normally rated at torque and pressure levels corresponding to about a 10% leakage through the motor.

PDMs operate between recommended maximum and minimum flow rates. The maximum flow rate is based on a limiting linear velocity of the rotor. The

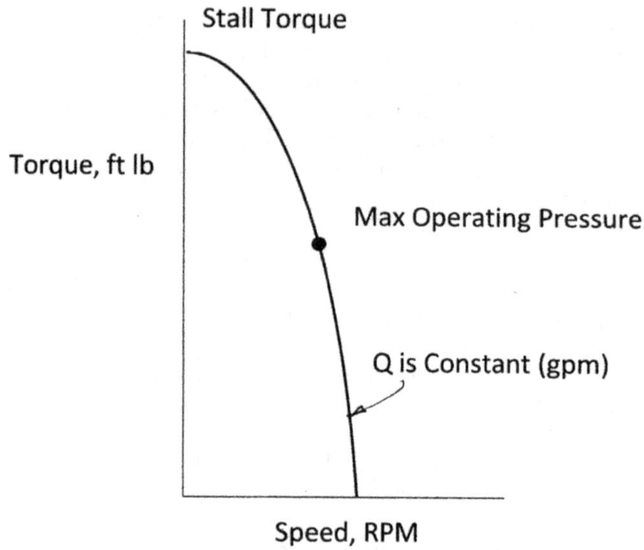

Figure 5-3. Typical performance diagram for Moineau type motors.

rotor is constantly undergoing accelerations which generate forces on rubber undulations in stators. This could cause premature failure of the rubber. Minimum flow rate is based on inefficiency, which is typical at low flow rates. Based on these guidelines, laboratory performance data can be approximated as shown in Fig. 5-4. The numbers are typical of a 1:2 PDM. The performance of PDMs with multiple lobes deviates somewhat from this model.

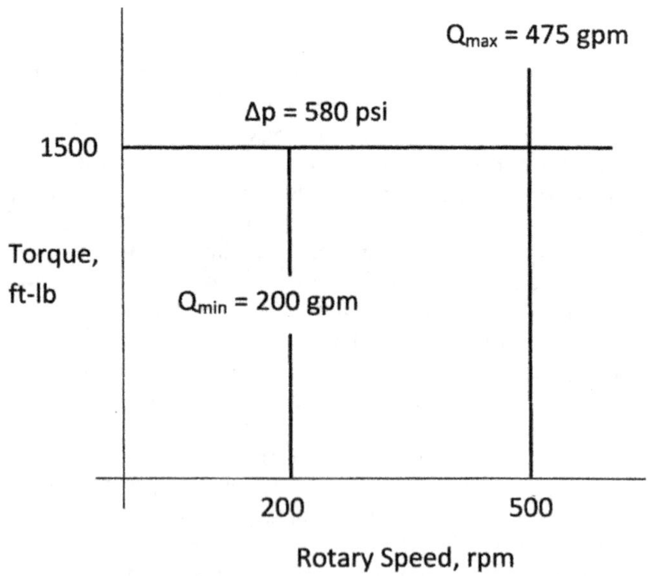

Figure 5-4. Idealized performance diagram for Moineau motors.

It is important to note that maximum recommended operating torque and pressure levels are not stall conditions. Stall torque is approximately 50% higher than the recommended operating torque.

The capability to monitor performance of PDMs from the surface is a distinct advantage. Pressure drop across these motors can be monitored by standpipe pressure and output speed can be monitored by pump flow rate.

It has long been recognized that out of the total mechanical power (~8000 hp) required to operate a drilling rig, only a small fraction is used at drill bits. For example, under normal rotary drilling operations, bit horsepower is

$$P_{\text{bit}} = \frac{2\pi TN}{33{,}000}\,\text{hp} \tag{5-24}$$

$$P_{\text{bit}} = \frac{2\pi(1500)80}{33{,}000} = 22.8\ \text{hp}$$

The output speed of a 1:2 PDM is in the range of 400 rpm. Corresponding output power is

$$P_{\text{bit}} = \frac{2\pi(1500)400}{33{,}000} = 114\ \text{hp}$$

Bit power is increased by a factor of 4. This increase in bit power comes at a cost because motors cost (C_{pdm}) is added into footage cost calculations. However, PDM cost is offset by increased rate of penetration (ROP).

One of the issues facing the use of PDMs was the need for a companion drill bit which could operate at the higher rotary speeds. Roller cone drill bits operated successfully up to 150 rpm. PDC drill bits satisfy this requirement and can operate in the 400-rpm range.

Positive displacement motors come in different sizes and lobe numbers (Fig. 5-5). In general, the higher the lobe numbers the higher the torque and the lower the speed. A 1:2 combination rotor to stator configuration creates the highest speed. A 5:6 lobe configuration, for example, performs at a higher torque and lower speed.

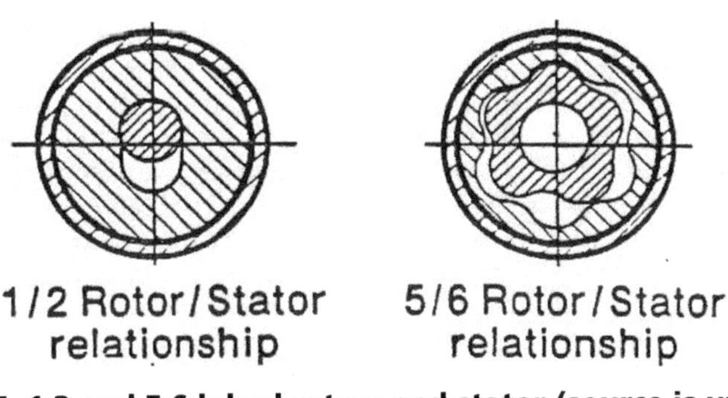

Figure 5-5. 1:2 and 5:6 lobed rotary and stator (source is unknown).

Sometimes, it is desirable to rotate the entire drillstring simultaneously with a PDM. Drillstring rotation disorients the tool face when PDMs with tilted bearings are used. It also reduces drillstring friction. The question concerns the amount of mechanical power delivered to the drill bit. Since mechanical power depends on both rotational speed and torque, consider the speed first. Bit speed is the absolute speed of the motor housing plus the rotary speed of the drill bit relative to the housing. So if the drillstring rotates at 60 rpm, and the output speed of the motor (based on flow rate) is 400 rpm, then absolute bit speed is 460 rpm.

If motors operate within the recommended torque range, total bit speed is the direct sum of the two speeds. When motor torque reaches stall torque as determined from performance tests, the rotor speed relative to the housing is zero and bit speed is due only to pipe rotation; the motor is locked and forms a direct drive between bit and rotary table. It is only when bit torque goes above stall torque that the motor housing overtakes the rotor. However, this is an adverse loading condition on the motor and should be avoided.

The torque that gets to the drill bit is controlled by the motor. Bit torque can be no greater than the motors allows. Assuming zero friction, reactive motor torque gets transmitted back to the rotary table so bit and surface torque are the same (for this idealized example). In this case, mechanical power at the drill bit is the sum of input power at the surface plus power generated by the motor.

5.1.5.8 Downhole Drilling Turbines

Downhole drilling turbines develop torque fundamentally differently from positive displacement motors. Torque is developed by fluid momentum while torque in a PDM is developed by direct application of fluid pressure. Output speed of a turbine depends directly on output torque while output speeds of PDMs depend on flow rate and is independent of torque. Maximum power is developed at a particular rotor speed and herein lies a monitor/control issue. Monitoring and operating turbines at their optimum power speed is necessary for best use of drilling turbines. Since turbine torque is the result of fluid momentum, torque output depends on both drilling mud density and flow rate.

Each turbine stage has a fixed set of blades (stator) and a rotating set of blades (rotor). Stator blades direct drilling fluid onto rotor blades where output torque is generated. Output torque, output power and overall pressure drop of a turbine is a direct multiple of the number of stages in the turbine. Tiraspolsky [4] gives a comprehensive description of drilling turbines used in Russia.

When the rotor blade is stationary, fluid momentum changes the most, and the torque is the greatest. When the rotor blade is moving, fluid momentum is not changed as much so torque drops off. The faster the rotor turns, the lower the torque generated by the rotor blades. At a particular high speed of rotor rotation, fluid leaving the stator blades cannot catch the rotor blade so its momentum does not change. This particular speed is the runaway speed, and at this speed, no torque is developed on the rotor blades.

In practice, rotational speed is not set in turbines; output speed responds to applied torque or bit weight. Torque is applied and rotational speed automatically adjusts to the torque. Both stall torque and runaway speed change with mud weight and flow rate. Turbines are rated at maximum power. Whether or not maximum power is developed by downhole turbines depends on the ability to monitor and control bit torque.

The mechanics of turbine power is explained by considering momentum changes across the control volume, abcd (Fig. 5-6). The rate of change of fluid momentum across cd and ab requires a force vector, $\overline{R}$ [5].

$$\overline{R} = Q_m(\overline{V}_2 - \overline{V}_1) \tag{5-25}$$

where

$\overline{V}_1$ — absolute velocity of fluid leaving stator blade boundary ab
$\overline{V}_2$ — absolute velocity of fluid leaving rotor blade boundary cd
Q_m — mass flow rate through control volume
$\overline{R}$ — resultant force on fluid in control volume

The component of this vector in the tangent direction is

$$R_\theta = Q_m(U - 2V\cos\beta) \tag{5-26}$$

or

$$R_\theta = Q_m\left(U - \frac{2u}{\tan\beta}\right) \tag{5-27}$$

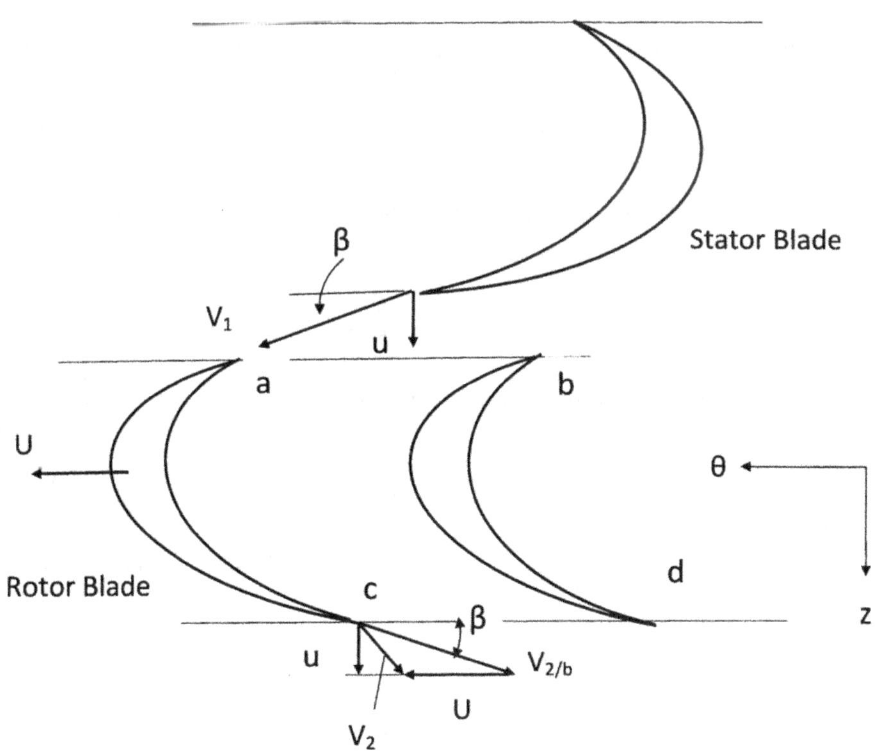

Figure 5-6. Rotor and stator turbine blades.

Since this is the force acting on the fluid in the control volume, the fluid force applied to the blade is

$$F_\theta = Q_m\left(\frac{2u}{\tan\beta} - U\right) \tag{5-28}$$

Output torque is produced by this force. Total torque generated by one turbine stage is

$$T = rQ_m\left(\frac{2u}{\tan\beta} - U\right) \tag{5-29}$$

where Q_m is now the total mass flow rate through the turbine and U is the tangent speed of the rotor and r is the average radial distance. Also

$$u = \frac{(1-\varepsilon)Q}{A\lambda} \tag{5-30}$$

where

ε — coefficient of fluid loss around outside of rotor
λ — contraction coefficient due to flow area occupied by blade thickness

In consideration of these factors, the torque generated by one turbine stage is

$$T = r\rho Q\left[\frac{2(1-\varepsilon)Q}{A\lambda\tan\beta} - U\right] \tag{5-31}$$

where Q_m has been replaced by ρQ. From this expression, the maximum torque occurs when $U = 0$ or the rotor blade is stalled.

$$T_{max} = r\rho\frac{2(1-\varepsilon)Q^2}{A\lambda\tan\beta} \tag{5-32}$$

On the other hand, output torque is zero when

$$U = \frac{2(1-\varepsilon)Q}{A\lambda\tan\beta} \tag{5-33}$$

This velocity is the runaway speed of the turbine. When the blade tip angle decreases from 30 deg to 20 deg, maximum torque and runaway speed increase by a factor of 1.59.

Output mechanical horsepower (Power = 2π TN) is formulated in terms of rotor/stator design parameters, drilling fluid density and flow rate.

Starting with

$$T = r\frac{\gamma}{g}Q\left[\frac{2(1-\varepsilon)Q}{A\lambda\tan\beta} - U\right] \tag{5-34}$$

and

$$U = 2\pi N_r = \frac{u}{\tan\beta} = \frac{(1-\varepsilon)Q}{A\lambda\tan\beta} \tag{5-35}$$

Power output is defined by

$$P_{out} = \frac{\gamma}{g}Q\left[\frac{2(1-\varepsilon)Q}{A\lambda\tan\beta} - U\right]U \tag{5-36}$$

Maximum power occurs at

$$\frac{d(P_{out})}{dU} = \frac{2(1-\varepsilon)Q}{A\lambda\tan\beta} - 2U = 0 \tag{5-37}$$

or

$$U = \frac{2(1-\varepsilon)Q}{A\lambda\tan\beta} \tag{5-38}$$

which gives

$$P_{max} = \frac{w}{g}\frac{(1-\varepsilon)^2}{A\lambda\tan\beta}Q^3 \tag{5-39}$$

Pressure drop across one turbine stage is determined from the Bernoulli equation.

$$\frac{\Delta p}{\gamma} = \frac{P_{max}}{QW} \tag{5-40}$$

So

$$\Delta p = \frac{\lambda}{g}\frac{(1-\varepsilon)^2}{A\lambda\tan\beta}Q^2 \tag{5-41}$$

The above equations predict the performance for drilling turbines. If performance information is known for one set of operating parameters, the performance at other operating parameters can be calculated using

$$N_2 = \frac{Q_2}{Q_1}N_1 \tag{5-42}$$

$$T_2 = \left(\frac{Q_2}{Q_1}\right)^2\left(\frac{\gamma_2}{\gamma_1}\right)T_1 \tag{5-43}$$

$$\Delta p_2 = \left(\frac{Q_2}{Q_1}\right)^2\left(\frac{\gamma_2}{\gamma_1}\right)\Delta p_1 \tag{5-44}$$

$$HP_2 = \left(\frac{Q_2}{Q_1}\right)^3\left(\frac{\gamma_2}{\gamma_1}\right)HP_1 \tag{5-45}$$

These formulas are based on the preceding derivation.

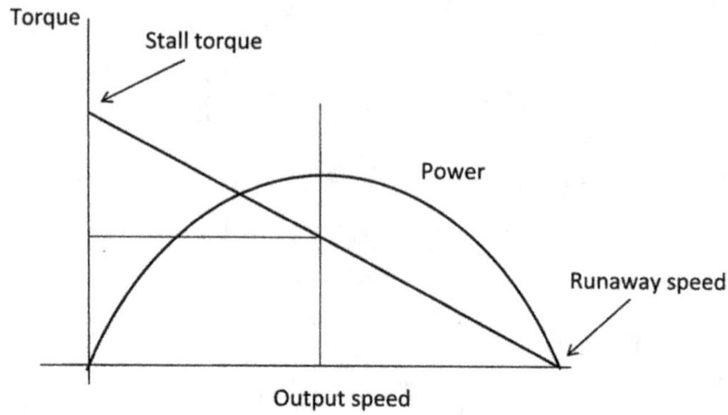

Figure 5-7. Performance curves for drilling turbines.

The torque and rotary speed are related graphically as a straight line (Fig. 5-7). Both stall torque and runaway speed change with mud weight and flow rate. Mechanical power delivered by turbines can be derived from the torque-speed line as follows:

$$T = T_{st} - \frac{T_{st}}{N_{rs}} N \tag{5-46}$$

where

T_{st} — stall torque
N_{rs} — runaway speed

The mechanical output power is

$$P = \frac{TN}{5255} \text{ hp} \tag{5-47}$$

when Eq. (5-46) is substituted in to this equation, the result is a parabola that reaches a maximum value at ½ stall torque and ½ runaway speed (Fig. 5-7).

$$P = \frac{T_{st}}{5255} N\left(1 - \frac{N}{N_{rs}}\right) \tag{5-48}$$

Because turbine speed depends on applied torque, mechanical power delivered by turbines depends on applied torque. When applied torque changes, output speed changes and power is delivered at particular combinations of torque and speed. A means of monitoring turbine speed is necessary to make maximum use of turbine power under drilling operations.

5.1.5.9 Measurement While Drilling Tools

Measurement while drilling (MWD) tools measure downhole parameters and transmit this information back to the surface by means of mud pulse telemetry.

This is done without disturbing drilling operations. MWD is used in directional drilling to monitor hole direction and inclination. They also measure and transmit other downhole parameters, such as force, torque, pressure, and temperature.

These tools include electronics, mechanical devices, and sensors, all of which must perform in the harsh drilling environment. Reliability is of utmost importance. The tool can be self-monitoring and relay dynamic loading caused by sever vibrations and impact environment.

5.1.5.9.1 Directional Drilling

Important directional drilling parameters are hole inclination and direction. These parameters are needed to calculate well path, and tool face orientation for hole corrections (Fig. 5-8).

There are different methods for predicting the location of station 2 based on data measured at station 1 and station 2. The tangent method provides a simple way to calculate incremental changes in *xyz* coordinates.

$$\Delta x = \mathrm{CL} \sin\alpha \sin\varepsilon \tag{5-49}$$

$$\Delta y = \mathrm{CL} \sin\alpha \cos\varepsilon \tag{5-50}$$

$$\Delta z = \mathrm{CL} \cos\alpha \tag{5-51}$$

where

CL — closure distance
α — hole inclination
ε — hole direction

This method gives an approximate location of station 2 because the direction of the closure line is not exactly the same as the direction of the unit vector of the well path at station 1 and the path between station 1 and station 2 is usually curved. The smaller the closure distance the more accurate the projected

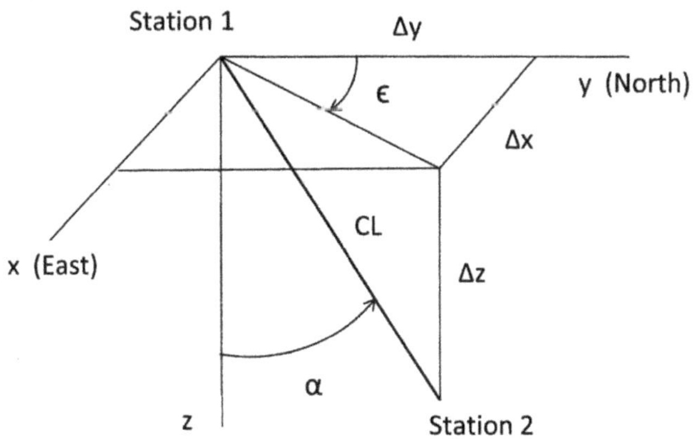

Figure 5-8. Incremental steps between two survey stations.

location of station 2 by the tangent method. Other methods are average angle, radius of curvature, minimum curvature.

As a directional well progresses, decisions are made on whether to make a correction run or stay with a given bottom hole assembly. The present assembly may not be performing as expected. The real dilemma arises when a bit is still green and cannot be used as a reason for tripping out of the hole. The more hole drilled with an ineffective assembly, the shorter the remaining distance for a hole correction. It is common practice to project ahead after each survey to make sure the hole can be turned to the target, still keeping the well path within specified limits. Hole corrections are made by properly orienting the tool face in a required plane of drilling. There are various techniques for calculating this tool face orientation. The method given below is the vector method (Fig. 5-9).

The objective for the formulation is to determine angle λ, which is measured from the high side of the hole to the desired plane of drilling. The plane of drilling contains the closure line, represented by vector, $\bar{a}_T$, and tangent vector at station 2, $\bar{a}_2$. Unit vector, $\bar{b}$, is perpendicular to the unit tangent vector, $\bar{a}_z$ and points toward the high side of the hole. Unit vector, $\bar{c}$, is perpendicular to the desired plane of drilling or the plane containing the oriented tool face. These unit vectors are defined by

$$\bar{a}_2 = (\sin\alpha_2 \sin\varepsilon_2)\bar{i} + (\sin\alpha_2 \cos\varepsilon_2)\bar{j} + \cos\alpha_2 \bar{k} \tag{5-52}$$

$$\bar{b}_2 = (\cos\alpha_2 \sin\varepsilon_2)\bar{i} + (\cos\alpha_2 \cos\varepsilon_2)\bar{j} - \sin\alpha_2 \bar{k} \tag{5-53}$$

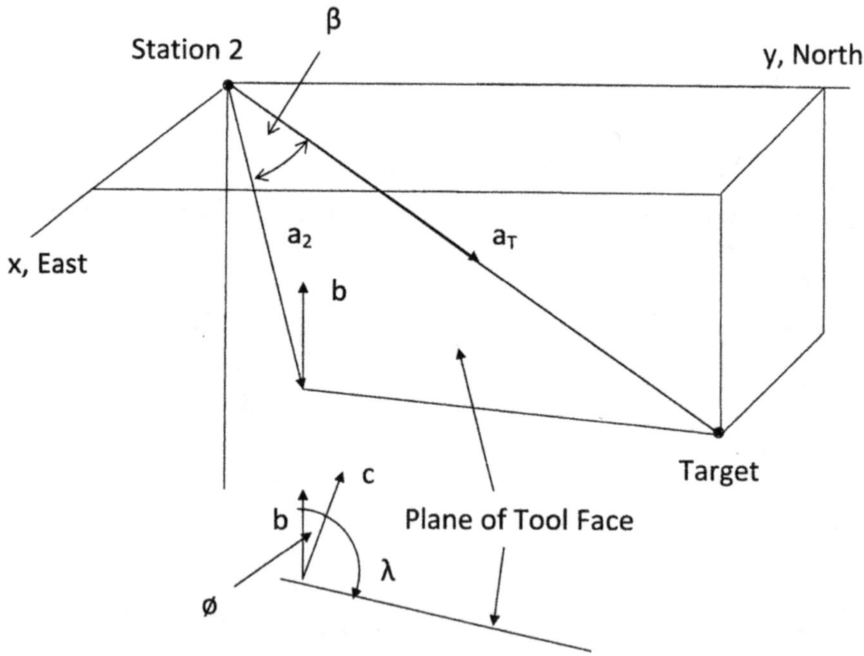

Figure 5-9. Vector method for determining tool face orientation.

$$\bar{c}\sin\beta = \bar{a}_2 x \bar{a}_T \tag{5-54}$$

where $\bar{i}, \bar{j}, \bar{k}$ are unit vectors in x, y, z directions.

The angle, β, is the projected dog leg angle. The actual dog leg angle required to turn the hole toward the target will be greater than β because the hole cannot be turned instantaneously. The desired tool face orientation referenced from the top side of the hole is

$$\lambda = 90 \pm \phi \tag{5-55}$$

where φ is determined from

$$\cos\phi = \bar{b} \bullet \bar{c} \tag{5-56}$$

Historically, these calculations were made over a relative large increment of the well path. With the use of rotary steerable tools, these measurements and hole corrections are made instantaneously, allowing precise navigation toward the target.

5.1.5.9.2 Vibration Monitoring

Mechanical vibrations in and around BHAs can be very severe and can damage tools. Vibration control is essential. Frequency tuning is the cause of severe vibrations. MWD can monitor vibrations, identify the source and lead to steps necessary to reduce or eliminate these dynamic forces.

5.1.5.10 Stabilized Assemblies

Stabilizers are special tools that centralize drill collars at specific locations to establish a desired side load to drill bits. Stabilizer blades are roughly two feet long and allow fluid to flow upward in the annulus with minor flow restriction. Stabilized assemblies can build, hold, or drop hole angle, depending on stabilizer number and spacing [6].

5.1.5.10.1 Building Assemblies

Building assemblies develop an upward side force on drill bits. A near bit stabilizer acts as a fulcrum point for the collar weight to pry against (Fig. 5-10a). The mechanical advantage of this assembly is greatest when the first stabilizer is close to the bit. A second stabilizer is positioned some distance to establish a large free span distance to increase the strength of building assemblies. Positioning of the second stabilizer depends on collar size. Stiffer or larger drill collars will allow a longer free span distance.

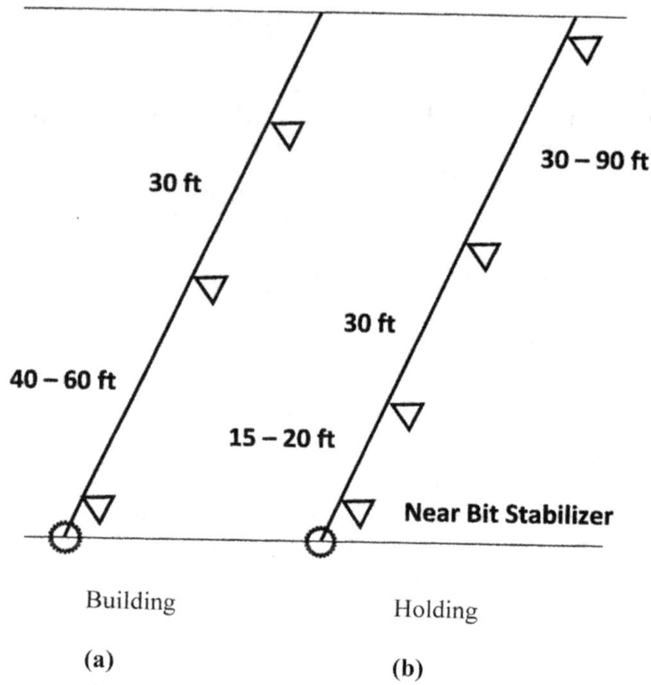

Figure 5-10. Bottom hole assemblies.

5.1.5.10.2 Holding Assemblies

Stabilizers are positioned within drill collars so that relatively small side forces (usually less than 200 lbs) are developed at the bit. When drill bits having long shanks or large gauge protection areas are used, it is desirable to minimize the tilt angle with stabilizer placement to alleviate bit binding. A typical holding assembly is illustrated in Fig. 5-10b. Calculations show that hole angle directly affects the magnitude of the side force for each of these assemblies. Side forces increase with hole angle.

5.1.5.10.3 Dropping Assemblies

Unlike building and holding assemblies, dropping assemblies have no near bit stabilizers (Fig. 5-11). The drill bit stands alone and is pushed downward by the weight of drill collars suspended between the bit and the first stabilizer which is roughly 60 ft from the bit. The first stabilizer in this case acts much like a hinge, allowing the suspended collars to swing downward like a pendulum.

To make maximum use of the pendulum assembly, the freely suspended length should be as long as possible. If the first stabilizer is located too far from the bit, the suspended collars can sag too much, touch the inside of the well bore and lose the strength of the pendulum. Stiffer or larger drill collars weigh more and allow a greater suspended distance for a stronger dropping assembly.

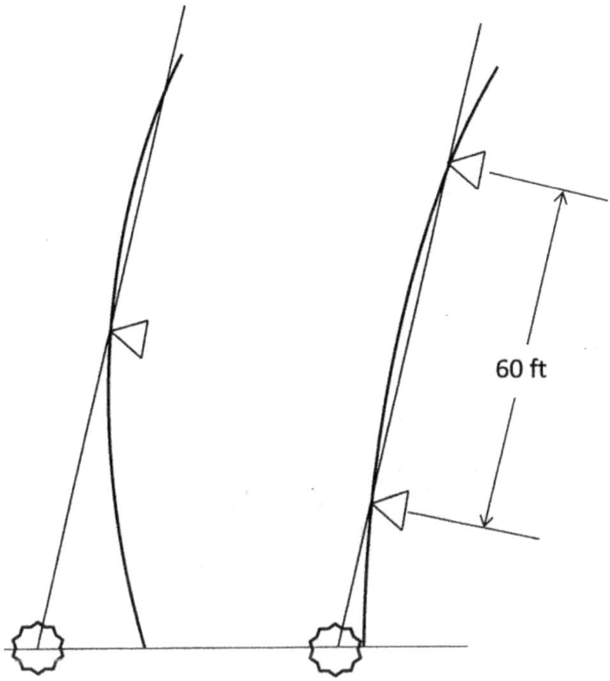

Figure 5-11. Pendulum assemblies.

Dropping Assemblies for Soft Formations—Special types of dropping assemblies are used in drilling vertical wells (see Fig. 5-11a). The 60 pendulum means a single stabilizer is placed some 60 ft from the bit. The drill bit is forced downward by the upward deflection of the drill collars above the stabilizer. This causes the collars below the stabilizer to swing back to the vertical. This assembly is especially effective in soft to medium hard formations because the threshold sidewise drilling force is low in these formations. The standard pendulum assembly generates small side forces on the frill bit in near vertical holes. Bit weight is usually very low (~ 10,000 lbs).

Dropping Assemblies for Hard Formations—The 60-ft pendulum is slow to respond in hard formations because of low allowable bit weights. An alternative is a 30 to 90 pendulum (Fig. 5-11b), which allows higher weight on bit and thus a higher rate of penetration; the distance between the first and second stabilizer is 60 ft. WOB could be near 50,000 lbs. In addition the higher bit weight tends to make the bit drop hole angle even faster.

The drill collars above the second stabilizer lie on the low side of the hole due to gravity. This produces an upward bow in the drill collars between the first and second stabilizers which forces the drill bit sideways toward the vertical. The greater the weight on bit, the greater the side force, which moves the bit toward the vertical. This assembly typically holds hole inclination within 2 ½ deg from vertical. It begins the corrective action around 2 deg inclination.

As an example consider the following case where

Hole size — 8.5 in.
Hole inclination — 4 deg.
Drill collar size — 6 ¾ OD × 2.8124 ID (83.27 lb/ft)
WOB — 50,000 lbs
Mud weight — 10 ppg (BF = 0.847)

Drill collar displacements and bit side force was determined for this dropping assembly using the method explained in Chapter 3. The deflection of the drill collars from drill bit to upper tangent point is shown in Fig. 5-12. Drill collar deflections between drill bit and the first stabilizer are magnified in Fig. 5-13. Zero deflection points are located at the drill bit and two stabilizers. The deflection at the upper tangent point is –0.875 in. (8.5 to 6.75)/2. Reaction force at the drill bit is 39 lbs, while the action forces at the first and second stabilizer are 290 and 352 lbs, respectively. The reaction at the upper tangent point is 157 lbs. Note that the bit side force is relatively small but positive. This is significant since WOB was 50,000 lbs.

5.1.5.11 Drill Collars

Drill collars, here, refers only to the tubular. Its primary function is to apply force to the drill bit.

Drill collar size and length depend on hole size, formation hardness, and directional control. Because of the relative size of drill collars in comparison with drill pipe, drill collars tend to dictate the overall dynamics of BHAs and often times the entire drillstring.

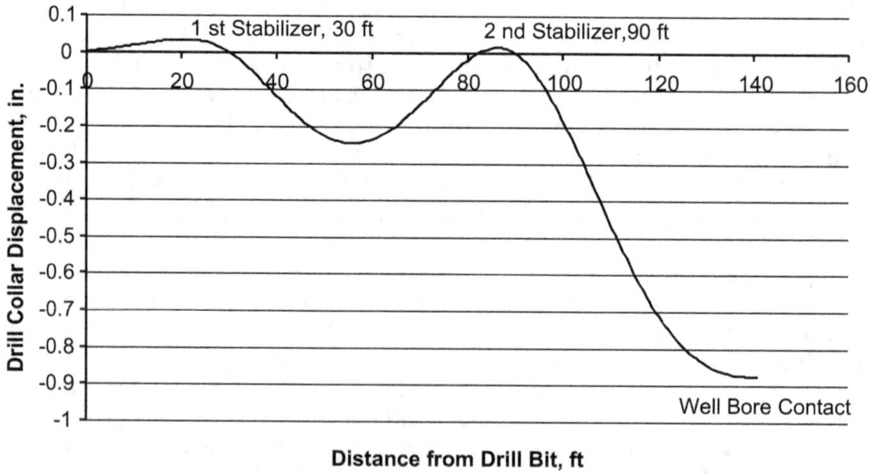

Figure 5-12. Drill collar deflection for 30–90 dropping assembly (hole inclination is 4 deg).

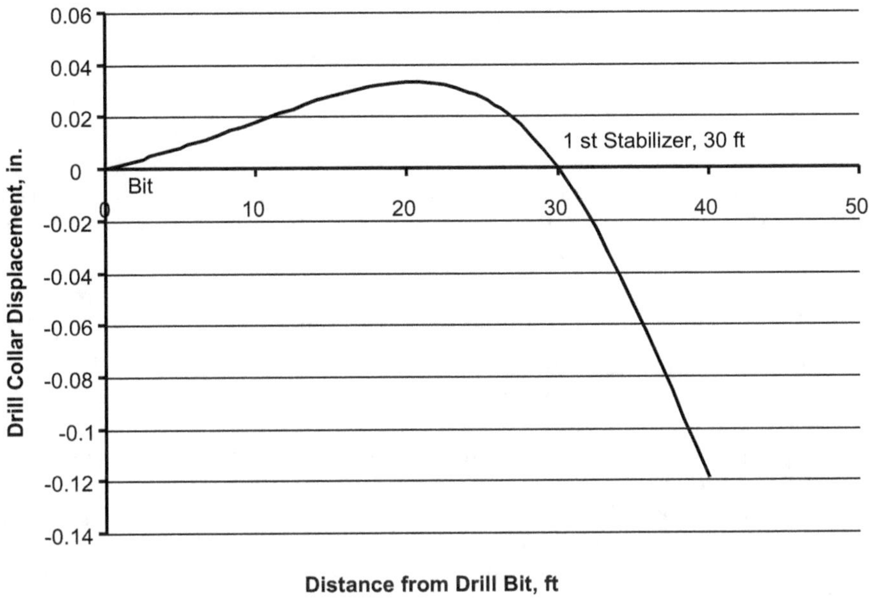

Figure 5-13. Drill collar deflection near drill bit and first stabilizer.

Make-up torque for drill collars is limited by to a stress level of 62,500 psi in the box or pin, which ever is weaker [7]. An important aspect of rotary shouldered drill collar connections is the relative bending stiffness of the pin and box. If the box is too thin, the pin can force high local bending in the box and cause early failure. If the pin is too small the box becomes the dominant member. Experience has shown that a bending strength ratio of 2.50:1 balances the bending strength of box and pin. As drill collars wear and the outside diameter of a box reduces, the bending strength ratio (BSR) may reduce to a level that is undesirable. Therefore, BSRs slightly higher than 2.50:1 may be selected, especially in abrasive formations in anticipation of reductions in BSR.

Drill collars are produced in two grades of steel according to Table 5-2. Drill collars are made of high strength steel because connections are machined directly into the drill collar body.

The length of drill collars depends on the expected force to be applied to the drill bit (WOB). Because of high compressive drill bit forces, the drill collar section buckles and produces bending at the connections. Make-up torque must be

Table 5-2. Physical properties of drill collars [7].

Drill collar OD range (in.)	Minimum yield psi	Minimum tensile strength psi
3 ¼ thru 6 7/8	110,000	140,000
7 thru 10	100,000	135,000

high enough to keep the shoulders engaged in order to protect the threads from cyclic bending and fluid erosion.

5.1.5.12 Drill Pipe

The purpose of drill pipe is to support the entire drillstring under a variety static and dynamic conditions. Specifications for the well bore and drill pipe selections are important. Typical specifications are:

- Maximum hook load accounting for over pull
- Maximum expected torque
  - Accounting for friction and well path
- Well path plan and expected Dog-leg severity with location
- Anticipated dynamic loading including frequency and amplitude
  - Axial mode
  - Torsion mode

The performance capabilities and operating guidelines are documented [7], and it is not the purpose here to repeat this well-established data. However, it is appropriate to review a few loading situations that can affect the performance of drill pipe.

5.1.5.12.1 Drill Pipe Stress Components

Figure 5-14 shows different types of stress components that can be generated in the body dof drill pipe. This three dimensional state of stress is typical directly below the rig floor. As an example, consider the following load condition directly below the rig floor.

F_x = 250,000 lbs	(hook load; including margin of overpull)
T = 20,000 ft-lbs	(surface torque)
p = 3,000 psi	(standpipe pressure)

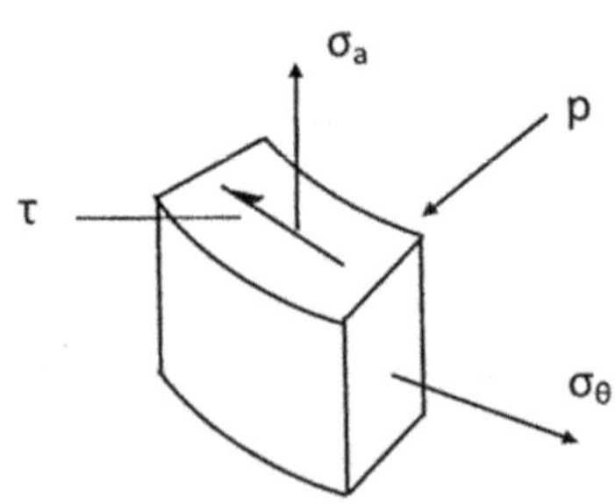

Figure 5-14. Stresses on a pipe element.

Furthermore assuming 5 ½ (19.2 lb/ft) drill pipe having wall thickness of 0.304 inches. The $\frac{r}{t}$ ratio is 8.546 which is somewhat lower than $\frac{r}{t} \geq 10$ generally specified for thin wall pressure vessels [8]. Cross section area is 4.9624 in.^{2} Polar sectional modulus is 12.221 in.^{3}

OD = 5 ½ in.
ID = 4.892 in.
r_0 = 2.75 in.
t = 0.304 in.
A = 4.9624 in.^{2}
S = 12.221 in.^{3} (polar section modulus)

Then stress components on the outside surface are

$$\sigma_a = \frac{F_x}{A} = 50{,}380 \text{ psi} \quad \text{(axial stress)} \tag{5-57}$$

$$\tau = \frac{Tr_0}{J} = \frac{T}{S} = 19{,}640 \text{ psi} \quad \text{(shear stress)} \tag{5-58}$$

$$\sigma_\theta = \frac{r_0}{t} p = 27{,}140 \text{ psi} \quad \text{(hoop stress)} \tag{5-59}$$

Stress components on the inside surface are the same except there is a radial component

$$\sigma_r = -3{,}000 \text{ psi}$$

Using the above numbers, principal stresses components [8] on the outside surface are:

$$\sigma_{p1} = 61{,}580 \text{ psi}$$

$$\sigma_{p2} = 15{,}940 \text{ psi}$$

$$\sigma_{p3} = 0$$

Principal stresses on the inside surface are

$$\sigma_{p1} = 61{,}580 \text{ psi}$$

$$\sigma_{p2} = 15{,}940 \text{ psi}$$

$$\sigma_{p3} = -3{,}000 \text{ psi}$$

Assuming the von Mises criteria of failure, material yielding takes place when

$$\sigma_f \geq \sigma_{yld} \tag{5-60}$$

Substituting for the assumed loading, von Mises stress [8] on the outside surface is

$$\sigma_f^2 = \sigma_{p1}^2 - \sigma_{p1}\sigma_{p2} + \sigma_{p2}^2 \quad \text{(2D case)} \tag{5-61}$$

$$\sigma_f = 55{,}360 \text{ psi}$$

and von Mises stress on the inside surface is

$$2\sigma_f^2 = (\sigma_{p1} - \sigma_{p2})^2 + (\sigma_{p2} - \sigma_{p3})^2 + (\sigma_{p3} = \sigma_{p1})^2 \quad \text{(3D case)} \tag{5-62}$$

$$\sigma_f = 57{,}500 \text{ psi}$$

Assuming a pipe grade of G105, the factor of safety on the outside surface is

$$FS = \frac{105{,}000}{55{,}360} = 1.9$$

The FS on the inside surface is

$$FS = \frac{105{,}000}{57{,}500} = 1.83$$

Bending stresses are not included here, but could be important farther down the well bore.

5.1.5.12.2 Bending at Drill Pipe/Drill Collar Interface

This transition from drill pipe to drill collars is an area of frequent drillstring failures [7] and is particular vulnerable when drill bits are rotated off bottom. This condition can create high bending when the transition is located in dog legs. Heavy weight drill pipe is sometime used in this transition region to alleviate this problem.

Bending moments and stresses are, however, quite nominal under buckled conditions in a straight vertical hole. To illustrate, assume that drill pipe is allowed to buckle as shown in Fig. 3-2 (Chapter 3). This data applies to the case of α = 4000 and $\beta = \beta_{cr}$. Figure 3-2 shows that

$$\frac{ML^2}{EIY_0} = 200 \quad \text{(first buckle mode)} \tag{5-63}$$

and

$$\frac{ML^2}{EIY_0} = 250 \quad \text{(second buckle mode)} \tag{5-64}$$

Assuming 5 in. (19.5 lb/ft) drill pipe

ID= 4.276 in.
I = 14.269 in.^{4}
E = 30 (10)^6 psi
L = 5,000 ft
Y_0 = 2.5 in.

The bending moment for the first and second modes are:

$$M_1 = 200\frac{EIY_0}{L^2} = 200\frac{30(10)^6 14.269(2.5)}{5000^2}\left|\frac{1ft}{12in}\right|^3 = 4.95\, ft-lb$$

$$M_2 = \frac{250}{200}4.95 = 6.19\, ft-lb$$

Corresponding bending stresses are

$$\sigma_{B1} = 10.42 \text{ psi}$$

$$\sigma_{B2} = 13.02 \text{ psi}$$

These bending stress levels are quite low and support the practice of sometimes running drill pipe in compression in horizontal drilling. However, bending in this transition zone becomes significant in high curvature situations [9].

5.1.5.12.3 Effect of Dog Legs on Bending Stress

Bending caused by local dog legs can be significant. Consider the mathematical model (Fig. 5-15) showing tool joints lying flat against the inside of a curved

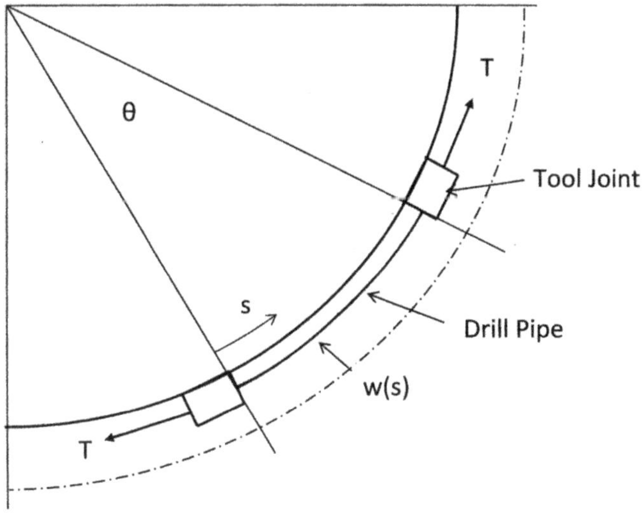

Figure 5-15. Drill pipe bending between tool joints.

well bore. A drill pipe section of length, L, is assumed to be under tension, T. This condition creates high localized bending at each tool joint as explained below.

The curvature of the displaced beam is expressed in terms of beam deflection by (see Timoshenko and Woinowsky-Krieger [10, pp503]

$$\frac{1}{\rho} = \frac{1}{R} + \frac{w}{R^2} + \frac{d^2w}{ds^2} \tag{5-65}$$

Assuming Euler bending

$$M = EI\frac{1}{\rho} \tag{5-66}$$

and combining eqs 5-65 and 5-66 with eqs 3-207, 208, 209 gives

$$\frac{d^4w}{ds^4} - \lambda^2\frac{d^2w}{ds^2} = \zeta \tag{5-67}$$

where

$$\lambda^2 = \frac{T}{EI} - \frac{1}{R^2}$$

$$\zeta = \frac{T}{REI}$$

The solution to Eq. (5-67) is

$$w(s) = -\frac{1}{2}\chi s^2 + As + B + C\cosh\lambda s + D\sinh\lambda s \tag{5-68}$$

where

$$\chi = \frac{\varsigma}{\lambda^2}$$

Boundary conditions are

$$w(0) = w(L) = \Delta$$

$$\frac{dw}{ds}(0) = \frac{dw}{ds}(L) = 0$$

where

$$\Delta = \frac{1}{2}(d_H - d_{TJ})$$

The four boundary conditions can be casts into

$$\begin{bmatrix} 0 & 1 & 1 & 0 \\ L & 1 & \cosh\lambda L & \sinh\lambda L \\ 1 & 0 & 0 & \lambda \\ 1 & 0 & \lambda\sinh\lambda L & \lambda\cosh\lambda L \end{bmatrix} \begin{Bmatrix} A \\ B \\ C \\ D \end{Bmatrix} = \begin{Bmatrix} \Delta \\ \Delta + \frac{1}{2}\chi L^2 \\ 0 \\ \chi L \end{Bmatrix} \tag{5-69}$$

Once A, B, C, D have been determined, local curvature and bending stress can be determined from Eqs. (5-65) and (5-66). First and second derivatives are

$$\frac{dw}{ds} = -\chi s + A + C\lambda\sinh\lambda s + D\lambda\cosh\lambda s \tag{5-70}$$

$$\frac{d^2w}{ds^2} = -\chi + C\lambda^2\cosh\lambda s + D\lambda^2\sinh\lambda s \tag{5-71}$$

As an example consider the following drill pipe and hole size.

d_H = 9 in.
d_P = 5 in. (19.50 lb/ft; ID = 4.276 in.)
A_c = 5.2746 in.^{2}
d_{TJ} = 7 in. OD
E = 30 (10^6) psi
I = 14.269 in.^{4}
L = 30 ft.

What is of interest here is the magnitude of the bending stresses near tool joints. Stress magnitude depends on hole curvature (1/R) and tension throughout the pipe joint. This tension is assumed to be constant over the pipe joint.

Figure 5-16 shows how bending stress varies over a 30-ft pipe joint when the radius of hole curvature is R = 1000 ft. Tension is assumed to be 100,000 lbs. The highest bending occurs at the tool joints as expected. The magnitude of maximum bending at s = 0 (and s = 30 ft) is

$$M_{max} = EI\frac{1}{\rho}\bigg|_{s=0} \tag{5-72}$$

$$M_{max} = EI\left[\frac{1}{R} + \frac{\Delta}{R^2} + \left(-\chi + C\lambda^2\right)\right] \tag{5-73}$$

Figure 5-17 shows how this maximum bending stress is affected by hole radius and tension. Total stress is the algebraic sum of bending and axial stresses. In practice, an early dog-leg, where tension is highest, is not good.

These bending stresses are cyclic and herein lay the problem. Using the previous example, the stress condition contains an average component and a cyclic component.

$$\sigma_{dyn} = \sigma_{ave} + \sigma_{cyc}$$

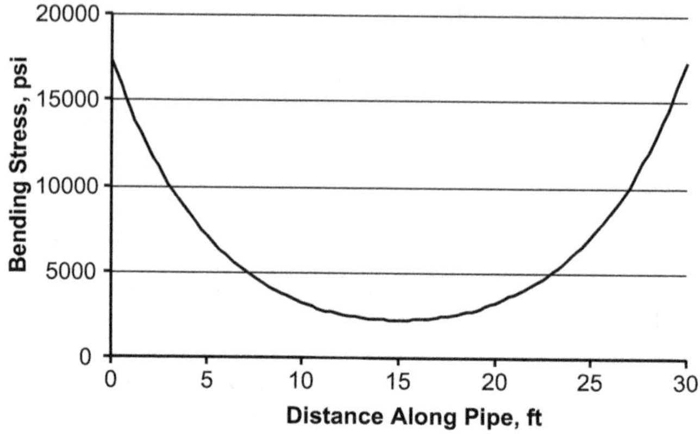

Figure 5-16. Bending stress along a section of 5 in. drill pipe (*R* = 1000 ft, *T* = 100,000 lbs).

where

$$\sigma_{ave} = 18{,}958 \text{ psi}$$
$$\sigma_{cyc} = 17{,}500 \text{ psi}$$

Fatigue tests in cases of complete reverse of stress indicate that the endurance limit of steels ranges between 35% and 60% of the ultimate strength [11,12]. For example, the material properties for SAE 4340 (heat treated) are [8]

$$\sigma_{ave} = 132{,}000 \text{ psi}$$
$$\sigma_{ult} = 150{,}000 \text{ psi}$$
$$\sigma_{end} = 76{,}000 \text{ psi}$$

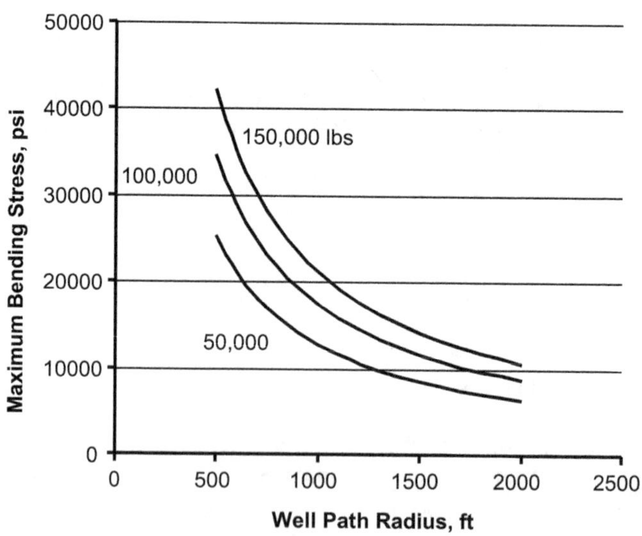

Figure 5-17. Bending stress versus tension and well path radius.

Table 5-3. Endurance limits for drill pipe grades [13]*.

Strength	API minimum yield strength (ksi)	API yield strength maximum (ksi)	API tensile strength minimum	Average Tensile strength of test samples (ksi)	Endurance limit	
					Minimum test value	Median test value
E75	75	105	100	123	30	32
X95	95	125	105	132	32	35
G105	105	135	115	144	34	38
S135	135	165	145	167	36	40

* Youngstown Sheet and Tube Company, 1969 ASME Conference, Tulsa, Oklahoma.

The endurance limit is 50% of the ultimate strength. Endurance limit depends on many factors such as, surface roughness, size, temperature, and stress concentration.

Fatigue tests conducted by Morgan and Roblin [13] show that endurance limits for four API grades of drill pipe are at the lower percentage level (Table 5-3). Their tests preserved the surface and size effects of these pipes.

Average stress also affects the endurance limit and has to be considered in drill pipe fatigue analysis. The modified Goodman diagram is commonly used to account for mean stress effects [12]. Figure 5-18 shows the Modified Goodman diagram for API Grade X95. The square dot gives the location of the stress condition (18.9,17.7) for the above example. This stress condition is below the solid line and no fatigue failure is expected.

Dog legs, especially high up where tension is high, can generate high stress reversal as indicated above. Lubinski [9] gives graphs defining limits on DLS standard grades and sizes of drill pipe.

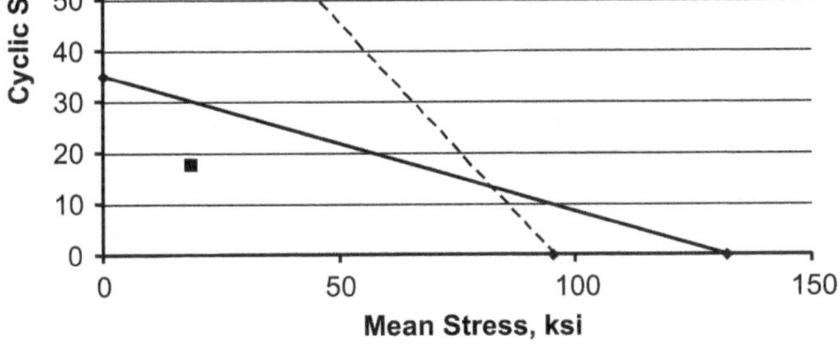

Figure 5-18. Modified Goodman diagram for API Grade X95.

5.1.5.12.4 Tool Joints and Make up Torque [7]

Each drill pipe is made up of a tubular, which has forged upsets at each end. Tool joints are welded onto these upsets. The yield strength of tool joint material is 120,000 psi. Since the first tool joint in 1910, improvements have been made in metallurgy, design, and thread form [14]. Round thread profiles became an API standard in 1928.

Tool joints are, in general, stronger than the main body of drill pipe tubular. Failure of tool joints is usually associated with improper make-up torque. If make-up torque is too low, the shoulders may not seal the inside fluid pressure. Inadequate contact pressure in the shoulders can lead to a wash-out through the threads and shoulder or allow high bending stresses in the pin.

Make up torque is based on the screw jack formula

$$T = \frac{F}{12}\left[\frac{p}{2\pi} + \frac{R_t f}{\cos\theta} + R_s f\right] \tag{5-74}$$

where

T — torque, ft-lbs
p — lead of threads, in.
R_t — average mean radius of threads, in.
f — coefficient of friction (~ 0.08)
R_s — mean radius of shoulder, in.
θ — ½ of included angle of thread
F — contact force between mating shoulders, lbs

The screw jack formula shows that total make-up torque is distributed among three areas. One component of the torque drives the mating shoulders together creating the contact force. If there were no friction in the connection all of the applied torque would create this force which, in this special case, is determined by

$$2\pi T(12) = Fp \tag{5-75}$$

The second component of torque is the torque required to overcome friction in the threads. The third torque component overcomes friction in the shoulder.

The tensile strength of a tool joint is based on a 120,000-psi stress level at the cross-sectional area at the root of the thread located 5/8 in. from the shoulder in either pin or box, whichever is weakest. In the case of a new NC50 (6 3/8 in. × 3 ¾ in.) tool joint connection, the tensile load to yield the pin is 939,000 lbs. The tool joint is therefore stronger in tension than 4 ½ in. (16.6) Grade E drill pipe which yields at a 330,558-tension load. This force level would be generated by a make-up torque of 37,700 ft-lbs [7].

The make-up force or internal contact force between the two mating shoulders directly affects the structural integrity and pressure sealing capacity of the tool joint. Consider, for example, the torque components in a new NC50 (6 3/8 in. × 3 ¾ in.) tool joint using the following set of numbers:

p = 0.25 in.
f = 0.08
θ = 30 deg
R_t = 2.385 in.
R_s = 2.922 in.

Substituting these numbers into Eq. (5-74) gives

$$T = \frac{F}{12}\left[\frac{0.25}{2\pi} + \frac{2.385(0.08)}{0.866} + (2.922)0.08\right]$$

$$T = \frac{F}{12}(0.0398 + 0.2203 + 0.233)$$

$$T = \frac{F}{12}(0.4931)$$

This calculation shows that only 8% of make up torque forces the shoulders together. The remaining 92% overcomes friction. The coefficient of friction of thread dope is therefore critical. Make-up torque recommendations [7] are based on a coefficient of friction of f = 0.08.

If tool joints are made up too tight, the pin could be over stretched. If tool joints are made up too low, threads can be exposed to washouts or bending fatigue. In practice, tool joints are generally made up too low. According to one drilling engineer, "if tool joints are made up too low, they will not make up downhole. On the other hand, drill collar connections will make up downhole. If drill collar connections are difficult to break out, it usually means they were made up too low initially."

During directional drilling operations, drill pipe is usually limited by rotating torque and not by direct pull. One practice is to allow rotary torque to reach 80% of make-up torque.

5.2 HYDRAULICS OF ROTARY DRILLING

The main purpose of the circulating system is to clean cuttings from underneath drill bits and carry them back to the surface for analysis and disposal. In recent years downhole motors have placed additional demands on the hydraulic system. Drilling mud also cools drill bits and lubricates bearing surfaces throughout drillstrings. When downhole motors or turbines are used, hydraulic power (pressure and flow rate) is extracted from fluid flow.

Total fluid pressure can be viewed as the sum of hydrostatic pressure and hydrodynamic pressure. Hydrostatic mud pressure affects many aspects of drilling including well control, drillstring mechanics and drilling rate. Hydrostatic pressure is the result of gravity and pressure varies with depth and mud density according to

$$p = \gamma L \tag{5-76}$$

In terms of typical oil field units

$$p = \frac{7.48\gamma L}{144} = 0.052\gamma L \text{ (psi)} \tag{5-77}$$

where

γ — ppg
L — ft

Hydraulic pressure is the first line of defense against a blowout.

Hydrodynamic pressure pushes drilling mud down and around the drillstring. Total input pressure is party consumed by friction. The remaining pressure is used to clean drill bits and power downhole motors. Mud pumps are literally the heart of the circulating system.

5.2.1 Mud Pumps

Two types of oil field mud pumps are triplex and duplex. Triplex pumps are single acting and have three pistons. Duplex pumps are double acting and have two pistons. Pumping units contain both pump and power source and are usually skid mounted. The power end is driven by electric motors. Older versions were driven by diesel engines by means of a direct drive.

Mud pumps have two distinct segments: a fluid end and a power end (Fig. 5-19). The fluid end contains: fluid manifold, intake and exhaust valves, pistons, and

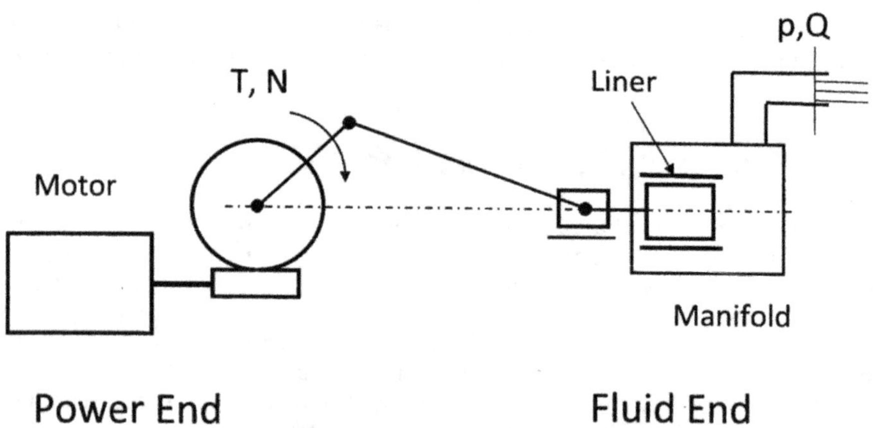

Figure 5-19. Schematic of mud pumps.

liner. The power end is a mechanical drive containing a slider crank mechanism, which converts rotary motion and torque to linear piston motion and force.

Mud pumps are positive displacement pumps. Volume output is directly related to the number of strokes of the crank. Output capacity is normally expressed in terms of volume per stroke. Volume output depends on liner size and stroke length. Actual flow rate output depends on volumetric efficiency of the pump and suction intake.

Fluid ends are limited by fluid pressure in the manifold. Triplex pumps are limited to a maximum discharge pressure of approximately 5500 psi. Higher manifold pressures can cause fatigue cracks and premature fatigue failure.

Power ends are limited by force transmitted from the piston through the slider crank mechanism. Bearings usually limit the maximum allowable force on the piston and this force depends on fluid pressure and piston area. Allowable fluid pressure in large liners is low (high volume). Allowable pressure in small liners is typically high (small volume).

While pressure and volume output change with liner size, power input remains the same. Hydraulic horsepower is related to pressure and flow rate by

$$\text{Power} = pQ \tag{5-78}$$

Equation 5-78 is adjusted to horsepower units by

$$\text{HHP} = pQ\left|\frac{144\ \text{in}^2}{1\ \text{ft}^2}\right|\left|\frac{1\ \text{ft}^3}{7.48\ \text{gal}}\right|\left|\frac{1\ \text{min}}{60\ \text{sec}}\right|\left|\frac{1\ \text{hp}}{550\ \text{ft-lb/sec}}\right|,\ \text{hp}$$

$$\text{HHP} = \frac{pQ}{1714},\ \text{hp} \tag{5-79}$$

where

HHP — hydraulic horsepower, hp
p — fluid pressure, psi
Q — volume flow rate, gpm

Performance of mud pumps is measured in terms of hydraulic horsepower output. Mud pumps produce maximum hydraulic horsepower with different combinations of output pressure and flow rate (Fig. 5-20). The constant horsepower curve is established from Eq. (5-79).

The rectangle that touches the horsepower curve represents pressure-flow rate capability the pump. The horizontal line indicates the maximum operating pressure set by the manufacturer. The Contractor usually operates somewhat below this maximum pressure. Flow rate can be varied by changing strokes per minute.

5.2.2 Parasitic Losses

When drilling mud is pumped through a drilling system, energy is lost due to fluid friction. This friction loss is taken from the hydraulic horsepower and can be viewed as parasitic. Losses also occur in surface equipment, inside and out

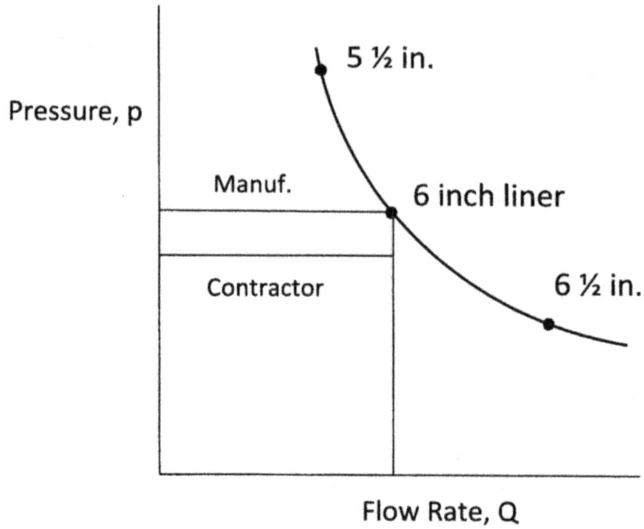

Figure 5-20. Pump performance versus liner size.

of the drillstring. Energy loss across drill bit nozzles is not considered parasitic. Parasitic losses are substantial and must be considered in designing the overall circulating system. A major portion of the total hydraulic horsepower input by the mud pumps is lost to friction. The remaining part is used for bottom hole cleaning and power for downhole motors or turbines.

Two rheological fluid models common used to describe the physical properties of drilling mud are Bingham and Power Law models. These models can be used to predict pressure losses under laminar flow conditions. However, when flow is turbulent the shear stress-strain rate curves converge, and each model will predict approximately the same pressure drop.

It is common practice, when flow is turbulent, to predict pressure losses using the empirical Darcy-Weisbach equation [15]

$$\frac{\Delta p}{L} = f \frac{\gamma}{D} \frac{V^2}{2g} \tag{5-80}$$

where

f — friction factor
V — average velocity
D — diameter of flow area
γ — density of drilling mud

In oil field terms

$$\frac{\Delta p}{L} = f \frac{\gamma V^2}{93{,}000D} \tag{5-81}$$

and

Δp — pressure drop, psi
L — pipe length, ft
f — friction factor
V — average fluid velocity, fps
D — diameter of flow area, in.
γ — density of drilling mud, ppg

The Blasius approximation for the turbulent flow friction factor depends on Reynolds number.

$$f = \frac{a}{N_R^b} \tag{5-82}$$

For weighted muds flowing through drillstring tubulars, this relation is approximately [16]

$$f = \frac{0.046}{N_R^{0.2}} \tag{5-83}$$

The Reynolds number in oil field terms is

$$N_R = \frac{15.47\gamma VD}{\mu} \tag{5-84}$$

where the viscosity in turbulent flow is also approximated by

$$\mu = \frac{PV}{3.2} \tag{5-85}$$

where PV is plastic viscosity. By substitution

$$N_R = \frac{49.5\gamma VD}{PV} \tag{5-86}$$

Combining these expressions with the Darcy-Weisback equation gives

$$\frac{\Delta p}{L} = 2.27(10^{-7})\frac{PV^{0.2}\gamma^{0.8}V^{1.8}}{D^{1.2}} \tag{5-87}$$

Since by the continuity of flow

$$V = \frac{24.51Q}{D^2} \tag{5-88}$$

The pressure loss inside pipe [16,17] is therefore

$$\frac{\Delta p}{L} = 7.7(10^{-5})\frac{PV^{0.2}\gamma^{0.8}Q^{1.8}}{D^{4.8}} \tag{5-89}$$

where

D — ID of pipe, in.
PV — plastic viscosity, cp
γ — mud weight, ppg
Q — flow rate, gpm
L — pipe length, ft
Δp — pressure drop, psi

and pressure loss in the annulus is [16]

$$\frac{\Delta p}{L} = 7.7(10^{-5})\frac{PV^{0.2}\gamma^{0.8}Q^{1.8}}{(D_h - D_p)^3(D_h + D_p)^{1.8}} \tag{5-90}$$

5.2.3 Nozzle Selection

Cleaning and removing cuttings from underneath drill bits is enhanced by pressure drop across bit nozzles. Bit nozzles transform available energy into kinetic energy which is discharged under the bit for cleaning and cuttings removal. Flow rate selection and nozzle size are important for making the best use of available hydraulic horsepower.

Bit pressure drop depends on total flow area (TFA) of all nozzles in a drill bit. Applying Bernoulli's energy equation between sections 1 and 2 (see Fig. 5-21) gives

$$\frac{p_1}{\gamma} + \frac{V_1^2}{2g} = \frac{p_2}{\gamma} + \frac{V_2^2}{2g} \tag{5-91}$$

Since V_1 is small by comparison with V_2

$$p_1 - p_2 = \Delta p = \frac{\gamma}{2g}V^2 \tag{5-92}$$

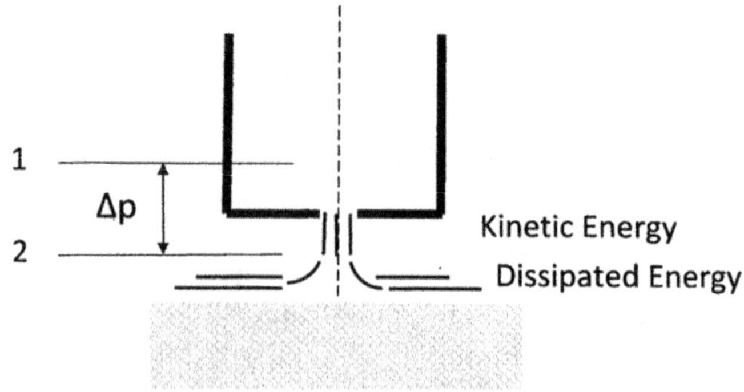

Figure 5-21. Schematic of Fluid flow through bit nozzles.

In oil field units

$$\Delta p = \frac{7.48\gamma}{2(32.2)(144)} V^2 \tag{5-93}$$

$$\Delta p = \frac{\gamma V^2}{1240} \tag{5-94}$$

where

Δp — psi
V — fps
γ — ppg

Applying a nozzle coefficient of 0.95 [16]

$$\Delta p = \frac{\gamma V^2}{1120} \tag{5-95}$$

Substituting the continuity equation

$$V = \frac{0.32Q}{A_n} \tag{5-96}$$

into Eq. (5-95) gives

$$\Delta p = \frac{\gamma Q^2}{10{,}938 A_n^2} \tag{5-97}$$

where

Δp — psi
Q — gpm
γ — ppg
A_n — total nozzle area, sq. in.

It is desirable to use up available horsepower for best bottom hole cleaning. This is accomplished by selecting the best flow rate and then solving for total nozzle flow area.

Dividing bit hydraulic horsepower $\left(\text{HHP} = \frac{\Delta p Q}{1718}\right)$ by the cross-sectional areal of the hole give hydraulic horsepower per square inch of hole or HSI. An HSI of 5 to 7 is considered a high level of hydraulic horsepower across these bit types.

5.2.4 Annular Velocity Requirements

Fluid flow in the annulus carries bit cuttings back to the surface for geological evaluation and disposal. While the upward movement of the drilling mud is lifting the cuttings, gravity tends to pull the cuttings back down the hole and cause them to slip relative to the fluid flow. A minimum annular velocity is therefore required to adequately clean the wellbore. If annular velocity is high, fluid turbulence can develop and possibly erode the wellbore.

Formulas for calculating slip velocity and minimum flow rate have been developed. However, a general guide commonly used to select annular velocities is

$$\mathrm{AV} = \frac{11800}{D_{\mathrm{h}}\gamma} \tag{5-98}$$

where

AV — recommended annular velocity, ft/min
D_{h} — hole diameter, in.
γ — drilling mud density, ppg

As an example, assume a hole diameter of 10 in. and drill mud of 12 ppg, then

$$\mathrm{AV} = \frac{11800}{10(12)} = 98\ \mathrm{ft/min}$$

Typical annular flow rates are around 120 ft/min.

5.2.5 Available Hydraulic Horsepower

Availability of hydraulic horsepower at the drill bit is explained by use of Fig. 5-22. The pump line represents power delivered by the mud pump assuming it operates at rated pressure. Hydraulic output increases linearly with flow output. System pressure losses vary exponentially with flow rate as discussed previously. The system loss curve represents parasitic losses predicted at different flow rates. The difference between these two curves represents the amount of hydraulic horsepower available for bit cleaning (or drilling motors). The best flow rate for optimum bit cleaning corresponds to the peak power point. This is the basis for nozzle selection.

According to Eq. (5-89) parasitic loss is proportional to flow rate as given by

$$\Delta p = CQ^{1.8} \tag{5-99}$$

where C is a constant accounting for all other variables in Eq. (5-89). For convenience, the following calculations are based on a total parasitic loss of 800 psi at a flow rate of 400 gpm. The constant C, in this case is

$$C = \frac{800}{400^{1.8}} = 0.0166$$

Based on this assumption and for the sake of simplicity, parasitic losses are calculated by

$$\Delta p = 0.0166Q^{1.8}$$

Hydraulic horsepower lost due to fluid friction is then expressed by

$$\mathrm{HHP_{loss}} = 0.0166\Delta pQ^{2.8} \tag{5-100}$$

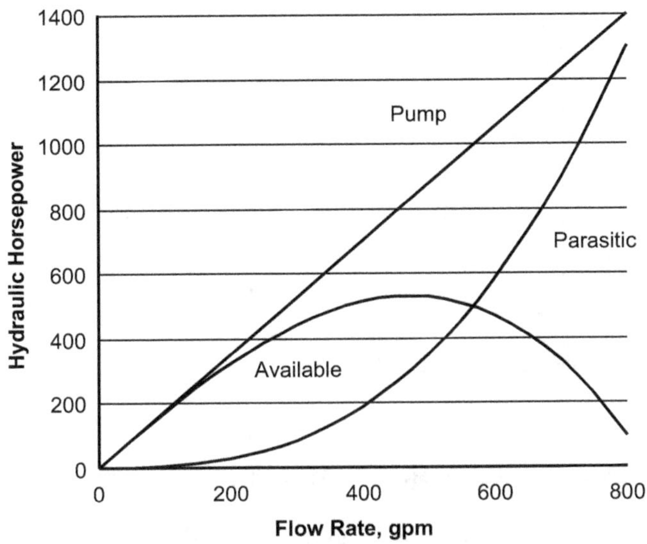

Figure 5-22. Available hydraulic horsepower at the drill bit.

Using an allowable mud pump pressure of 2800 psi and Eq. (5-100) an interesting set of horsepower curves (Fig. 5-22) is developed showing pump input and parasitic losses at different flow rates. The difference between these two curves represents the amount of horsepower available at the bottom end of the drillstring for bit cleaning and drilling motors or turbines.

The choice of flow rate also depends on the average velocity of the drilling fluid in the annulus. Annular velocity must be high enough to overcome the downward slip velocity of the cuttings due to gravity. In unstable formations, fluid turbulence is a consideration in selecting flow rate. In practice, the choice of flow rate is a compromise between adequate bit cleaning and cuttings removal from the well bore.

5.2.6 Power Demands of Downhole Motors

Power consumption of PDMs has to be considered in designing drilling hydraulics. Part of the available horsepower (see Fig. 5-23) is consumed by the motor and reduces the amount of power available to clean the drill bit. Typical PDM power consumption (based on Fig. 5-4) is shown in Fig. 5-23 along with the available HHP curve to show how much power is consumed by a PDM. The difference between to two curves represents the HHP available for bit cleaning. This difference is the basis for selecting nozzle size for best bottom-hole cleaning. This type of diagram is useful for determining best flow rate, too. Flow rate selection also has to consider annular velocity requirements.

Successful application of downhole turbines depends on the amount of hydraulic horsepower available at the bottom of the drillstring. Hydraulic analysis of the total circulating system is critical. Figure 5-24 shows an available hydraulic power curve and turbine power consumption curve; assuming the

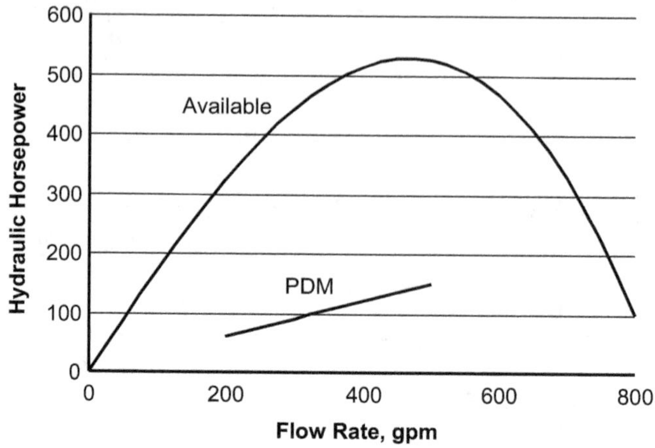

Figure 5-23. Typical PDM power consumption.

turbine operates at maximum power out put. Note that the power consumption curve is basically different from the PDM power curve.

Turbines usually operate at high flow rates to make best use of the turbine's power capability. This usually means that all of the available hydraulic horsepower is consumed by turbines and there is little or no hydraulic horsepower left for bit cleaning. This condition is represented by the intersection of the turbine power with the available power curve. For this reason, open drill bits (no nozzles) are often used with turbines. In this case, bit cleaning and cooling is accomplished by high flow rates.

The composite hydraulics diagram (Fig. 5-25) gives the available hydraulic horsepower at depth-in and depth-out and shows the amount of power consumed by both turbines and positive displacement motors. The annular velocity line is an essential part of the diagram because the final choice of flow rate,

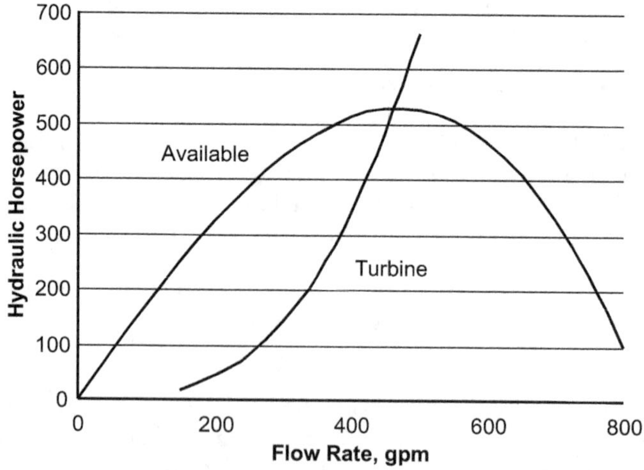

Figure 5-24. Power consumption of drilling turbine.

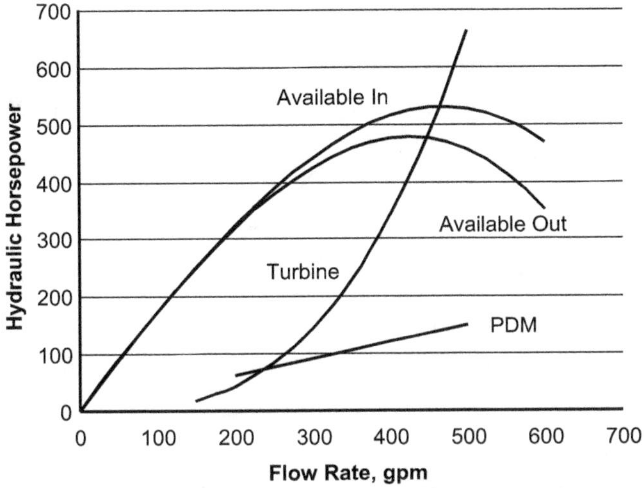

Figure 5-25. Composite hydraulic horsepower diagram.

Q, is usually a compromise between bottom hole hydraulic horsepower and adequate annular velocity for hole cleaning.

5.3 OPTIMUM DRILLING PRACTICE

It is perhaps impossible to achieve a mathematical optimum in drilling operations because of multiple variables that affect drilling rate and cost. However, if attention is given to key aspects of drilling, the result is the very best that can be achieved with current drilling technology. Good drilling practice emphasizes

- drill bits
- weight on bit (WOB)
- rotational speed
- hydraulics
- pressure balance
- mud properties
- bottom hole assembly

Paying attention to these factors usually leads to maximum rate of penetration (ROP) and minimum footage cost and is commonly called optimum drilling practice.

5.3.1 Drill Bit Selection

Bit selection is based to a large extent on statistical data accumulated for each bit type. Footage cost can be estimated from

$$C = \frac{C_{\text{rig}}(t_{\text{t}} + t_{\text{d}}) + C_{\text{bit}}}{\text{ROP} * t_{\text{d}}} \tag{5-101}$$

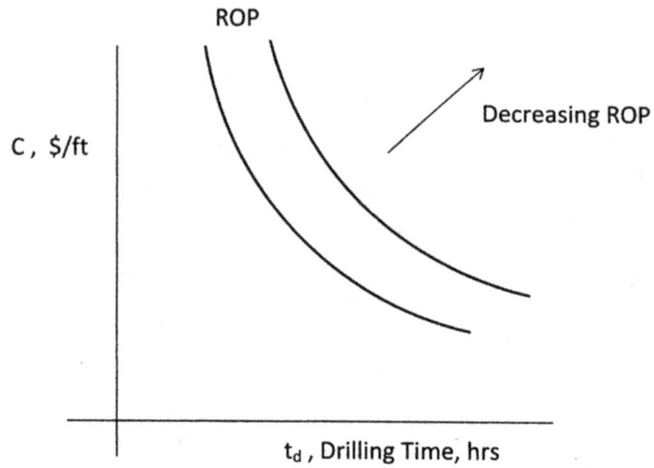

Figure 5-26. Footage cost versus drilling time and ROP.

where

C — footage cost ($/ft)
C_{rig} — rig rate ($/hr)
C_{bit} — bit cost ($)
t_{t} — trip time (hr)
t_{d} — drilling time (hr)

Equation (5-101) can be displayed graphically as shown in Fig. 5-26.

This figure show that for a given rate of penetration, footage cost steadily reduces with on-bottom drilling time, which depends on bit life and reliability of other downhole tools. Multiple cost curves can be drawn on the same diagram where each curve corresponds to a given average rate of penetration. The cost curves shaft upward as average rate of penetration reduces, indicating footage cost increases. Conversely, if average rate of penetration increases, footage cost reduces. With this type of diagram, one can follow or track footage cost by dividing footage drilled by drilling time and locating the corresponding footage cost on the diagram. The footage cost at each time step is the footage cost if the bit were tripped out of the hole at that particular time. If ROP is constant, footage cost would reduce along a given ROP curve. If average ROP reduces with drilling time, the footage cost point moves from one ROP cure to a lower ROP curve.

Footage cost is directly affected by rig rate and rate of penetration. Equation (5-101) can be rearranged to compare the benefits of one drill bit over another, considering bit cost and ROP. Economic curves can show when one bit breaks even and out performs another. Downhole motors and turbines can be added to Eq. (5-101) as discussed earlier.

5.3.2 Bit Weight and Rotation Speed

These two factors directly affect rate of penetration and bit life [18,19]. In the early 1960s, Maurer [20,21] developed a perfect cleaning theory and showed that if chips or cuttings are completely removed from their craters, ROP increased with the square of bit weight. His perfect cleaning theory has been verified in drilling situation that have nearly perfect cleaning such as raised boring and horizontal shaft boring. In reality, bit cleaning is not perfect because actual rates of penetration fall short of the perfect cleaning bench mark. Over balance pressure conditions and high viscosity are factors.

According to Maurer's theory [20], rate of penetration should increase directly with rotary speed. In practice, this does not occur either because of the reasons mentioned above.

The difference in rate of penetration between perfect cleaning and normal cleaning is even greater when downhole motors and turbines are used because each operate at much higher speeds. There is much incentive to improve bit cleaning when PDMs and turbines are used. Many attempts have been made to formulate the effects of bit weight and rotary speed on rate of penetration. A common formula is

$$\mathrm{ROP} = \mathrm{WOB}^a \mathrm{RPM}^b \tag{5-102}$$

where $a = 2$ and $b = 1$ are for perfect cleaning while $a \le 2$ and $b \le 1$ are for imperfect cleaning.

5.3.3 Optimum Hydraulics for Drill Bit Cleaning

Hydraulics refers to fluid flow and fluid pressures throughout the circulating system. Fluid energy and impact forces are required to clean drill bits and to remove cuttings from underneath the bit. Good drilling rate depends on it. Available hydraulic power was discussed earlier and is a drilling factor that can be mathematically optimized.

Hydraulic Horsepower

The available hydraulic horsepower at the bottom of the drillstring is

$$\mathrm{HHP}_a = p_p Q - C Q^{2.8} \tag{5-103}$$

Horsepower is a maximum when

$$\frac{\mathrm{dHHP}_a}{\mathrm{d}Q} = p_\mathrm{p} - 2.8 C Q^{1.8} = 0 \tag{5-104}$$

But system losses are

$$p_s = C Q^{1.8} \tag{5-105}$$

Combining Eq. (5-105) and Eq. (5-104)

$$p_s = \frac{1}{2.8} p_\mathrm{p} \tag{5-106}$$

which means available HHP is maximum when parasitic losses are about 1/3 of maximum allowable pump pressure. Noting that system losses at different flow rates can be expressed by

$$\frac{p_2}{p_1} = \left(\frac{Q_2}{Q_1}\right)^{1.8} \tag{5-107}$$

or

$$\frac{Q_2}{Q_1} = \sqrt[1.8]{\frac{p_2}{p_1}} \tag{5-108}$$

then

$$Q_{opt} = Q_1 \sqrt[1.8]{\frac{p_{max}}{2.8 p_1}} \tag{5-109}$$

If system losses are known to be, say, $p_1 = 444$ psi at $Q_1 = 300$ gpm and the maximum allowable pump pressure $p_p = 2800$ psi, then the optimal flow rate is $Q_{opt} = 450$ gpm. Under this condition, $p_a = 2000$ psi and $Q_{opt} = 450$ gpm. This was the drilling condition under which Fig. 5-22 was developed. Note that the maximum available hydraulic horsepower occurs at a flow rate of 450 gpm.

5.3.4 Impact Force

The impact force of drilling mud striking the hole bottom is a measure of the fluid's ability to blast chips or cutting off of the formation directly under a bit [22]. The magnitude of this fluid impact depends on fluid velocity leaving the bit nozzles and mud weight. Impact force is the result of fluid momentum change. The fluid momentum vector is turned 90 deg from the axis of nozzles. An impact force is required to turn the fluid 90 deg the fluid strikes the formation.

From impulse-momentum

$$F = Q_m v = \frac{\gamma}{g} Q v \tag{5-110}$$

From Bernoulli's energy equation

$$\frac{1}{2}\frac{v^2}{g} = \frac{\Delta p}{\gamma} \tag{5-111}$$

$$v = \sqrt{\frac{2\Delta p g}{\gamma}} \tag{5-112}$$

Combining Eqs. (5-110) and (5-112) gives

$$F = \frac{\gamma}{g} Q \sqrt{\frac{2\Delta p g}{\gamma}} \tag{5-113}$$

or

$$F = Q\sqrt{\frac{2\gamma\Delta p}{g}} \tag{5-114}$$

In terms of oil field units

$$F = Q\sqrt{\frac{\gamma\Delta p}{3008}} \tag{5-115}$$

where

F — hydraulic impact force, lbs
Q — circulation rate, gpm
γ — mud weight, ppg
Δp — bit pressure drop, psi

Both impact force and hydraulic horsepower are shown in Fig. 5-27, assuming parasitic pressure loss is equal to $p_s = 0.0166Q^{1.8}$ and all available pressure is absorbed across the bit; there are no downhole motors. Note that while bit hydraulic horsepower reaches a maximum (472 hp) at a flow rate of Q = 450 gpm (corresponding to a bit pressure drop of 1800 psi), hydraulic impact force reached a maximum at a flow rate of Q = 600 gpm (corresponding to a bit pressure drop of 1337 psi).

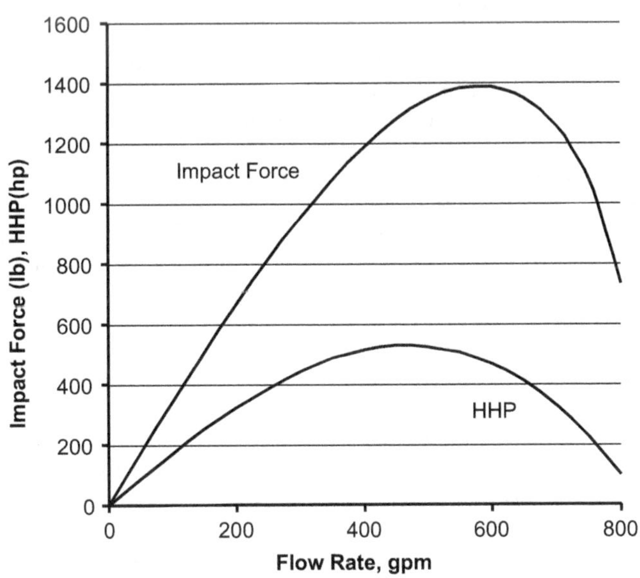

Figure 5-27. Comparison of impact force and hydraulic horsepower.

The condition for which impact force is maximized is determined as follows. From Eq. (5-113)

$$F = Q\left[\frac{2\gamma}{g}(p_p - p_s)\right]^{\frac{1}{2}} \tag{5-116}$$

$$F = \left[\frac{2\gamma}{g}(p_p Q^2 - CQ^{3.8})\right]^{\frac{1}{2}} \tag{5-117}$$

Differentiating F with respect to Q and setting the expression equal to zero gives

$$p_s = \frac{2}{3.8} p_p \tag{5-118}$$

which states that impact force is maximum when system pressure loss is about ½ the allowable pump pressure.

Whether it is best to maximize hydraulic or impact force has been debated in the past. It does appear that hydraulic horsepower is preferable where bit cleaning and formation cleaning are needed simultaneously, and impact force is preferable where bottom hole formation cleaning is needed such as in excessive chip hold down cases. Fortunately, when hydraulic horsepower is maximized, impact force is 95% of its maximum value; and when impact force is maximized, hydraulic horsepower is 95% of its maximum value. It is therefore difficult to distinguish which cleaning parameter is more effective.

5.3.5 Pressure Balance

The pressure difference between mud column and formation pressure significantly affects drilling rate of penetration. Wellbore pressure is the primary means of controlling formation pressure while drilling. Yet, too much hydrostatic mud pressure can breakdown formations and slow drilling rates [23,24]. Balanced pressure drilling means that the hydrostatic pressure is adjusted to match formation pressure. Underbalanced drilling occurs when hydrostatic pressure is less than formation pressure. Under this condition, drilling rate is the highest. This effect is sometimes used to sense when a reservoir has been reached; the rate of penetration suddenly increases and can be sensed immediately at the surface.

5.3.6 Drilling Mud Properties

Drilling mud is the lifeblood of the rotary drilling operation. Its composition and physical properties directly impact rate of penetration. Its density (mud weight), rheological properties (viscosity and gel strength) and chemistry also affect penetration rate.

Mud viscosity slows down drilling rate, a cleaning effect. Studies show that drilling rate decreases inversely with the square of viscosity [25]. If mud viscosity doubles, drilling rate reduces by one fourth.

5.3.7 Bottom Hole Assembly Design

Bottom hole assembly design affects drilling cost by allowing desired bit weights and rotary speeds to be applied to drill bits for best ROP without causing hole deviations. Walker [6] points out that an optimum BHA is one whose stabilizer placements simultaneously provides:

- a zero resultant force angle at the bit
- the minimum tilt angle at the bit
- the minimum iron in the BHA
- minimum drill collar deflection

This is accomplished within the bottom 150 ft of a drill string.

REFERENCES

1. Dareing, Don W.: *Engineering Design and Problem Solving*, Tennessee Valley Publishing Company, 2010.
2. http://en.wikipedia.org/wiki/Progressive_cavity_pump.
3. Robello Samuel G. and Ken J. Saveth, "Optimal Design of Progressing Cavity Pumps (PCP)" JERT, Dec 2006 Vol 128, Issue 4, 275 (5 pages).
4. Tiraspolsky, W., *Hydraulic Downhole Drilling Motors*, Gulf Publishing Company, 1985.
5. Hebbeler, R. C.; *Engineering Mechanics: DYNAMICS*, 11th ed. Pearson-Prentice Hall, 2007.
6. Bruce H. Walker; "Some Technical and Economic Aspects of Stabilizer Placement", *Journal of Petroleum Technology*, June 1973.
7. API RP 7G; Recommended Practice for Drill Stem Design and Operating Limits, American Petroleum Institute.
8. Riley, William F., Sturges, Leroy D., and Morris, Don W.; Mechanics of Materials, 6th ed., John Wiley & Sons, 2007.
9. Lubinski, Arthur; "Maximum Permissible Dog-Legs in Rotary Boreholes", Journal of Petroleum Technology, Feb, 1961.
10. Timoshenko, S. and Woinowsk-Krieger, S.; *Theory of Plates and Shells*, McGraw-Hill, 1959.
11. H. O. Fuchs and R. I. Stephens, *Metal Fatigue in Engineering*, Wiley, NY, 1980 (69–71).
12. Shigley, Joseph Edward and Larry D. Mitchell, *Mechanical Engineering Design*, McGraw-Hill, 1983 (p275).
13. Morgan, R. P. and Roblin, M. J., "A Method for the Investigation of Fatigue Strength in Seamless DrillPipe," ASME Conference, Tulsa, Oklahoma, September 22, 1969.
14. Brantly, J.E., 1971, *History of Oil Well Drilling*, Gulf Publishing Company; also see *History of Petroleum Engineering*, API, 1961.
15. Fox, Robert W., Alan T. McDonald and Philip J. Pritchard; *Introduction to Fluid Mechanics* (6th ed.), John Wiley & Sons, 2004.
16. Moore, Preston L., *Drilling Practice Manual*, PennWell Books, Tulsa, Oklahoma, 1974.

17. Gatlin, Carl, *Petroleum Engineering Drilling and Completion*, Prentice-Hall Inc, Englew3ood Cliffs, N. J. (1960).
18. Woods, H.B. and Galle, E. M., "Constant Bit Weight and Rotary Speed" *Oil and Gas Journal*, Oct. 6, 1958.
19. Woods, H.B. and Galle, E.M., "Bit Weight and Rotary Speed", *Oil and Gas Journal*, Nov 14 & 21, 1963.
20. Maurer, W. C., "The Perfect-Cleaning Theory of rotary Drilling", *Journal of Petroleum Technology, AIME*, **225**, November 1962, pp. 1270–1274.
21. Maurer, W. C., "Bit Tooth-Penetration Under Simulated Borehole Conditions", *Journal of Petroleum Technology, AIME,* **234**, December 1965, pp. 1433–1442.
22. Kendall, H.A. and Goins, W. C. "Design and Operation of Jet-bit Programs for Maximum Horsepower, Maximum Impact Force and Maximum Jet Velocity", *Trans., AIME,* 1960.
23. Murray, A. S. and Cunningham, R.A., "Effect of Mud Column Pressure on Drilling Rates", *Trans. AIME,* **204**, 1955, pp. 196–204.
24. Cunningham, R.A. and Eenink, J.G., "Laboratory Study of Effect of Overburden, Formation and Mud column Pressure on Drilling Rate of Permeable Formations", *Trans. AIME,* **216**, 1959, pp. 9–17.
25. Eckel, J.R., "Microbit Studies of the Effect of Fluid Properties and Hydraulics on Drilling Rate", paper SPE 2244, *SPE Annual Conference*, 1968.

Chapter 6

SELECTED TOPICS IN MARINE RISER DESIGN

Effective tension is one of the subtle aspects of riser performance. This term contains direct pull within riser pipes and fluid densities inside and outside. Effective tension affects everything: bending, buckling, natural frequencies, and vibration response. Other factors that directly affect riser behavior are: 1) actual tension resulting from direct top pull and gravity; 2) side loading caused by current, waves, and rig offset and motion; and 3) drilling mud density. These factors should be considered early in configuring riser design. Quasi-static analysis is fine for this. A dynamic analysis can follow. Modifications to surface tension may be required since it affects natural frequencies. Vortex-induced vibrations (VIV) should also be checked during the design phases and monitored during operations. As in any design, initial designs are modified to account for unexpected vibrations.

6.1 MARINE DRILLING RISERS

Reference [1] gives excellent guidance in the design and operation of drilling risers. Three operating situations are discussed.

1. Drilling mode
2. Non-Drilling mode
3. Riser Disconnect mode

A few additional comments are offered here.

6.1.1 Drilling Mode

The drilling mode refers to normal drilling operations. There is some discretion on wall thickness in each case, depending on operating and environmental loads as well as BOP size. Flex/ball joint are used at the top and bottom to allow structural flexibility while restricting bend angle across the joint. This angle is controlled to allow passage of downhole tools. Rotational stiffness of flex joints range between 10,000 and 30,000 ft-lb/deg of rotation [1]. Mean value of dog leg severity (DLS) across flex/ball joints is 2 deg with a 4-deg maximum. DLS is controlled by top tension and flex/ball joint rotation stiffness. Flex/ball joint stiffness values are factored into boundary conditions need in mathematical models.

As an example consider the following case:

Riser size – 22 in. OD – 20.75 in. ID
Area – 41.97 in.^{2}
w = 142.81 lb/ft
Riser length – 1500 ft
Ball joint at top ($K_\phi = 0$)
Flex/ball joint at bottom (K_ϕ = 50,000 ft-lb/deg)
Maximum rig offset (δ = 50 ft)
Bottom tension (F_B = 100,000 lbs)
$\gamma_i = 104.72$ *lb/ft*^{3} (14 ppg)
$\gamma_o = 64$ *lb/ft*^{3}

These numbers yield
$\alpha = 1289.9$
$\beta = 173.4$

6.1.1.1 *Effective Tension and Riser Buckling*

Since tensioning is a key factor in controlling bending, it is important to determine the pull capacity of the riser body. This number depends on yield strength and is easy to determine. Choosing X65, for our example, the yield force ($F_{yld} = \sigma_{yld} A_c$) is 65,000 × 41.97 = 2,728,029 lbs. This is an outside number to keep in mind. This number also sets the force requirement for motion compensation.

Based on the drilling program and expected mud weights, check the pull requirement corresponding to lateral buckling of the riser. For our example and assuming a maximum mud weight of 14 ppg, then the critical buckling force (F_B) based on Eq. (4-8) is 17,270 lbs. This means that the corresponding top pull force is total air weight plus this buckling force or T_{top} = 1500(142.81) + 17,270 = 231,485 lbs. This is a good reference number because it represents the very minimum tension at the surface. This is well within the maximum yield tension of 2,728,029 lbs. A larger force may have to be applied to control bending caused by ocean side loads. Buoyant modules are sometimes used in very deep waters to reduce the load on motion compensators and axial stress in the riser pipe. Multiple buoys may be used for this purpose.

6.1.1.2 *Rotation Across Flex/Ball Joints*

Deflection and bending calculations were made for a 22 in. × 20.75 in. riser have a length of 1500 ft with a rig offset of 50 ft. Tension at the lower end is F_B = 100,000 lbs. The bottom end is attached to a flex/ball joint. Three different values of K_ϕ are used in this example. Only the deflections over the lower 150 ft (ς = 0.1) are shown in Fig. 6-1. Above this point, the riser deflection bends back to the rig offset displacement point. The corresponding bending moment at the lower end is 159,000 ft-lbs (see Table 6-1).

In each case, there is a point of inflection above the flex joint. Plus (+) moment is created by the flex/ball joint. Total rotation across a flex/ball joint is the sum of the effects of various side loads. Calculations for side loading are given in Chapter 4.

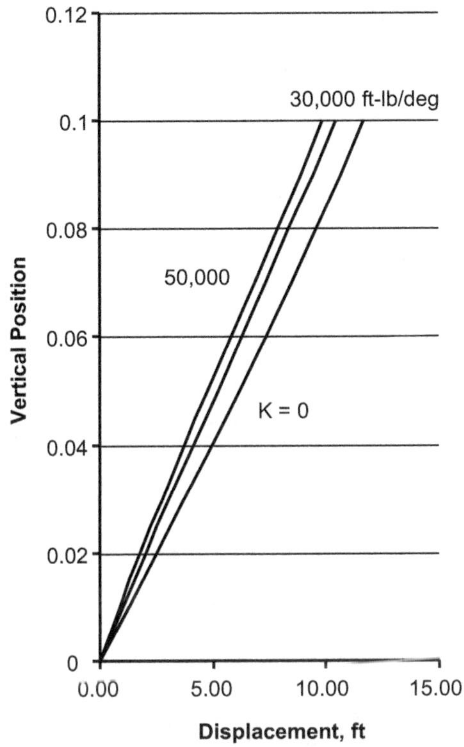

Figure 6-1. Effect of flex/ball joint stiffness on riser displacement.

Forces on well head equipment are important too. The shear force in the riser pipe is determined from

$$V = \frac{dM}{dx} - F_x \frac{dy}{dx} \tag{6-1}$$

The pull force on the well head is F_B. Top pull force can be determined from

$$F_{top} = F_B + wL \tag{6-2}$$

Table 6-1. Rotation across flex ball joint.

Torsion stiffness (ft-lb/deg)	a_1 (ft)	Angle (deg)	Bending moment (ft-lbs)	Bending stress (ksi)*
0	126.36	4.83	0	0
10,000	114.55	4.38	43,800	2.41
20,000	104.76	4.00	80,000	4.40
30,000	96.51	3.69	110,700	6.09
40,000	89.46	3.42	136,800	7.53
50,000	83.38	3.18	159,000	8.75

* $\sigma_b = \frac{Mc}{I}\left|\frac{12\text{in}}{\text{ft}}\right|$, c = 11 in., I = 2399 in.[4]

The bending moment on the well head is

$$M(0) = K_\phi \Delta\theta$$

where

$$\Delta\theta = \frac{dy}{dx}(0)$$

Loading to the well head for the K_ϕ = 50,000 ft-lb/deg example is calculated below. The riser analysis produced the following coefficients, which are used below:

$$a_0 = 0$$
$$a_1 = 83.38$$
$$a_2 = 370.79$$
$$a_3 = -3040.31$$

Each coefficient has units of length.

Pull force

$$F_B = 100{,}000 \text{ lbs}$$

Rotation across flex/ball joint

$$\Delta\theta = \frac{dy}{dx}(0) = \frac{a_1}{L}, \text{ radians}$$

$$\Delta\theta = \frac{83.38}{1500} = 0.0556 \text{ radians}$$

or

$$\Delta\theta = 3.18 \text{ deg}$$

Bending moment

$$M(0) = EI\frac{d^2y}{dx^2}(0) = \frac{EI}{L^2}\frac{d^2y}{d\zeta^2}(0)$$

$$M(0) = \frac{EI}{L^2}2a_2$$

$$M(0) = \frac{29(10)^6 2399}{1500^2}2(370.79)\left|\frac{1\text{ft}^2}{144\text{in}^2}\right|$$

$$M(0) = 159{,}236 \text{ ft-lb}$$

This number agrees with the product of $\Delta\theta$ and K_ϕ given above.

$$M(0) = \Delta\theta K_\phi = 3.18(50{,}000) = 159{,}000 \text{ ft-lbs}$$

Shear force

$$V(0) = EI\frac{d^3y}{dx^3}(0) - F_B\frac{dy}{dx}(0)$$

$$V(0) = \frac{EI}{L^3}\frac{d^3y}{d\zeta^3}(0) - \frac{F_B}{L}\frac{dy}{d\zeta} \qquad (6\text{-}3)$$

$$V(0) = \frac{EI}{L^3}(6a_3) - \frac{F_B}{L}(a_1)$$

$$V(0) = \frac{29(10)^6 2399}{1500^3} 6(-3040.31)\left|\frac{1\text{ft}^2}{144\text{in}^2}\right| - \frac{10^5}{1500}(83.38)$$

$$V(0) = -8{,}170 \text{ lbs}$$

The minus sign means that shear applied to the bottom of the riser is to the left while the shear force applied to the well head is to the right.

6.1.2 Non-Drilling Mode

This is a precautionary mode in preparation for a possible well shut down and disconnect, if necessary. Once the well is shut in, drilling mud may be displaced with sea water to help stabilize the riser.

6.1.2.1 Replacing Drilling Mud with Sea Water

Replacing drilling mud with sea water increases effective tension without actually pulling the riser more at the surface. Table 6-2 shows the effect of mud weight on the critical buckling force at the lower end. Riser length is assumed to be 1500 ft and the bottom is structurally fixed. When drilling mud is replaced by sea water, the buckling force at the lower end is a compressive force of

Table 6-2. Critical buckling force (F_{Bcr}); drilling riser size (22 in. OD × 20.75 in. ID) riser length is 1500 ft.

Mud Weight, ppg	Buckling Force, kips
8.56	−95.00
10	−65.88
11	−45.37
12	−24.66
13	−3.77
14	17.27
15	38.46
16	59.77

(–) 95.0 kips. On the other hand, if mud weight is, say, 14 ppg, the riser will buckle with a bottom force of (+) 17.27 kips tension.

These numbers were calculated using (from Chapter 4)

$$F_{Bcr} = -\phi_{cr}\left(\frac{EI}{144}\right)^{\frac{1}{3}}(w - A_o\gamma_o + A_i\gamma_i)^{\frac{2}{3}} - (LA_o\gamma_o - LA_i\gamma_i) \tag{4-8}$$

Drilling mud density has a huge effect on riser behavior through effective tension. Effective tension distribution for different mud weights are shown in Fig. 6-2. High effective tension reduces bending while low effective tension moves riser pipes toward instability. This graph was determined from

$$T_{eff} = \left[(F_B + wx) + (L - x)\left(A_0\gamma_0 - A_i\gamma_i\right)\right] \tag{6-4}$$

Each line corresponds to a bottom tension of F_B = 100,00 lbs. The effective tensions at the bottom are not 100,000 lbs because of mud weight and sea water density. The top tension is the same for each case and is equal to

$$T_{top} = 100{,}000 + 1500(142.81) = 314{,}215 \text{ lbs}$$

Since the bottom tension is 100,000 lbs and greater than the critical buckling forces given in Table 6-2, each case is stable.

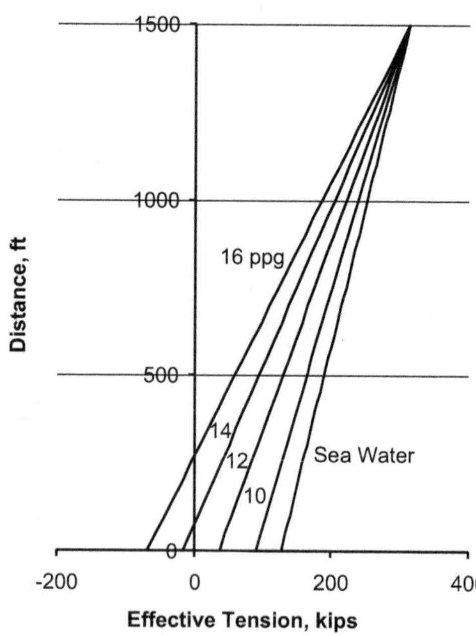

Figure 6-2. Mud weight and effective tension.

6.1.2.2 *Effect of Drilling Mud on Bending Stress*

The effect of mud weight on riser bending stress is profound. This is clearly shown in the comparison of the two response predictions of Figs. 6-3a and 6-3b. The conditions for these two calculations are

L = 1500 ft
Riser Pipe = 22 in. OD × 20.75 in. ID
F_B = 100,000 lb
γ_0 = 64 lb/ft³ (sea water)
Rig offset = 50 ft

Only the mud weight is changed from 12 ppg to sea water density. Actual internal tension within the riser is unchanged; bottom tension is 100,000 lb in each case. When drilling mud is replace by sea water effective tension increases as shown in Fig. 6-2. The top and bottom tensions are the same for both cases.

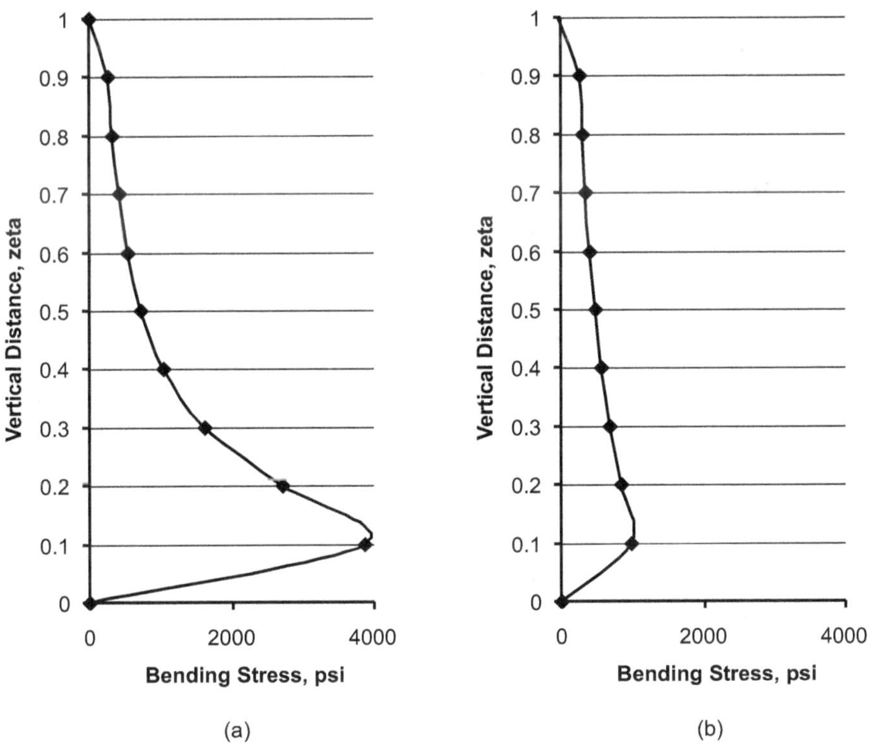

Figure 6-3. Comparison of bending stress when drilling mud is replaced with sea water. (a) 12 ppg drilling mud, (b) sea water.

Replacing drilling mud with sea water greatly increases the effective tension and stability against buckling. It also reduces bending stresses along the riser as shown in Fig. 6-3. For this example, the critical buckling force corresponding to 12 ppg mud is 24.66 kips internal compression at the lower end (Table 6-2). When the drilling mud is replaced by sea water, the critical buckling force is 95.0 kips compression. The riser is much more stable under this condition. It also means that bending stresses cause by side loading due to ocean currents and waves are significantly reduced. The bending stress curves in Fig. 6-3 correspond to 100 kips (tension) at the bottom end.

If environmental conditions create excessive stresses under this condition, then the riser should be disconnected.

6.1.3 Disconnect Mode

The disconnect mode means the lower end of the riser is disconnected from the well head. If a severe storm is imminent then the riser may be removed completely and stacked on the drill ship. Sometimes a riser may be left suspended below the drill ship. In this case, the riser is highly unconstrained and responds only to ocean waves and currents and gravity. Ocean waves could excite one of its natural modes of vibration. Vortex-induced vibrations are also possible and should be checked. In addition, vessel heaving could excite axial modes. Natural lateral modes of a freely hanging pipe can be determined using data given in Chapter 2 (Case 12). The special case of riser response with a BOP stack hanging freely at the lower end is discussed in Appendix 6A.

6.2 MARINE PRODUCTION RISERS

Production risers are designed to stay on location over the production lifetime of a reservoir. Each riser is structurally pinned to the production platform and attached to mudline equipment by means of tapered flex joints or flex/ball joints to reduce bending stresses while accommodating production tubing. Cumulative fatigue damage over the expected lifetime of production risers should be evaluated. Critical structure areas are: 1) end connections and 2) zones within the riser body that may experience high cyclic fatigue stresses caused by vortex-induced vibrations.

6.2.1 Top-Mounted Tapered Flex Joint

Bottom-mounted tapered flex joints were considered in Chapter 4 based on Dareing [2]. Deflections for a composite riser are given (Fig. 6-4). Input data for this example are:

Riser pipe = 22 in. OD × 20.75 in. ID
L = 1500 ft
Mud weight = 12 ppg
Bottom pull = 100,000 lbs
Tapered Flex Joint geometry (see Table 4-2)

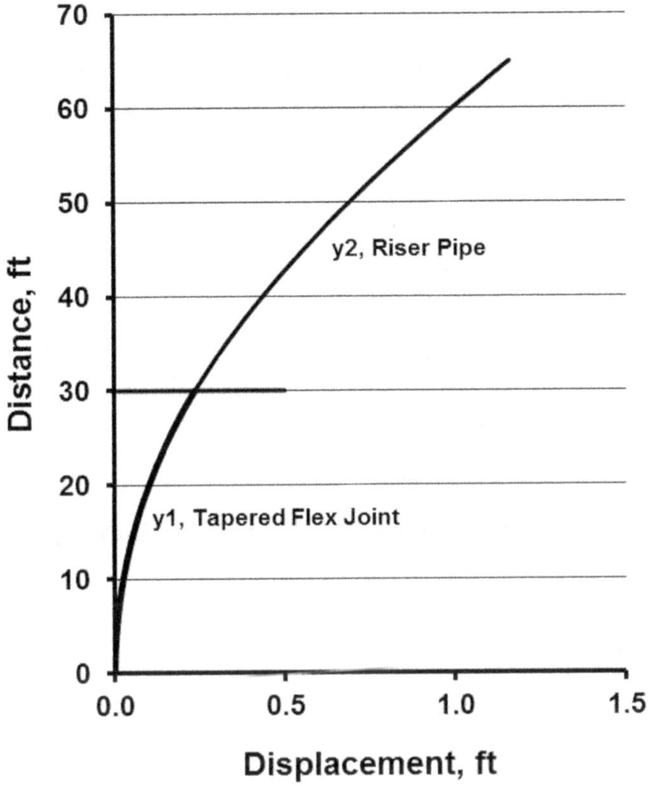

Figure 6-4. Bending of tapered flex joint at bottom location.

On occasion, BOP stacks are top mounted, and this imposes a different structural condition on riser pipes. The situation considered here is a top-mounted BOP stack with a tapered stress joint positioned directly below the BOP stack [3]. The arrangement is illustrated in Fig. 6-5. Bending moment and stress distribution along the stress joint can be substantial when the BOP platform is tilted. This rotation creates bending and a shear force at the top end of the stress joint as indicated.

The differential equation of bending for the flex joint (identified by the subscript number 1) is

$$EI(x)\frac{\mathrm{d}^2 y_1}{\mathrm{d}x^2} - Ty_1 = -M_0 + S_o x \tag{6-5}$$

The differential equation of bending for the riser pipe (identified by the subscript number 2) is

$$EI\frac{\mathrm{d}^2 y_2}{\mathrm{d}x^2} - Ty_2 = -M_0 + S_o x \tag{6-6}$$

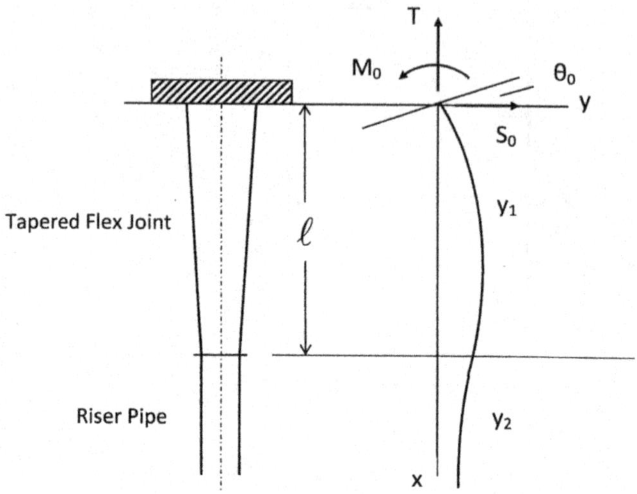

Figure 6-5. Tapered stress joint and coordinate system.

The moment of inertia is constant below the stress joint. In this case, two solutions are sought, one for the tapered flex joint and one for the riser pipe.

The solution to the bending equation (Eq. (6-6)) for the riser pipe is

$$y_2(x) = C_3 e^{\beta x} + C_4 e^{-\beta x} + \frac{M_0}{T} - \frac{S_0}{T}x \tag{6-7}$$

where

$$\beta = \sqrt{\frac{T}{\mathrm{EI}_p}}$$

The solution to the bending equation (Eq. (6-5)) for the tapered flex joint is given below. The differential equation of bending in this case is

$$\frac{\mathrm{EI}_p}{a^2}(h-x)^2\frac{d^2y_1}{dx^2} - Ty_1 = S_0x - M_0 \tag{6-8}$$

The solution is written as

$$y(x) = y_P(x) + y_C(x) \tag{6-9}$$

where the particular solution is

$$y_P(x) = \frac{M_0}{T} - \frac{S_0}{T}x \tag{6-10}$$

The complementary solution is determined as before making the total solution

$$y_1(x) = C_1(h-x)^{D_1} + C_2(h-x)^{D_2} + \frac{M_0}{T} - \frac{S_0}{T}x \tag{6-11}$$

Equation (6-11) defines lateral deflections in the tapered flex joint section.

Equations (6-7) and (6-11) contain six unknown constants, C_1, C_2, C_3, C_4, M_0, and S_0, which are determined from boundary conditions.

At the top of the riser system, the deflection is taken as zero, while the slope is equal to that of the surface platform when not at horizontal. The boundary conditions for the top of the tapered joint are

$$y_1(0) = 0 \tag{6-12a}$$

$$\frac{dy_1}{dx}(0) = \theta_0 \tag{6-12b}$$

which give

$$0 = C_1 h^{D_1} + C_2 h^{D_2} + \frac{M_0}{T} \tag{6-13}$$

$$\theta_0 = -D_1 C_1 h^{(D_1-1)} - D_2 C_2 h^{(D_2-1)} - \frac{S_0}{T} \tag{6-14}$$

At the interface of the flex joint and pipe sections, the deflections, as well as the slopes, of the two separate sections are equal. The boundary conditions at the intersection of the stress joint and pipe are

$$y_1(\ell) = y_2(\ell) \tag{6-15a}$$

$$\frac{dy_1}{dx}(\ell) = \frac{dy_2}{dx}(\ell) \tag{6-15b}$$

These two boundary conditions produce

$$C_1 (h-l)^{D_1} + C_2 (h-l)^{D_2} + \frac{M_0}{T} - \frac{S_0}{T} l = C_3 e^{\beta l} + C_4 e^{-\beta l} + \frac{M_0}{T} - \frac{S_0}{T} l \tag{6-16}$$

$$-D_1 C_1 (h-l)^{(D_1-1)} - D_2 C_2 (h-l)^{(D_2-1)} - \frac{S_0}{T} = \beta C_3 e^{\beta l} - \beta C_4 e^{-\beta l} - \frac{S_0}{T}. \tag{6-17}$$

At some depth, it is assumed that both the deflection and the slope of the lower section of pipe are both zero. L is initially chosen as a large number, say 500 ft. Hence, the boundary equations at the lower end of the system are

$$y_2(L) = 0 \tag{6-18a}$$

$$\frac{dy_2}{dx}(L) = 0 \tag{6-18b}$$

which give

$$0 = C_3 e^{\beta L} + C_4 e^{-\beta L} + \frac{M_0}{T} - \frac{S_0}{T} L \tag{6-19}$$

$$0 = \beta C_3 e^{\beta L} - \beta C_4 e^{-\beta L} - \frac{S_0}{T} \tag{6-20}$$

The boundary condition equations converted to matrix form yield the following 6×6 matrix system.

$$\begin{vmatrix} h^{D_1} & h^{D_2} & 0 & 0 & \frac{1}{T} & 0 \\ -D_1 h^{(D_1-1)} & -D_2 h^{(D_2-1)} & 0 & 0 & 0 & \frac{-1}{T} \\ (h-l)^{D_1} & (h-l)^{D_2} & -e^{\beta l} & -e^{-\beta l} & 0 & 0 \\ -D_1 (h-l)^{(D_1-1)} & -D_2 (h-l)^{(D_2-1)} & -\beta e^{\beta l} & \beta e^{-\beta l} & 0 & 0 \\ 0 & 0 & e^{\beta L} & e^{-\beta L} & \frac{1}{T} & \frac{-L}{T} \\ 0 & 0 & \beta e^{\beta L} & -\beta e^{-\beta L} & 0 & \frac{-1}{T} \end{vmatrix} * \begin{vmatrix} C_1 \\ C_2 \\ C_3 \\ C_4 \\ M_0 \\ S_0 \end{vmatrix} = \begin{vmatrix} 0 \\ \theta_0 \\ 0 \\ 0 \\ 0 \\ 0 \end{vmatrix} \tag{6-21}$$

Solution of this matrix equation yields values for C_1, C_2, C_3, C_4, M_0, and S_0. Tension applied to the top of the stress joint, slope at the top of the stress joint, pipe geometry, and pipe physical properties are known. The deflections of the tapered flex joint and the suspended pipe can now be defined through Eqs. (6-7) and (6-11).

Bending moment expression is the same for both stress joint and pipe.

$$M(x) = Ty(x) + S_0 x - M_0 \tag{6-22}$$

The bending stress in the flex joint and pipe are

$$\sigma_1(x) = \frac{M(x)c}{I(x)} \tag{6-23}$$

and

$$\sigma_2(x) = \frac{M(x)c}{I_p} \tag{6-24}$$

The moment of inertia used to calculate bending stress in the tapered flex joint is taken as the parabolic approximation for $I(x)$.

$$I(x) = I_p \left(\frac{h-x}{a}\right)^2 \tag{6-25}$$

where

$$h = \frac{\ell}{1 - \sqrt{\dfrac{I_p}{I_{SJ}}}}$$

and

$$a = h\sqrt{\frac{I_p}{I_{sJ}}}$$

Total stress can be determined, for both sections of the system, as the sum of the bending stress and the axial stress. Axial stress for each section must be calculated using the appropriate respective area (Table 6-3).

$$\sigma_{1T}(x) = \frac{M(x)c}{I(x)} + \frac{T}{A} \tag{6-26}$$

$$\sigma_{2T}(x) = \frac{M(x)c}{I_p} + \frac{T}{A_p} \tag{6-27}$$

In order to show the effectiveness of tapered flex joints, consider a top-mounted case with and without a tapered flex joint. The riser has the following specifications:

- Riser tubular size = 13 3/8 in. OD × 12.414 in. ID
- Top pull force = 1500 kips
- Rotation of fixed end is 2 deg from the horizontal
- Minimum yield strength of the material is 110 ksi
- Stress joint is a one piece straight tapered tubular

The numbers show that the assumed profile is nearly straight and the moment of inertia for the assumed shape differs from that of the straight profile by about 3.5% at the most.

Without a tapered flex joint, the bending stress distribution is shown in Fig. 6-6.

The axial stress in this case is a constant 77,062 psi over the effected span. When added to the bending stress, at the top cross section, the maximum stress is about 157,000 psi. This level exceeds the yield strength (110,000 psi) of the riser pipe.

When a flex joint having 15 in. OD base which tapers down to 13 3/8 in. OD over a 15 ft length, the bending stress drops dramatically to a maximum of only

Table 6-3. Comparison of wall thickness and moment of inertia over a 15-ft flex joint (15" OD at base; 13 3/8" OD at interface with riser; ID is 12.414").

	Wall thickness, *t*, in.		Moment of Inertia, I, in⁴	
Location, *x* ft	**Parabolic**	**Straight**	**Parabolic**	**Straight**
0	1.293	1.293	1319.3	1319.3
1	1.234	1.239	1242.0	1248.3
2	1.175	1.185	1167.1	1178.8
3	1.117	1.131	1094.5	1110.8
4	1.060	1.076	1024.2	1044.3
5	1.003	1.022	956.3	979.3
6	0.946	0.968	890.7	915.7
7	0.891	0.914	827.4	853.6
8	0.836	0.860	766.4	792.8
9	0.782	0.806	707.8	733.5
10	0.729	0.751	651.5	675.5
11	0.677	0.697	597.6	618.8
12	0.626	0.643	546.0	563.4
13	0.576	0.589	496.7	509.4
14	0.527	0.535	449.7	456.6
15	0.481	0.481	405.1	405.1

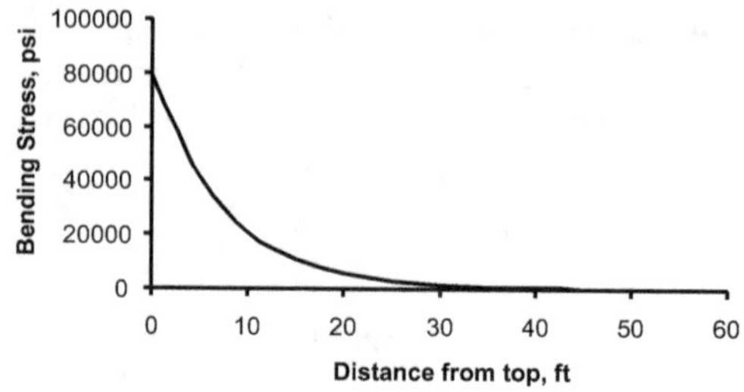

Figure 6-6. Bending stress without tapered flex joint.

40,000 psi. The total stress including axial stress component is also shown in Fig. 6-7. In this case, the maximum stress (100,000 psi) is at the flex joint/riser interface. This stress level is below the yield strength of the riser body.

A comparison of calculations made using the tapered flex joint equations and using multiple straight sections is given in Table 6-4.

6.2.2 Optimizing Tapered Flex Joints

The objective of the process is to configure the stress joint so that it is as short as possible (for cost and operational considerations) and maintain maximum stresses within acceptable levels. Design questions area follows: What is the best base OD? What is the best stress joint length?

Consider stress joint having a length of 15. Figure 6-8 shows how total stress distribution varies over the 15-ft length for several different base OD's. The numbers show that a base OD of 14.2 in. gives a stress balance between the base and smaller end of the stress joint. Similar calculations are given for a 30 ft long stress joint [3].

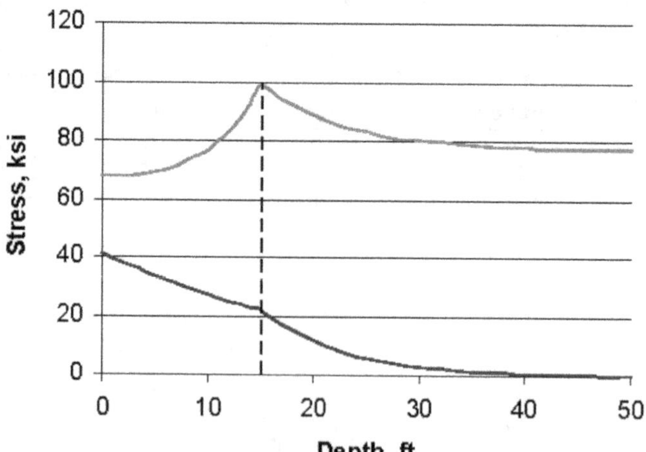

Figure 6-7. Stress distribution in 15 flex joint.

Table 6-4. Comparison of Parabolic Flex Joint Predictions with Multiple Linear-Step Element Predictions (15" OD at base; 13 3/8" OD at interface with riser; ID is 12.414") 15 ft Flex Joint.

	Bending stress, ksi		Total stress, ksi	
Location, *x* ft	Parabolic	Linear step	Parabolic	Linear step
0	41.17	39.27	68.11	66.21
1	39.63	37.46	67.98	65.69
2	38.12	35.90	68.01	65.64
3	36.64	34.64	68.22	65.82
4	35.18	33.05	68.62	65.93
5	33.76	31.50	69.25	66.26
6	32.38	30.00	70.14	66.83
7	31.02	28.47	71.32	67.68
8	29.71	27.01	72.83	68.85
9	28.44	25.58	74.73	70.42
10	27.22	24.18	77.08	72.45
11	26.06	22.81	79.97	75.04
12	24.96	21.48	83.49	78.35
13	23.94	20.19	87.76	82.55
14	23.02	18.94	92.96	87.91
15	22.21	17.76	99.28	94.82

Note that when the diameter is reduced to the OD of the riser tubular, the maximum stress becomes the stress corresponding to no stress joint. With no stress joint the maximum stress in the riser pipe would be about 150 ksi. These numbers how that stress levels at the stress joint/riser interface are best controlled by stress joint length, while the stress levels at the stress joint base are controlled by base OD.

Stress calculations for both 15 ft and 30 ft stress joints show that in each case, the stress curves intersected at a point near the end of the stress joint. The location of this point in each diagram is independent of the base diameter. Even the stress curve for a bare riser without a stress joint passes through this point. The intersection point represents stress levels and location of stresses in a bare riser. Since these points are near the end of stress joints, they could be useful in estimating stress joint lengths for achieving acceptable stress levels in the riser. The base OD would dictate stresses in the stress joint.

In reality, the final stress joint configuration is established by a series of trade-offs between, acceptable stress level, operational considerations, time and scheduling, costs, etc.

6.3 BENDING OF BUNDLED TUBULARS ATTACHED TO PRODUCTION RISERS

Production tubing is sometimes fixed to the main body of marine production risers as shown in Fig. 6-9. Tubulars are held in a given pattern within or outside of the riser with spacers holding them in place relative to the inside of the riser. Over a given section of a riser pipe, there may be several spacers, which cause each pipe to move with the riser. Each tube then deflects with the riser but is constrained by the spacers, which cause each tube to behave as a multi-spanned beam column.

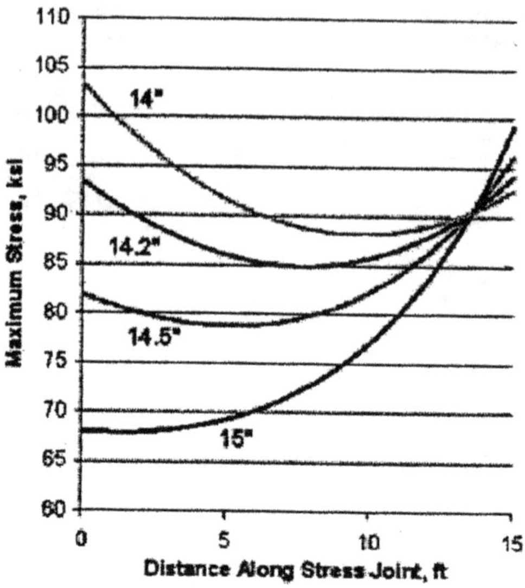

Figure 6-8. Stress curves for a 15-ft tapered flex joint [2].

It is important to note that the local curvature of each attached tube is not the same as the main riser pipe. Local bending in the attached tubes can be much higher than that of the riser pipe. The basic question here relates to the bending stresses within the production tubing as it is held in place spacers and as the riser moves from side to side.

It is assumed that riser motion has been determined from a separate analysis and that riser deflections at each spacer are imparted to each tube [4]. Deflections, bending and stresses between each span of a tube are needed so that critical points and stresses can be identified and quantified in the tube. Each tube is treated as a multi-spanned beam column where the lateral displacement of each spacer is known.

6.3.1 Method of Solution

The solution to this type of problem normally involves four constants of integration for each span. For n spans, the constants of integration are found by solving 4n algebraic equations simultaneously. If n = 5 then the number of constants to be determined is 20. Local deflections and stresses can be determined once these constants are found,. The calculation procedure follows the method of transfer matrix [5]. This approach requires the solution to only two algebraic equations. Appendix 3A gives a step by step example of this method as applied to stabilized drill collars. The same procedure applies here as well.

A multi-spanned tube (beam) is depicted in Fig. 6-10. Support numbers and span numbers are identified. Support positions are known from a separate riser analysis.

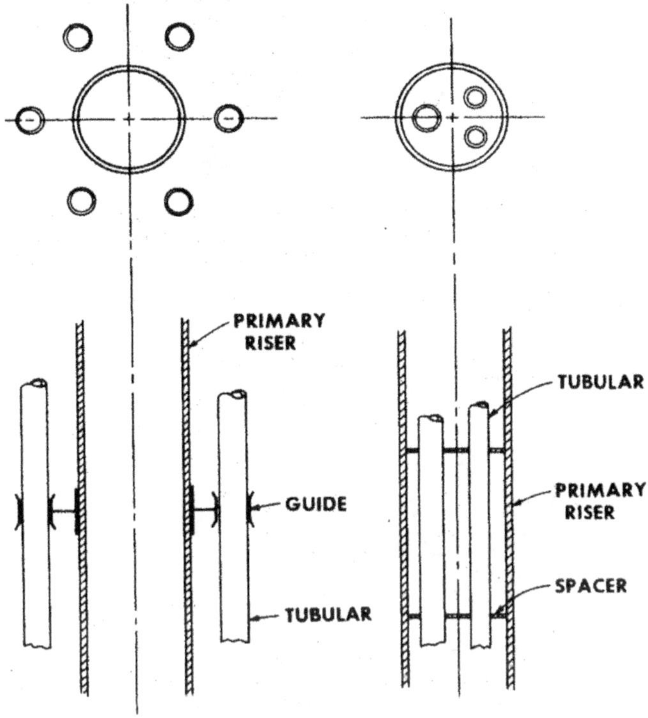

Figure 6-9. Bundled tubing attached to marine risers.

The bending between each spacer is depicted in Fig. 6-11a. The bending equation over each span is derived from the differential element shown in Fig. 6-11b.

Following the sign convention shown Fig. 6-11b, the slope, θ, moment, *M* and shear, *V* are

$$\theta = \frac{dy}{dx} \tag{6-28}$$

$$M = -EI\frac{d^2y}{dx^2} \tag{6-29}$$

$$V = -EI\frac{d^3y}{dx^3} + T\frac{dy}{dx} \tag{6-30}$$

Combining these equations, gives the differential equation of bending.

$$EI\frac{d^4y}{dx^4} - T\frac{d^2y}{dx^2} = 0 \tag{6-31}$$

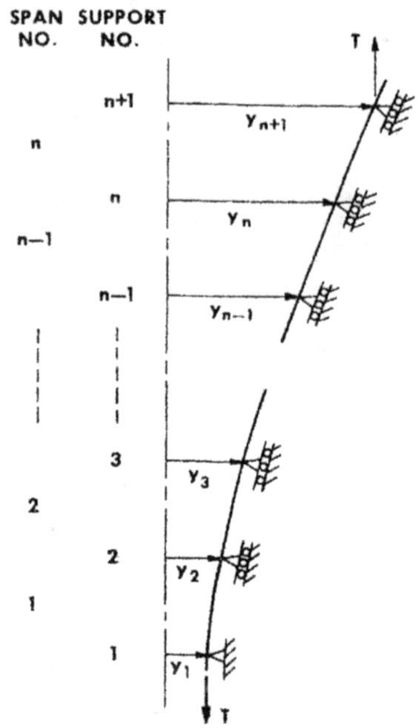

Figure 6-10. Schematic of multi-spanned continuous beam.

Here tension in the tube is assumed to be constant over a given span. This tension should be viewed as "effective tension" which can account for actual tension plus fluid pressure in and around the tube.

Assuming the tube is in tension ($T > 0$), the solution to Eq. (6-31) is

$$y = A_i x + B_i + C_i \cosh \lambda x + D_i \sinh \lambda x \tag{6-32}$$

where

$$\lambda = \sqrt{\frac{|T|}{EI}}$$

It follows that

$$\theta = A_i + \lambda(C_i \sinh \lambda x + D_i \cosh \lambda x) \tag{6-33}$$

$$M = -T(C_i \cosh \lambda x + D_i \sinh \lambda x) \tag{6-34}$$

$$V = TA_i \tag{6-35}$$

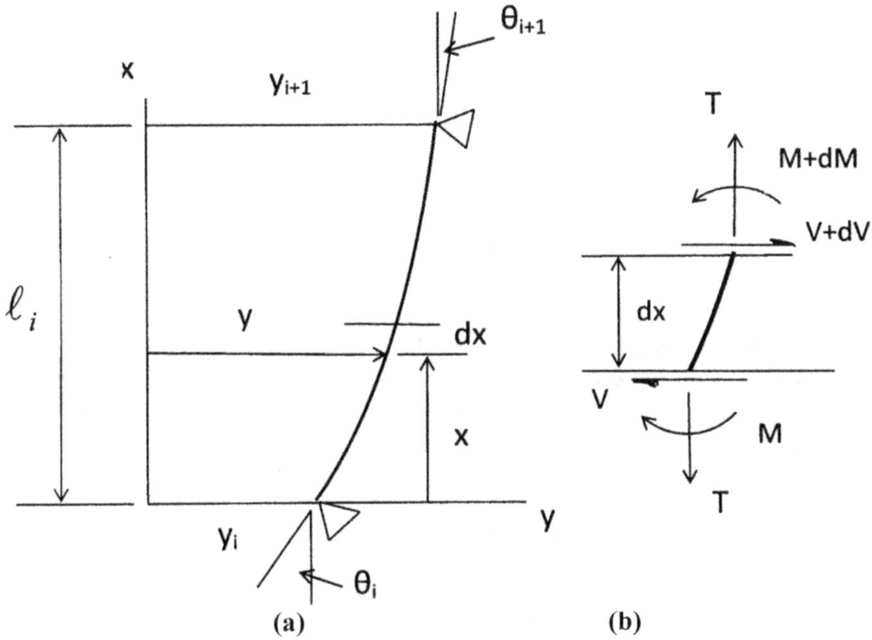

Figure 6-11. Freebody of differential element.

Applying boundary conditions, $y = y_i$ at $x = 0$ and $y = y_{i+1}$ at $x = \ell$ we obtain

$$A_i = \psi_i - \frac{\cosh \lambda\ell_i - 1}{\ell_i} C_i - \frac{\sinh \lambda\ell_i}{\ell_i} D_i \tag{6-36}$$

$$B_i = y_i - C_i \tag{6-37}$$

where

$$\psi_i = \frac{(y_{i+1} - y_i)}{\ell_i}$$

Note that the subscript i attached to A, B, C, D, ψ, L refers to those quantities of the ith span, while the subscript i fixed to y, θ, M refers to the respective values of the ith support.

When considering the deflection distribution over a given san, $x = 0$ is located at the lower end of the span. Therefore, by setting $x = 0$ in Eqs. (6-33) and (6-34) and combining the resultant equation with Eq. (6-36) gives

$$\begin{pmatrix} \theta_i \\ M_i \end{pmatrix} = \begin{bmatrix} \frac{1-\cosh\lambda\ell_i}{\ell_i} & \lambda - \frac{\sinh\lambda\ell_i}{\ell_i} \\ -T & 0 \end{bmatrix} \begin{pmatrix} C_i \\ D_i \end{pmatrix} + \begin{pmatrix} \psi_i \\ 0 \end{pmatrix} \tag{6-38}$$

By inversion

$$\begin{pmatrix} C_i \\ D_i \end{pmatrix} = [\alpha]_i \left(\begin{pmatrix} \theta_i \\ M_i \end{pmatrix} - \begin{pmatrix} \psi_i \\ 0 \end{pmatrix} \right) \tag{6-39}$$

where

$$[\alpha]_i = \frac{1}{T\left(\lambda - \dfrac{\sinh \lambda \ell_i}{\ell_i}\right)} \begin{bmatrix} 0 & -\left(\lambda - \dfrac{\sinh \lambda \ell_i}{\ell_i}\right) \\ T & \dfrac{1 - \cosh \lambda \ell_i}{\ell_i} \end{bmatrix} \tag{6-40}$$

Next formulate expressions for θ_{i+1} and M_{i+1} by letting $x = \ell$ in Eqs. (6-33) and (6-34), and combine the resultant equations with Eqs. (6-36) and (6-39). This gives the following recursion equation.

$$\begin{pmatrix} \theta_{i+1} \\ M_{i+1} \end{pmatrix} = [\beta]_i [\alpha]_i \left(\begin{pmatrix} \theta_i \\ M_i \end{pmatrix} - \begin{pmatrix} \psi_i \\ 0 \end{pmatrix} \right) + \begin{pmatrix} \psi_i \\ 0 \end{pmatrix} \tag{6-41}$$

where

$$[\beta]_i = \begin{bmatrix} \dfrac{1 - \cosh \lambda \ell_i}{\ell_i} + \lambda \sinh \lambda \ell_i & \dfrac{-\sinh \lambda \ell_i}{\ell_i} + \lambda \cosh \lambda \ell_i \\ -T \cosh \lambda \ell_i & -T \sinh \lambda \ell_i \end{bmatrix} \tag{6-42}$$

According to Eq. (6-41), $\begin{pmatrix} \theta_i \\ M_i \end{pmatrix}$ at the ith support can be expressed in terms of $\begin{pmatrix} \theta_1 \\ M_1 \end{pmatrix}$ at the first support, i.e.,

$$\begin{pmatrix} \theta_i \\ M_i \end{pmatrix} = [\gamma]_i \begin{pmatrix} \theta_1 \\ M_1 \end{pmatrix} + (\delta)_i \text{ for } i = 1, 2, \ldots n+1 \tag{6-43}$$

where

$$[\gamma]_1 = \begin{bmatrix} 1 & 0 \\ 0 & 1 \end{bmatrix}$$

and

$$(\delta)_1 = \begin{pmatrix} 0 \\ 0 \end{pmatrix}$$

The matrices $[\gamma]_{i+1}$ and $(\delta)_{i+1}$ can be related to $[\gamma]_i$ and by $(\delta)_i$ substituting Eq. (6-43) into Eq. (6-41). The results are

$$[\gamma]_{i+1} = [\beta]_i [\alpha]_i [\gamma]_i \tag{6-44}$$

$$(\delta)_{i+1} = [\beta]_i [\alpha]_i \left((\delta_i) - \begin{pmatrix} \psi_i \\ 0 \end{pmatrix} \right) + \begin{pmatrix} \psi_i \\ 0 \end{pmatrix} \tag{6-45}$$

Thus starting with $i = 1$, $[\gamma]_i$ and $(\delta)_i$ can be successively obtained from Eqs. (6-44) and (6-45) since $[\beta]_i$, $[\alpha]_i$ and $[\psi]_i$ are known. Finally

$$\begin{pmatrix} \theta_{n+1} \\ M_{n+1} \end{pmatrix} = [\gamma]_{n+1} \begin{pmatrix} \theta_1 \\ M_1 \end{pmatrix} + (\delta)_{n+1} \tag{6-46}$$

This matrix equation contains two linear equations with M_1 and M_{n+1} as variables. Eliminating M_{n+1} we have

$$M_1 = \frac{\theta_{n+1} - (\gamma_{1,1})_{n+1}\theta_1 - (\delta_1)_{n+1}}{(\gamma_{1,2})_{n+1}} \tag{6-47}$$

After the value of M_1 is obtained, θ_i and M_i are obtained by using Eq. (6-43). The integration constants, C_i and are D_i obtained by using Eq. (6-39), while Ai and Bi are calculated using Eqs. (6-36) and (6-37). Next the deflection, slope, moment and shear at any section can be obtained. Finally, the reaction at the ith support can be obtained from the difference of the shears on each side of a given support.

6.3.2 Example Calculation

Consider the tubing bundle shown nested within a riser as shown in Fig. 6-12.

Further assumptions are as follows:

- pipe size is 4 in. OD × 0.262 in. wall
- riser section is 50 ft long
- each pipe is supported by six spacers; there are five 10 ft spans
- internal pipe pressure is 5,000 psi
- each of the nested pipes is free to expand by means of a slip joint

The pipe is offset by 3 in. from the neutral axis of the composite riser. In this case, tubular tension, T, is comprised of only effective tension produced by internal pressure.

Assuming a seal diameter, d_s, of 4.375 at the sliding end,

$$T = -\frac{\pi}{4} d_s^2 p = -75{,}165 \text{ lb}$$

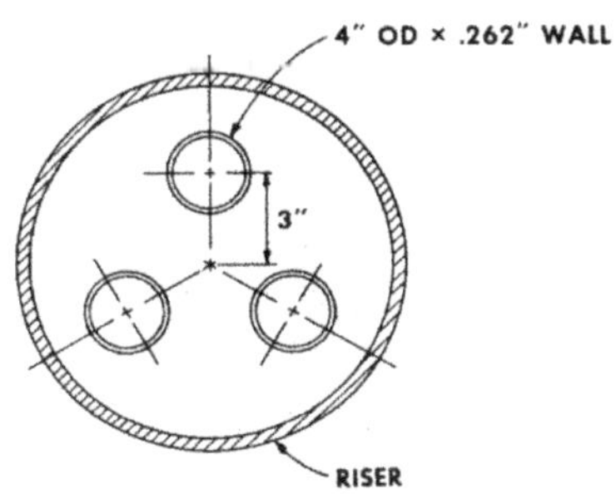

Figure 6-12. Internal bundling of production tubing.

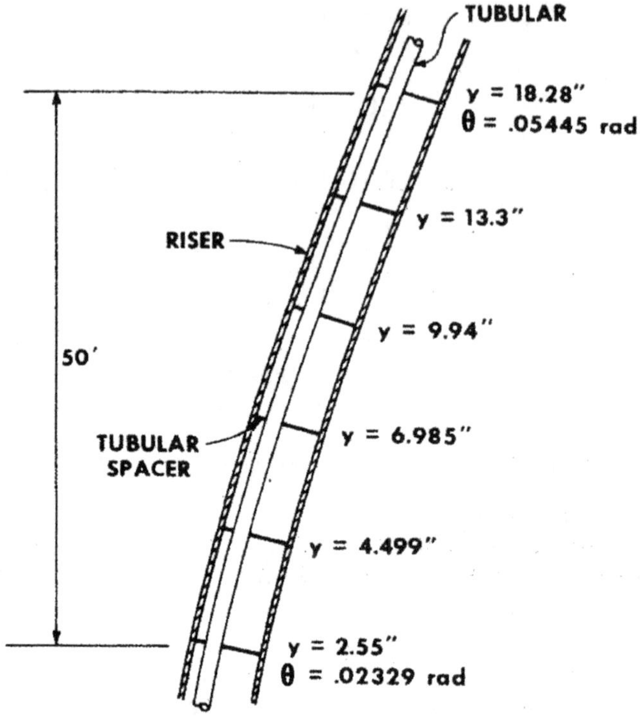

Figure 6-13. Displacement imposed on a production tube.

Since effective tension is minus ($T < 0$) then the differential equation of bending becomes

$$\mathrm{EI}\frac{\mathrm{d}^4 y}{\mathrm{d}x^4} + T\frac{\mathrm{d}^2 y}{\mathrm{d}x^2} = 0$$

Its solution changes slightly from the previous case ($T > 0$). When effective tension is minus (T < 0)

$$y = A_i x + B_i + C_i \cos\lambda x + D_i \sin\lambda x$$

The solution proceeds the as before [4].

It has been determined through a separate analysis that the main riser pipe is deflected as shown in Fig. 6-13.

Bending moments and stress were determined following the procedure discussed above and in ref [4]. The total stress, which has bending and axial components, is shown in Fig 6-14. Stress distribution shows significant difference in bending stress over each of the five spans.

Reference [4] also discusses the case where each end of a given tube is fixed and effective tension is plus.

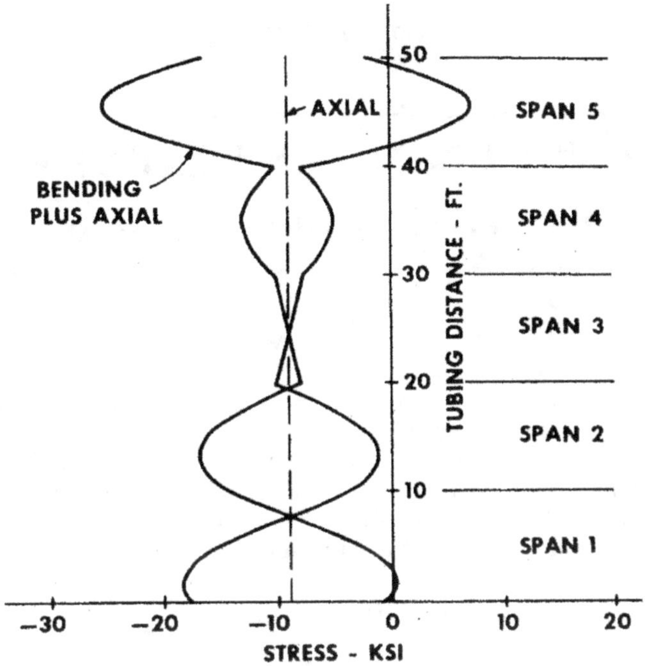

Figure 6-14. Stress distribution in a tubular.

Appendix 6A

BOP SUSPENDED AT LOWER END OF DRILLING RISER

Disconnecting a riser with a significant mass attached at the lower end, such as a BOP stack is considered here. The weight and angular moment of inertia impacts static and dynamic responses. Consider, for example, the natural modes of vibration of a riser with a BOP attached at the bottom end as shown in Fig. 6-15. The BOP stack is disconnected in this case.

The differential equation for the natural frequency of the riser body is

$$EI\frac{\partial^4 y}{\partial x^4} - \frac{\partial}{\partial x}\left\{\left[(F_B + wx) + (L-x)(A_0\gamma_0 - A_i\gamma_i)\right]\frac{\partial y}{\partial x}\right\} - m\frac{\partial^2 y}{\partial t^2} = 0 \quad (6\text{-}48)$$

The boundary conditions are

$$y(L,t) = 0 \quad (6\text{-}49a)$$

$$\frac{\partial^2 y}{\partial x^2}(L,t) = 0 \quad (6\text{-}49b)$$

$$EI\frac{\partial^2 y}{\partial x^2}(0,t) = M_0 \quad (6\text{-}49c)$$

$$EI\frac{\partial^3 y}{\partial x^3}(0,t) - F_B\frac{dy}{dx}(0,t) = V_0 \quad (6\text{-}49d)$$

A freebody analysis of the BOP stack gives

$$\sum M_G = I_G\alpha \quad (6\text{-}50a)$$

$$\sum F_y = Ma_G \quad (6\text{-}50b)$$

where M is mass of BOP stack. Expanding Eq. (6-50a) gives

$$M_0 - \ell V_0 - d\frac{\partial y}{\partial x}(0,t)F_B = I_G\frac{\partial^3 y}{\partial x \partial t^2}(0,t) \quad (6\text{-}51)$$

Expanding Eq. (6-50b) gives

$$-V_0 = M\left(\frac{\partial^2 y}{\partial t^2}(0,t) - d\frac{\partial^2 y}{\partial x \partial t}(0,t)\right) \quad (6\text{-}52)$$

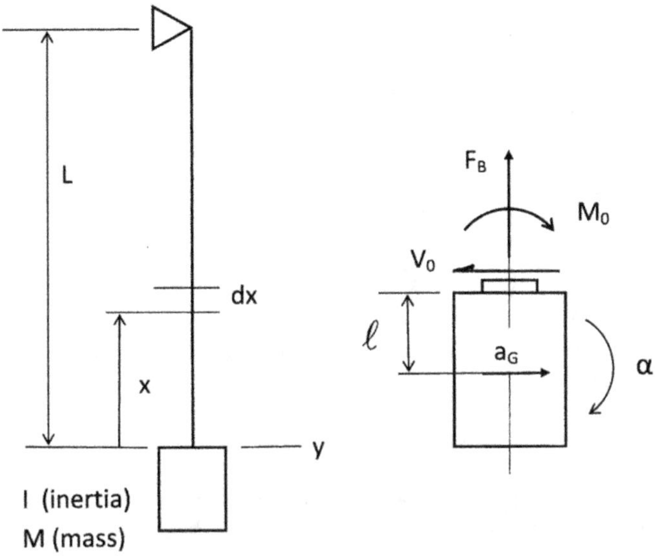

Figure 6-15. BOP stack suspended from riser.

Assuming the solution to Eq. (6-48) of the form

$$y(x,t) = Y(x)\sin \omega t \tag{6-53}$$

And substituting into Eq. (6-48) gives

$$\frac{d^4Y}{d\zeta^4} - \frac{d}{d\zeta}\left[(\beta + \alpha\zeta)\frac{dY}{d\zeta}\right] - \lambda^4 Y = 0 \tag{6-54}$$

where

$$\lambda^4 = \frac{m\omega^2 L^4}{\mathrm{EI}}$$

$$\zeta = \frac{x}{L}$$

The boundary conditions convert to

$$Y(L) = 0 \tag{6-55a}$$

$$\frac{d^2Y}{dx^2}(L) = 0 \tag{6-55b}$$

$$EI\frac{d^2Y}{dx^2}(0) = \omega^2\left[\frac{Wd}{g} - I_G\frac{dY}{dx}(0)\right] + Wd\frac{dY}{dx}(0) \tag{6-55c}$$

$$EI\frac{d^3Y}{dx^3}(0) - F_B\frac{dY}{dx}(0) = \omega^2\left[MY(0) - Md\frac{dY}{dx}(0)\right] \tag{6-55d}$$

The solution continues as described in Chapter 2 by expressing each coefficient, a_n, in terms of a_0, a_1, a_2, a_3 using

$$a_n = F_n a_0 + G_n a_1 + H_n a_2 + I_n a_3 \tag{6-56}$$

and setting up the characteristic equation as explained in Chapter 4. Rotational motion of the rigid mass could induce high bending stresses at the point of connection.

Appendix 6B

DRILL PIPE WHIRL WITHIN DRILLING RISERS

Marine risers sometime exhibit circular motion in the ocean. This motion could be due to external fluid forces. It also could be caused by internal forces associated with drill pipe whirl. This second scenario is discussed below.

The riser itself is rigidly attached at the top and bottom so backward, forward or synchronous whirling motions of the type discussed in Chapter 2 are not possible. However, the drill pipe may experience these whirling motions as illustrated in Fig. 6-16.

If the damping ratio is less than one (1), drill pipe whirl is unstable (see Fig. 2-20). Amplitude grows with time until contact is made with the inside surface of the riser. Assuming further that the velocity ratio is greater than one, the drill pipe would continue with backward whirl moving as a gear inside the riser. On the other hand, if the velocity ratio is less than one, the pipe would be unstable and whirl within the riser with forward whirl. Once contact is made between drill pipe and the inside of the riser, radial forces are developed against the riser causing circular motion of the riser center.

In this case, drill pipe whirl could be a source of excitation to the riser. The whirling frequency of drill pipe could excite one of the natural modes of the

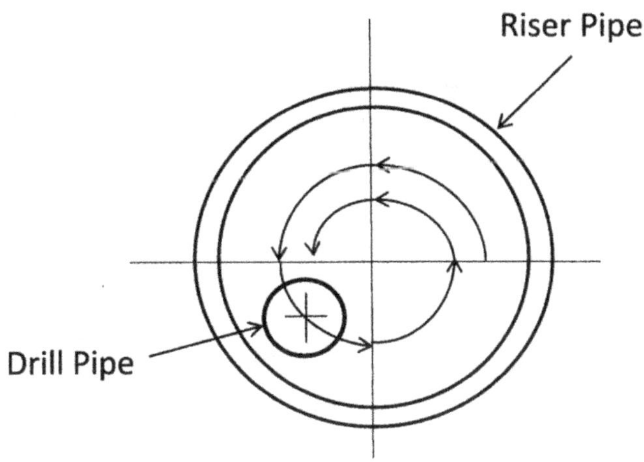

Figure 6-16. Drill pipe whirling within a marine riser.

riser. The solution to riser circular motion is to operate drill pipe at damping ratios greater than one.

Rotational Stability with Axial Thrust

A reasonable approach here is to assume average effective tension in drill pipe over the riser interval. This approach combines the effects of pipe weight, mud weight, and thrust force in the effective tension term and eliminates the variable coefficient. In essence, short beam theory could be used to predict the onset of drill pip whirl. The possible whirling and rotation stability of drill pipe within a marine riser, where effective tension is very high, might be predicted with short beam theory. Equation (2-200a, b) simplify to

$$EI\frac{\partial^4 x}{\partial z^4} - T\frac{\partial^3 y}{\partial z^3} - T_{\text{eff}}\frac{\partial^2 x}{\partial z^2} + \rho\left(\frac{\partial^2 x}{\partial t^2} - 2\omega\frac{\partial y}{\partial t} - \omega^2 x\right) + c\left(\frac{\partial x}{\partial t} - \omega y\right) = 0 \qquad \text{(6-57a)}$$

$$EI\frac{\partial^4 y}{\partial z^4} + T\frac{\partial^3 x}{\partial z^3} - T_{\text{eff}}\frac{\partial^2 y}{\partial z^2} + \rho\left(\frac{\partial^2 y}{\partial t^2} + 2\omega\frac{\partial x}{\partial t} - \omega^2 y\right) + c\left(\frac{\partial y}{\partial t} + \omega x\right) = 0 \qquad \text{(6-57b)}$$

where T is torque throughout the section and T_{eff} is the average of the effective tensions at top and bottom of the pipe interval.

$$T_{\text{eff}} = [F_B + w(L - z)] + zw_m$$

$$\left(T_{\text{eff}}\right)_{\text{ave}} = \frac{(T_{\text{eff}})_{\text{top}} + (T_{\text{eff}})_{\text{bottom}}}{2}$$

$$\left(T_{\text{eff}}\right)_{\text{ave}} = F_B + \frac{1}{2}LwBF \qquad \text{(6-58)}$$

The parameter, z, is measured downward from the top (see Fig. 2-18).

Boundary conditions at $z = 0$ (top) and $z = L$ (bottom) are assumed to be structurally pinned for the sake of simplicity.

$$x(0) = y(0) = 0 \qquad \text{(6-59a)}$$

$$x(L) = y(L) = 0 \qquad \text{(6-59b)}$$

$$EI\frac{d^2 x}{dz^2}(0) = T\frac{dy}{dz}(0) \qquad \text{(6-59c)}$$

$$EI\frac{d^2 y}{dz^2}(0) = -T\frac{dx}{dz}(0) \qquad \text{(6-59d)}$$

$$EI\frac{d^2x}{dz^2}(L) = T\frac{dy}{dz}(L) \tag{6-59e}$$

$$EI\frac{d^2y}{dz^2}(L) = -T\frac{dx}{dz}(L) \tag{6-59f}$$

As before, Eq. (6-57a) and (6-57b) are combined into a single equation using the complex function

$$U(z,t) = x(z,t) + iy(z,t) \tag{6-60}$$

The resulting differential equation is

$$\frac{\partial^4 U}{\partial \zeta^4} - i\Theta\frac{\partial^3 U}{\partial \zeta^3} - \beta\frac{\partial^2 U}{\partial \zeta^2} + \frac{\rho L^4}{EI}\left[\frac{\partial^2 U}{\partial t^2} + i\left(2\omega\frac{\partial U}{\partial t}\right) - \omega^2 U\right] + \frac{cL^4}{EI}\left(\frac{\partial U}{\partial t} + i\omega U\right) = 0 \tag{6-61}$$

where

$$\zeta = \frac{L - z}{L}$$

Using the method of separation of variables, the solution to Eq. (6-61) is expressed in the form

$$U(\varsigma,t) = u(\varsigma)\tau(t) \tag{6-62}$$

The two components $u(\zeta)$ and $\tau(t)$ are separated and equated to a complex constant defined as

$$\gamma = R_\gamma + iI_\gamma \tag{6-63}$$

From the separation of variables we obtain two separate ordinary differential equations.

$$\frac{d^4u}{d\zeta^4} - i\Theta\frac{d^3u}{d\zeta^3} - \beta\frac{d^2u}{d\zeta^2} - (R_\gamma + iI_\gamma)u = 0 \tag{6-64}$$

$$\frac{d^2\tau}{dt^2} + 2\left(\frac{c}{2\rho} + i\omega\right)\frac{d\tau}{dt} + \frac{EI}{\rho L^4}\left[R_\gamma - \Omega^2 + i\left(I_\gamma + \frac{c\omega L^4}{EI}\right)\right]\tau = 0 \tag{6-65}$$

where

$$\Theta = \frac{TL}{EI}$$

$$\beta = \frac{T_{\text{eff}}L^2}{EI}$$

$$\Omega^2 = \frac{\rho\omega^2 L^4}{EI}$$

The criterion of stability is the same as before (see Eq. (2-220)). Stability depends on the complex constants R_γ and I_γ [6].

Establishing R_γ and I_γ

The solution to Eq. (6-64) along with boundary conditions yields the required eigenvalue, $\gamma R_\gamma + iI_\gamma$. It would appear that the solution to Eq. (6-64) is straight forward since there is no variable coefficient. Assume a solution of the form

$$u = Ce^{a\varsigma} \tag{6-66}$$

where C and a are both imaginary numbers. Substituting into Eq. (6-64) gives

$$a^4 - i\Theta a^3 - \beta a^2 - \gamma = 0 \tag{6-67}$$

There are four imaginary roots (a_1, a_2, a_3, a_4) to this equation and the broader solution becomes

$$u(\zeta) = C_1 e^{a_1\zeta} + C_2 e^{a_2\zeta} + C_3 e^{a_3\zeta} + C_4 e^{a_4\zeta} \tag{6-68}$$

Assuming the roots to Eq. (6-67) have been found, and applying boundary conditions gives a square matrix whose determinate represents the characteristic equation. The real and imaginary parts of the characteristic equation must both be zero from $\gamma = R_\gamma + iI_\gamma$. Obtaining these results is not a simple algebraic exercise.

However, by assuming a solution of the form

$$u = Ce^{ia\zeta} \tag{6-69}$$

where C is an imaginary number but a is a real number gives a different outcome. The substitution of Eq. (6-69) into Eq. (6-64) and simplifying gives

$$a^4 - \Theta a^3 + \beta a^2 - \gamma = 0 \tag{6-70}$$

This equation shows that by assuming a to be real, γ is also real, which means that $I_\gamma = 0$ and the problem reduces to synchronous whirl.

Southwell and Dough [7] arrived at the same algebraic equation starting with the assumption of synchronous whirl. They were able to determine the roots using numerous trial hand calculations. Their results are represented by Eq. (2-171).

They also point out that Eq. (2-172), though not exact, gives a decent approximation. Equation (2-172) is repeated here for convenience. Note that the force, P, is compression (+) and the formula is based on short beam theory.

$$\frac{P}{P_{\text{Euler}}}+\left(\frac{M_{\text{t}}}{M_{\text{t,Greenhill}}}\right)^2+\left(\frac{\omega}{\omega_{\text{crit}}}\right)^2=1 \tag{2-172}$$

where

$$P_{\text{Euler}}=\frac{\pi^2}{L^2}EI$$

$$M_{\text{t,Greenhill}}=\frac{2\pi}{L}EI$$

$$\omega_{\text{crit}}^2=\frac{EI}{m}\left(\frac{\pi}{L}\right)^4$$

If Eq. 2-172 is written in terms of parameters used in the discussion of long pipe,

$$\frac{F}{F_{\text{cr}}}+\left(\frac{T}{T_{\text{cr}}}\right)^2+\left(\frac{\omega}{\omega_{\text{cr}}}\right)^2\leq 1 \tag{6-71}$$

Using the formulas developed in the previous chapters

$$F_{\text{cr}}=\phi_{\text{cr}}(EI)^{\frac{1}{3}}(w-w_m)^{\frac{2}{3}} \tag{3-19}$$

T_{cr} - see Table 2-7

$$\omega_{\text{cr}}^2=\frac{EI}{m}\frac{R_\gamma}{L^4} \tag{2-224}$$

Consider the following set of input.

Drill Pipe – 5 ½ in. (19.2 lb/ft)
E = 29,000,000 psi
I = 16.8 in.^{4}
m = 0.6056 slugs/ft
L = 1,000 ft
γ_m = 12 ppg (BF = 0.817)
Surface Torque = 15,000 ft-lb
Rotary Speed = 100 rpm
Hook Load = 200,000 lbs

The dimensionless parameter, α = 4,637, which is outside the numbers given in Table 2-8. Extrapolation, however, gives $R_\gamma \approx 10^4$. Also see Table 2-3 and note that $R_\gamma \approx \lambda^4$. Using these input numbers gives

F_{cr} = 3,094 lbs (compression)

T_{cr} = 56,900 ft-lbs (Figure 2-17)

ω_{cr}^2 = 0.056 1/sec^2 (0.2364 rad/sec)

For comparison, the critical rotary speeds for zero tension and linear tension are

$$\omega_{cr}^2 = \frac{EI}{m}\left(\frac{\pi}{L}\right)^4 \qquad \text{(zero tension)}$$

$$\omega_{cr}^2 = \frac{EI}{m}\left(\frac{10}{L}\right)^4 \qquad \text{(linear tension)}$$

The critical rotary speed for linear tension is higher as expected. Applying the input parameters gives

F = 183,903 lbs
T = 15,000 ft-lbs (given)
$\omega = \frac{2\pi 100}{60} = 10.47 = \text{rad/sec}$

Bringing the numbers together by Eq. (6-71) gives

$$\frac{-183,903}{3094} + \left(\frac{15,000}{56,900}\right)^2 + \left(\frac{10.472}{0.2364}\right)^2 =$$

$$-59.44 + 0.0695 + 1,961 >> 1$$

While this is a rather crude approximation it does point to the possibility of drill pipe instability within risers. A dominating factor is the low critical rotary speed, which over rides the stabilizing effect of tension. Applied torque is a minor factor.

REFERENCES

1. American Petroleum Institute Recommended Practice, API RP 16Q.
2. Dareing, D. W.; "Stress Balanced Tapered Columns in Marine Risers", Proceedings of the 37th Petroleum Mechanical Engineering Workshop and Conference (ASME), Dallas, Sept. 13–15, 1981.
3. Gordon, Cormack C. and Don W. Dareing, "Configuring Marine Riser Tapered Stress Joints Used In Top-Supported Applications", *ASME Journal of Energy Resources Technology*, Vol 126, No.3, Sept 2004.
4. Huang, T., Dareing, D. W., and Beran, W. T., "Bending of Tubular Bundles Attached to Marine Risers," *Trans. ASME, Journal of Energy Resources Technology*, Vol. 102, March 1980, pp. 24–29.
5. Pestal, I. E. C. and Leckie, F. A.; *Matrix Methods in Elastomechanics*, McGraw-Hill, 1963.
6. Huang, T. and Dareing, D. W., "Predicting the Stability of Long Vertical Pipe Transmitting Torque in a Viscous Medium," *Trans., ASME, Journal of Engineering for Industry*, Vol. 88, Series B, No. 2, May 1966, pp. 191–200.
7. R. V. Southwell and B. S. Dough; "On the Stability of a Rotating Shaft Subjected Simultaneously to End Thrust and Twist," British Association for the Advancement of Science, Vol. 8 (1921) pp. 345–355.

INDEX

www.ingramcontent.com/pod-product-compliance
Lightning Source LLC
Chambersburg PA
CBHW062116240626
47202CB00003B/79
9780791859995